New Plant Sources for Drugs and Foods
from The New York Botanical Garden Herbarium

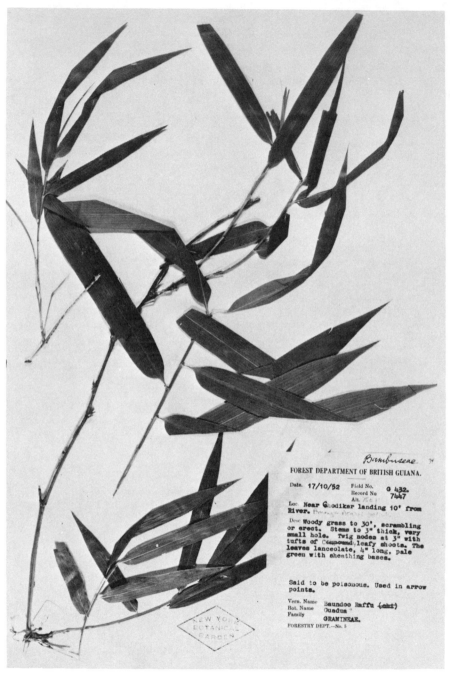

Herbarium specimen of a tropical woody grass (Poaceae), reported to be poisonous and used in arrow points in British Guiana. The plant is cited on p. 11 as number 75, *Guadua* indet. Photo courtesy of Alenka Remec.

New Plant Sources for Drugs and Foods from The New York Botanical Garden Herbarium

SIRI VON REIS AND FRANK J. LIPP, JR.

Harvard University Press
Cambridge, Massachusetts, and London, England
1982

Library of Congress Cataloging in Publication Data

Von Reis, Siri, 1931-
    New plant sources for drugs and foods from the New
York Botanical Garden Herbarium.

    Includes indexes.
    1. Medicinal plants — Catalogs and collections.
2. Plants, Edible — Catalogs and collections.   3. New
York Botanical Garden. Herbarium — Catalogs.   I. Lipp,
Frank J.   II. New York Botanical Garden. Herbarium.
III. Title.
QK99.A1V66   581.6′3   81-6280
ISBN 0-674-61765-7   AACR2

To Richard Evans Schultes: teacher, mentor, friend

Acknowledgments

My thanks extend to a number of people for making this volume possible. I am indebted to Dr. Howard S. Irwin, former Director of The New York Botanical Garden, for having allowed me to conduct the survey on which the catalogue is based. I am grateful also to Dr. Ghillean T. Prance, Vice President and Director of Research, for his support of the undertaking. Dr. Patricia K. Holmgren, Head Curator and Administrator of the Herbarium, was especially helpful, as were the members of her staff, to me and to my co-author, Frank J. Lipp, Jr., in our use of the collections. Mr. Lipp collaborated with me on the project and did the interpretive indexing. He trained, supervised, and worked with our assistants, John Ballman, Chester Chambers, Beth Dalal, Jacquelyn Kallunki, Rita Maroncelli, Abigail Miller, and Jessica Radloff. Others whom we wish to thank for assisting us in the Herbarium and with translations and interpretations of field notes are Dr. Rupert C. Barneby, Dr. Noel H. Holmgren, Dr. Tetsuo Koyama, Dr. James E. Luteyn, Dr. Scott A. Mori, and Eileen K. Schofield.

We received financial support from the Marstrand Foundation and from the Rockefeller Foundation. I particularly wish to thank the Rockefeller Foundation for having allowed me to spend a wonderful month as Scholar-in-Residence at Villa Serbelloni, working on this book.

Contents

New Plant Sources for Drugs and Foods
from The New York Botanical Garden Herbarium

Introduction

This book presents data relating to unusual drug and food plants. The format is the same as that of von Reis Altschul's *Drugs and Foods from Little-Known Plants: Notes in Harvard University Herbaria* (Harvard University Press, 1973). The notes and species are different, having been culled from another source, The New York Botanical Garden Herbarium. Together the two catalogues constitute a double resource of potentially useful species which, it is hoped, will be studied and evaluated before they are extinguished. Some of these plants have been partially domesticated by primitive societies that also are vanishing, along with their lore; many more exist in the wild and are likely to become extinct as natural environments are disturbed.

The National Research Council has reported that areas of tropical rain forest the size of Massachusetts are being destroyed each month. These forests are by far the richest associations on earth in numbers of plant species; but, unless there is a change in human nature, they will have disappeared by the turn of the century. Of the estimated four or five million living species of plants and animals, about three million are found in the tropics, and fewer than one-sixth of them have been identified and described (*Science,* April 25, 1980). Of that one-sixth, the mainly obscure four and a half thousand plants contained in this catalogue constitute only a tiny fraction.

The record assembled in this and the earlier catalogue was made on a basis of trust in the elements of empiricism in primitive medicine and nutrition. We believed that such a mass of folklore gathered in taxonomic order from a herbarium and published in the sequence in which it was found would reveal patterns of relationships between plants and their reputed usefulness. These relationships, or their absence, could direct investigators in the search for new medicines, foods, and related substances. The botanist interested in chemotaxonomy also might find clues to plant relationships in the catalogue.

One of the obstacles to locating and identifying plants referred to in the literature is the lack of voucher specimens (except in the annals of taxonomy, which usually are not concerned with uses of plants). Perhaps the

1

greatest value of these two catalogues is the existence of a voucher specimen for every entry. Because the data are firsthand and nearly every collection cited can be traced to its geographical source, the investigator has a head start. For exact localities and other details on any particular species, he still will need to seek out the herbarium sheets themselves.

SPECIES INCLUDED AND EXCLUDED

The data in this catalogue were gathered in a little over two years, starting in August 1975. The searchers scanned the labels of more than 3,000,000 sheets of dried specimens of angiosperms, or flowering plants, in The New York Botanical Garden Herbarium. Over 4500 species of possible interest were found and are presented here.

The orchid family was included in the search; it was omitted from the 1973 catalogue. The gymnosperms, pteridophytes, bryophytes, fungi, and algae, however, are excluded now as then, in the belief that they would yield few notes of interest and from a desire to limit the scope of the undertaking. During the search, unreckoned numbers of sheets were absent on loan, and some new acquisitions were not yet mounted; to that extent, this catalogue will not reflect completely the riches of its source. One may wonder, too, how many significant species are missed in a herbarium search because no field notes on their uses were recorded. The percentage represented therein could be fairly high, since botanical collectors are mainly taxonomists, whose concerns traditionally have not been with useful plants as such. Type specimens also were omitted from the catalogue since most of them are already in the record, albeit in journals of systematics. Similarly, species whose Latin epithets alone suggested medicinal or nutritional properties were not considered unless accompanied by corroborative or otherwise pertinent label data. Still other species were eliminated because of illegible handwriting or difficulties in interpretation.

Ordinarily, species were not included in the catalogue if they had appeared in J. C. Th. Uphof's *Dictionary of Economic Plants* (Lehre, Germany: J. Cramer, 1968); exceptions were made where a herbarium label added new aspects to uses or extended the geographical distribution of a previously known use. Similarly, duplications of collections found in the Harvard survey were eliminated as far as possible.

Where one collection of a species offered a note of interest, notations otherwise of little interest from other collections of the same species sometimes were included if they bore information which might extend the use of the species. Common names associated with a species having interesting uses occasionally were found applied to other species. Where this occurred, an attempt was made to include all species bearing that common name on the chance that some or all might possess similar properties. Common names or specific uses found in more than one collection of a given species within a country of moderate size are listed only once, as this compilation is not intended to record frequencies of occurrence. Any imbalance in geographical representation in the catalogue is a result of the fact that even large, cosmopolitan herbaria like that of The New York Botanical Garden specialize regionally, in this instance in the floras of tropical America and the United States. The coverage in the catalogue is influenced further by the collectors' depth of interest in applied botany.

Of the 399 families examined in the course of searching, 225—or more than half—figure herein, distributed among 1584 genera and 4571 species. The *Leguminosae (Mimosaceae, Caesalpiniaceae, Fabaceae)*, are the family best represented, making up approximately 10 percent of the species; the *Compositae (Asteraceae), Lauraceae, Rosaceae,* and *Rubiaceae* make up 4 percent each; *Euphorbiaceae, Labiatae (Lamiaceae),* and *Solanaceae,* 3 percent each; *Annonaceae, Ericaceae, Gramineae (Poaceae), Moraceae,* and *Verbenaceae,* 2 percent each; *Apocynaceae, Chrysobalanaceae, Clusiaceae, Flacourtiaceae, Myrtaceae, Polygonaceae, Sapindaceae,* and *Sapotaceae,* 1 percent each.

As in the Harvard survey, the two largest categories of plants are edible species (28 percent) and aromatic species (15 percent). Next are those used in gastrointestinal disorders (7 percent). Less numerous are species referred to on herbarium labels simply as "medicinal" or poisonous (6 percent each); those used to treat injuries, respiratory ailments, and skin diseases (3 percent each); those from which beverages are made, those used as febrifuges, those affecting the nervous system, and those used in connection with reproduction (2 percent each). All remaining notes hint at biodynamic properties of some kind.

Four indexes accompany this volume. Indexes to genera and families are presented for the use of investigators interested in a particular plant or plant group. The index to uses refers to special properties or employments. The index to common names may contain some erroneous entries caused by our difficulties in determining meanings in dialects and tribal languages. It is recommended that investigators not confine themselves to the indexes but thoroughly explore the text as well for subtleties of interpretation that otherwise might be missed.

EDITORIAL PRINCIPLES EMPLOYED

The use of this catalogue should not require a background in systematic botany or herbarium technique; but, for those who are interested in how certain problems were handled, the following remarks may be helpful. The species and notes belonging with them are arranged by family in the sequence obtaining in The New York Botanical Garden Herbarium, which is arranged after a slightly modified Engler and Prantl system.

Genera within each family are arranged alphabetically, a convenient system that, obviously, does not reflect natural relationships. The investigator should bear this fact in mind when comparing entries in this volume with those in the Harvard survey, the genera there being arranged in phylogenetic sequence.

Species are separated into ten geographical areas and then sorted alphabetically within each area, as in the herbarium. In the course of cataloguing, however, notes on a species which occurred in more than one geographical area were all brought together under the first entry of the species. This was done to facilitate the use of the catalogue.

Despite the partial use of alphabetical order in the system, similar uses among related plants, and within and among families, will be apparent, suggesting common properties. The most promising of such leads are reported uses shared, within a genus or species, by representatives which are widely separated geographically.

Family names throughout are spelled in accordance with the nomenclature used in The New York Botanical Garden Herbarium. Generic and specific names are spelled for the most part as written on the labels, except for obvious errors corrected with the help of J. C. Willis' *A Dictionary of the Flowering Plants and Ferns* (7th edition, Cambridge University Press, 1966) and *Index Kewensis*.

Throughout the text, specific and infraspecific epithets have been decapitalized in accordance with rules of the *International Code of Botanical Nomenclature* of 1978. In the interest of brevity, and because this book is not a taxonomic work, the botanical authorities have been omitted from all Latin binomials.

It may be helpful to show the format of each entry and to explain the various elements in detail:

(a) Catalogue no.   (b) Latin name of species   (c) Place of collection / (d) Collector and collector's no. / (e) Year collected / (f) Quotation of interest / (g) Common name   (h) Dialect

---

(a) Each species has been assigned a number (from 1 to 4571), for ease of handling and for indexing. The numbers serve no function outside the book.

(b) In all instances identifications of the plants are recorded as they were in the herbarium. Where there were discrepancies between the name on a sheet and that on the folder enclosing it, the name on the folder was used.

(c) Under each species, one or more collections are entered, each entry beginning with the name of the country or other political or geographical unit in which the collection was made. These names are presented as written on the labels. In some instances they are in archaic or foreign spelling, or designate countries or political entities that no longer exist. We have retained almost all these names because they can be used to estimate the ages of undated collections and to identify localities in countries whose borders have changed. For some older collections, where the exact country of collection could not be determined, we substituted the broad geographical area under which the specimen was filed. These areas, listed here in the se-

quence used in The New York Botanical Garden Herbarium, are capital-
ized in their entirety to distinguish them from countries, islands, and the like:

U.S. & CAN.	(includes Greenland)
MEX. & C. AM.	(includes Lower California or Baja California)
W.I.	(includes Bahamas and Bermuda)
S. AM.	(includes Trinidad and Tobago)
EUR.	(includes Iceland and European Russia)
AFR.	(includes Canary Is., Madagascar, and Seychelles Is.)
N. ASIA	(includes Russia east of the Urals)
TROP. ASIA	(includes Hainan)
AUS.	(includes New Zealand)
PACIFICA	(includes Hawaii)

(d) In each entry the name of the country is followed by that of the collector (or collectors) and his (or their) serial number. Where such numbers
were absent, denoted by the abbreviation "s.n.," herbarium numbers or
some combination of names and numbers were given, if available, to identify the collection. Note that a collector may not always sign his name in
the same way; hence, there may be variations of a name throughout the
text. We have attempted to include a name, usually of one or two initials
and the last name, complete enough to be identifiable without burdening
the text. Errors may occur in the spelling of names where handwriting was
difficult to decipher. We have referred to the available installments of *Index
Herbariorum, Part II* (1-4, A-M) for assistance with collectors' names.

(e) The year of collection is given exactly as it appears on the specimen
label, in the various forms of, for instance, 46, '46, ''46, 194, 946, and
1946. From such notations alone, the century in which a collection was
made may not always be ascertainable, and numbers assumed to be dates
occasionally may signify something else.

(f) The year is followed by the note or notes of possible interest to
medical or health sciences. Here we have quoted verbatim from herbarium
labels, including variations and errors in orthography, punctuation,
capitalization, and the like. We have not used *sic* because the need for it
arose too frequently.

(g) If there is a common name for a plant but this name does not appear
in the quotation relating to use, the name will follow as a new quotation.
Single quotes within double quotes denote double quotes on the label from
which the excerpt was taken. The abbreviation "N.v.," which appears often
in this portion of the entries, designates a vernacular name. Many species
have been included in this catalogue by virtue of the symbolism in their
common names. It was believed advisable to admit more, instead of fewer,
such species, to avoid missing something useful. Following this principle,
we have included aromatic plants, edible plants, plants used in magical
ceremonies, and so forth. Decisions to include or exclude material in

foreign tongues and local dialects were difficult to make. Some place names may have slipped in as common names, especially from old or illegible labels.

(h) Wherever available, information as to language, dialect, or locality of the name may be given, sometimes in parentheses or following the name. Occasionally the language of the name will be identified at the end of the entry, often in abbreviated form. In all instances the data are taken directly from the labels themselves.

The lack of uniform style in the entries arises from literal transcription of the data. Throughout, the aim has been to combine accuracy with clarity.

The investigator is advised that, if any given lead is traced back to the corresponding specimen in the herbarium and the plant is deemed worthy of further research, its identity should be confirmed by a taxonomic botanist familiar with the taxonomic group to which the specimen has been assigned. Confirmation is advisable because in all herbaria some plants have been identified only provisionally, pending scrutiny by monographers or other specialists.

The searches leading to this and the earlier catalogue were conceived as a means to bring to light new sources of food, medicine, and related materials. Notes in herbaria bearing on these subjects were of primary interest mainly because such notes were more plentiful than those touching on other uses. Future surveys, however, should consider plant uses of all kinds, since the discovery and preservation of all natural resources is increasingly urgent. Recovery of every kind of information will require only a little more effort than that already expended in massive and time-consuming surveys.

GNETACEAE

1   *Ephedra aspera* U.S.: Tex. / V. L. Cory 52959 / 1946 / " 'Mormon tea' "
Mexico: Coahuila / Stanford, Retherford & Northcraft 169 / 1941 / "heavily grazed by goats"

2   *E. californica* U.S.: Calif. / W. M. R. s. n. / 1927 / "Desert Tea"
U.S.: Calif. / F. W. Johnson 2191 / 1920 / " 'Desert Tea' "

3   *E. cutleri* U.S.: Ariz. / W. N. Clute 100 / 1920 / " 'Mormon Tea' "

4   *E. torreyana* U.S.: N.M. / H. Cutler 2076 / 1938 / "usually eaten by stock to less than 30 cm. high"

5   *E. viridis* U.S.: Nev. / Stretch s. n. / 65 / "used by Indians as a decoction for liver complaints"
U.S.: Nev. / Moore & Franklin 62 / 1937 / "Indian beverage"

6   *E. americana* Peru / F. R. Fosberg 27912 / 1947 / "said to be used to cure venereal diseases"

7   *E. alte* AFR. / Fouad I. Agric. Mus. HERB. 3/4 / 1935 / "Largely eaten by cattle"

8   *E. equisetina* China / F. N. Meyer 1095 / 1913 / "used by the Chinese as a medicine in urinary troubles"

9   *E. gerardiana* var. *saxatilis* Himalaya / W. Koelz 974 / 1930 / "used for cleaning teeth" / "Tsafad"
Himalaya / W. Koelz 562 / 1930 / "for cleaning teeth and use the ashes as snuff" / "chefrat"

10 *Gnetum urens* Panama / R. S. Williams 608 / 1908 / "said to be very poisonous" / " 'Canjura' "

11 *G. nodiflorum* Venezuela / B. & C. Maguire, Steyermark 53503 / 1962 / "Resin used as incense"
Brazil / J. Murça Pires 1190 / 1948 / "Frutos comestiveis (assados)" / " 'Itua' "

12 *G. paniculatum* Brazil / J. Murça Pires 807 / 1947 / "Cipó a beira d'agua; as castanhas assadas são comestiveis. Os Indios do Rio Icana extraem tapioca da semente" / " 'Curucuda' "

13 *G. schwackeanum* Brazil / J. M. Pires 788 / 1947 / "quando cortado expele u'a agua pouco adstringente que se bode beber. Fruto comestível quando assado e tem gôsto de batata. Nalgulmas partes, os indios preparam uma farinha da semente" / " 'Curucuda' "

14 *G. gnemon* Fiji / A. C. Smith 8590 / 1953 / "young leaves edible when cooked" / " 'Sukau' "
Fiji / A. C. Smith 8005 / 1953 / "young leaves edible, usually cooked with coconut milk" / " 'Mbele sukau' "
Fiji / A. C. Smith 7449 / 1953 / "young leaves edible" / " 'Mbele sukau' "

New Britain / J. H. L. Waterhouse 265 / 1934 / "Fruit and young leaves edible" / "Nula"

Philippine Is. / no collector 357 / "fts. used as substitute for coffee" / "Name in Tagbanua: 'bago' "

15 **G. gnemon** var. **domesticum** Fiji / A. C. Smith 6652 / 1947 / "young leaves cooked and eaten" / " 'Mbele sikau' "

Fiji / A. Tabualewa 15602 / 1941 / "Fruit edible" / " 'Masokau' "

Fiji / E. Ordonez 13615 / 1940 / "Fresh uncooked leaves eaten by Fijians"

Fiji / O. Degener 15157 / 1941 / "Fijians eat seed and use plant for firewood. Seruans do not eat leaves" / " 'Wasokau' (Serua)"

Fiji / Degener & Ordonez 14089 / 1941 / "young leaves and fruits eaten cooked"

16 **Gnetum** indet. Papua / C. E. Carr 11586 / 35 / "Fruit which is dull and is eaten by the natives" / "Native name KEM. Orikoro name RORO"

17 **Gnetum** indet. Indonesia / G. A. L. de Haan 1785 / 1950 / "Flowers young fruit and leaves used as a vegetable. Ripe fruit used for krupuk" / "Ganemo (Weda)"

PANDANACEAE

18 **Pandanus kirkii** Tanganyika / R. E. S. Tanner 3980 / 1958 / "aromatic"

19 **P. fascicularis** Sarawak / J. Purseglove P 5064 / "male flower . . . sweet scented"

APONOGETONACEAE

20 **Aponogeton dinteri** AFR. / A. & E. von Goldhammer 3485 / 66 / "Knollen am Grunde essbar"

JUNCAGINACEAE

21 **Triglochin maritima** Labrador / Gillett & Findlay 4930 / 1950 / ". . . acrid odor"

ALISMATACEAE

22 **Limnophyton obtusifolium** Tanganyika / R. E. S. Tanner 1117 / 1952 / "Sukuma: for aches anywhere; burnt & the preparation therefrom used by rubbing into small cuts made near the painful part . . . flowers . . . slightly aromatic" / "kisukuma: masobosobo"

23 **Sagittaria longiloba** Mexico / H. S. Gentry 4741 / 1939 / "Yaqui country" "Observed also in pots in Vicam pueblo"

24 *S. guayanensis* subsp. *lappula* China: Kwangtung / Lingnan (To & Ts'ang) 12720, 12729 / 1924 / "(crane vegetable)" / "Hok ts'oi"

LIMNOCHARITACEAE

25 *Limnocharis flava* Guatemala / P. Standley 74967 / 1940 / " 'Cebolla de chucho' "

HYDROCHARITACEAE

26 *Ottelia alismoides* China: Kwangtung / W. T. Tsang 26495 / 1934 / "Shui Kai Tsoi"

27 *Vallisneria americana* U.S.: Mass. / F. MacKeever N703 / 1963 / "wild or water-celery"

POACEAE

28 *Agropyron magellanicum* Falkland Is. / W. Davies BF 42 / 38 / "Sand oat grass"

29 *Agrostis scabra* U.S.: Mass. / F. C. Mackeever N5541 / 1961 / "Hair grass, fly-away grass, tickle grass"

30 *A. umbellata* Chile / Y. Mexia 7881 / 1936 / " 'Illusion' "

31 *Andropogon bicornis* Jamaica / W. Harris 370 / 1913 / " 'Sourgrass' "

32 *A. ceriferus* Jamaica / W. Harris s.n. / 1914 / " 'Lemon grass' "

33 *Andropogon* indet. Cuba / C. Lundell 15042 / 1953 / "aromatic"

34 *A. martini* Brazil / A. Chase 9302 / 1925 / "Entire plant has strong aromatic, peppery taste; tea is made of entire plant & given for colds; escaped from cultivation" / " 'Capim de São Jose' "

35 *A. citratus* Hawaiian Is. / O. Degener 11582 / 1937 / "Planted as spice in Filipino plantation camp"

36 *Aristida ternipes* Mexico / Y. Mexia 1053 / 1926 / " 'Aceitilla' "

37 *Aristida* indet. W. I.: Curaçao / Curran & Haman 23 / 1917 / "Jeerba die viento"

38 *A. arenaria* Australia: Queensland / S. Everist 2933 / 1947 / " 'Kerosene Grass' "

39 *Arundinella deppeana* Brit. Honduras / W. Schipp 791 / 1931 / "horses and other animals seem to avoid this grass"

40 *A. confinis* Jamaica / W. Fawcett 660 / 17 / " 'Sour grass', not eaten by cattle"

41 *Axonopus compressus* Jamaica / W. Harris s.n. / 1902 / " 'Pimento Grass' "

42   *Bambusa spinosa* Hainan / McClure & Merrill 8940 / 1923 / "seeds are eaten and considered to have medicinal value; young leaves which have not yet unfolded steeped to make 'tea' for medicine" / "Nai chuk"

43   *Bothriochloa barbinodis* Ecuador / W. Camp E-2507 / 1945 / " 'This grass not eaten much by the animals because it is sharp-flavored like red pepper' "

44   *B. pertusa* W. I.: St. Croix / M. Baker s.n. / 1929 / " 'Sour Grass' "

45   *Cathestecum erectum* Mexico / E. Palmer 270 / 1886 / "used medicinally" / " 'Gramma' "

46   *Cenchrus pauciflorus* Mexico / Dr. Gregg s.n. / 47 / "Said to be lactiferous, wherefore root tea drank by puerperal women" / "Zacate de Cadillo"

47   *Centotheca latifolia* Fiji Is. / O. Degener 15123 / 1941 / "For finger sickness called 'thava' Fijians cut finger lengthwise with bambo knife. The leaves (not stem) of this grass were mashed and wrapped in coconut cloth and their juice was pressed on sore" / " 'Mbitumbitu' (Serua)"

48   *Coix lachryma-jobi* Ecuador / W. Camp E-947 / 44 / "The flour is said to be a little heavy and needs to be mixed with wheat flour before it can be used for bread. It is said to be of excellent flavor" / " 'Trigo tropical' "

49   *Ctenium aromaticum* U.S.: Va. / Fernald & Long 6031 / 1936 / "base aromatic when bruised"

50   *Cymbopogon citratus* Brazil / Prance, Coelho & Monteiro 15022 / 1971 / "Cultivated by local residents for making tea" / "Capim santo"

51   *C. densiflorus* Tanganyika / Newbould & Harley 4319 / 59 / "The flowers are smoked alone or with tobacco by witch-doctors and cause dreams. It is said that these dream foretell the future"

52   *C. jwarancusa* India / Sri Ram s. n. / 1920 / " . . . a good crops use for Khash Tati at summer season & grazing for cattle"

53   *Cynodon dactylon* U.S.: Calif. / B. Trask s.n. / 1897 / " 'Devil-grass' "
El Salvador / S. Calderon 498 / 1922 / " 'Grama de gallina' "
Bermuda Is. / F. S. Collins 145 / 1913 / "common as pasture grass, sometimes causes staggers in cattle"
Peru / A. Samonamud 2211 / 1953 / " 'Grama dulce' "
Belgium / A. Cogniaux 43 / 1867 / "Les bestiaux et surtout les moutons mangent ses feuilles"

54   *Dactylis caespitosa* Falkland Is. / T. Morong 1341 / 1890 / "Cattle very fond of it" / "Called Tussock grass"

55   *Danthonia spicata* U.S.: N.Y. / E. H. D. 47 / 79 / " 'Wild Oat-Grass' "

56   *D. decumbens* Belgium / A. Cogniaux 127 / 1870 / "Excellent fourrage"

57   *Deschampsia nitida* Argentina / Y. Mexia 7941 / 1936 / "Said to be good feed"

58   *D. flexuosa* Belgium / A. Cogniaux 206 / 1871 / ". . . cependant les boeufs et les moutons la paissent avec plaisir"

59   *Digitaria sanguinalis* Belgium / A. Cogniaux 33 / 1867 / "Tous les bestiaux aiment et recherchent ce fourrage . . ."

60   *D. pole-evansii* S.W. Africa / R. Seydel 2715 / 1961 / "Gefressen (Okatussukanyi)"

61   *D. pruriens* New Guinea / L. Brass 5522 / 1933 / "considered good fodder"

62   *Diplachne muelleri* Australia: Queensland / S. L. Everist 2857 / 1947 / "Relished by cattle" / "Local name 'Sugar Grass'"

63   *Echinochloa polystachya* Brazil / A. Chase 7918 / 1924 / "I was told that cattle wade out to eat this; sometimes they are attacked by Piranhas (fish) and killed"

64   *Eleusine africana* S.W. Africa / R. Seydel 3889 / 64 / "saftiges Gras, Gefressen"

65   *Elymus innovatus* Alaska / Y. Mexia 2271 / 1928 / " 'Wild wheat' "

66   *E. virginicus* U.S.: Vt. / H. Banker 528 / '99 / "(Wild Rye)"

67   *Elyonurus tripsacoides* Colombia / J. C. Jacome s.n. / 1945 / " 'Paja amarga' "

68   *E. typica* Colombia / J. Idrobo 2443 / 1957 / "Los frutos farináceos, dulces . . ." / " 'Carrizo' "

69   *Eragrostis cilianensis* U.S.: Mass. / F. MacKeever N-904 / 1965 / "Skunk-, Snake-, or Stink-Grass"
Australia: Queensland / L. Smith 3046 / 1947 / "Local name Stink Grass"

70   *E. maypurensis* Mexico / Y. Mexia 1198 / 1926 / "Used for a tea"

71   *E. echinochloidea* S.W. Africa / R. Seydel 985 / 1957 / "gutes-Futtergras"

72   *E. omahekensis* S.W. Africa / R. Seydel / 1961 / "stark befressen"

73   *E. tef* Ethiopia / F. Meyer 7506 / 1961 / "The widely grown Ethiopian grain called 'Teff' ingredient of the national bread called 'injera'. Extensively cultivated . . ."

74   *Erianthus fulvus* Australia: Queensland / M. Koch 21 / no year / "Sugar grass. One of the most valuable grasses of the district"

75   *Guadua* indet. Brit. Guiana / For. Dept. Brit. Guiana G. 432 / 52 / "Said to be poisonous. Used in arrow points" / "Raffu"

76   *Heteropogon melanocarpus* U.S.: Fla. / A. Chase 281 / 1907 / "strongly lemon-scented"

77   *Hierochloë odorata* Canada: Québec / Frère Fabius 85 / 46 / "Vanilla Grass"
U.S.: N.D. / H. Bolley 196 / 1892 / "Vanilla-grass, Seneca-grass, Holy-grass"

Germany / L. Oberneder 920 / 1920 / "Wohlriechendes Mariengras"

78 *H. magellanica* Falkland Is. / Mrs. Vallentin 232 / 1910 / "Cinnamon grass"

79 *Hilaria cenchroides* Mexico / E. Palmer 197 / 1886 / " 'Gramma - used medicinally' "

80 *Holcus lanatus* Peru / C. Junge 2125 / 1937 / " 'Pasto miel' "
Germany / H. Wagner s.n. / no year / "Wolliges Honiggrass"

81 *H. spicigerum* Australia: Queensland / S. Blake 14267 / 1940 / "scented when bruised"

82 *Hordeum jubatum* Canada: Québec / Frère Fabius 493 / 46 / "Orge agréable"
U.S.: S.D. / T. Williams 197 / 1892 / "Wild Barley"
U.S.: Utah / H. Engelmann 24 / 1859 / " 'seeds gathered by Indians for food' "

83 *Ichnanthus panicoides* Venezuela / Aristeguieta & Lizot 7420 / 1970 / "El fruto tostado, estalla y se come. Indios Guaicas. (Yanomanö)"

84 *Ischaemum digitatum* New Guinea / Watson 18 / 54 / "AGARABI USE: medicine / "aBiémc?"

85 *Lasiacis oaxacensis* var. *oaxacensis* Brit. Honduras / P. H. Gentle 1769 / 1936 / " 'Rat Rice' "

86 *L. procerrima* Brit. Honduras / W. Schipp 785 / 1931 / "seeds much sought after by finches and other birds"

87 *L. sorghoidea* var. *sorghoidea* Brit. Honduras / P. Gentle 1910 / 1939 / " 'Wild rice' "

88 *L. divaricata* var. *austroamericana* Brazil / A. Chase 8619 / 1925 / ". . .birds eat spkets greedily"

89 *Leersia lenticularis* U.S.: Tex. / V. L. Cory 50008 / 1945 / " 'catchfly grass' "

90 *L. oryzoides* var. *oryzoides* Canada: Québec / Frère Fabius 544 / 46 / "Rice Cut-grass"

91 *Leptoloma cognatum* U.S.: Tex. / V. L. Cory 52475 / 1945 / " 'fall witch-grass' "

92 *Lolium multiflorum* U.S.: Mass. / F. MacKeever N836 / 1964 / "Italian Rye-Grass"
Brazil / Smith & Reitz 12947 / 1964 / "cattle feed" / " 'azevim' "

93 *Lophatherum gracile* China: Kwangtung / Lingnan (To & Tsang 12487, 12658) / 1924 / "Shan kai mai, (Pheasant rice)"

94 *Manisuris exaltata* Jamaica / C. Lewis s.n. / 1954 / ". . . with stinging hairs"
Jamaica / W. Harris 11256, 11387 / 1912 / " 'Rice-grass' "

95 *Milium effusum* U.S.: N.Y. / Torr. s.n. / ca. 1840 / "Millet Grass" Belgium / Lecoyer 204 / 1871 / "Son foin exhale une odeur agréable. Onl'emploie quelquefois pour aromatiser le tabac"

96 *Molinia coerulea* Belgium / E. Marchal 132 / 1870 / "Les pigeons recherchent beaucoup ses graines qui communiquent á leur chair un excellent fumet"

97 *Olyra latifolia* Brit. Honduras / P. H. Gentle 7116 / 1950 / " 'Wild Rice' "

98 *Oryza latifolia* Brit. Honduras / P. Gentle 3935 / 1942 / " 'wild rice' " Venezuela / Maguire & Wurdack 34755 / 1953 / "Stem cortex and epidermis used by Guaikas for cutting hair"

99 *O. perennes* Brazil / Cavalcante & Silva MG 32564 / 1967 / " 'aroz bravo' "

100 *O. oryzoides* Germany / L. Oberneder 959 / 1941 / "Wilder Reis"

101 *Oryzopsis asperifolia* U.S.: N.Y. / Torr. s.n. / ca. 1840 / "Large White-grained Mountain-rice"

102 *O. pungens* U.S.: N.Y. / Torr. s.n. / ca. 1840 / "Smallest Mountain-rice"

103 *O. racemosa* U.S.: N.Y. / Torr. s.n. / ca. 1840 / "Black-fruited Mountain-rice"

104 *Oxytenanthera brownii* Tanganyika / P. J. Greenway 8440 / 1949 / "used for making a bamboo wine. Cultivated . . . above 4,500 ft. alt. to c 7,000 ft. alt." / "Ulanzi (Kihehe)"

105 *Panicum capillare* Canada: Québec / G. P. Anderson s.n. / 1916 / "Old Witch Grass"

106 *P. clandestinum* U.S.: N.J. / H. N. Moldenke 1770 / 1931 / "Stems stinging . . ." U.S.: N.J. / H. N. Moldenke 10537 / 1938 / " 'Corn-grass' "

107 *P. fasciculatum* Brazil / G. & L. Eiten 3859 / 1962 / "n.v. 'Milhã' "

108 *P. ligularis* Brazil / A. Chase 11460 / 1930 / "(hairs irritating to skin)"

109 *P. mucromulatum* Brazil / A. Chase 8105 / 1924 / "hairs irritating to skin"

110 *P. pilcomayense* Brazil / A. Chase 10151 / 1929 / "Hairs . . . stinging"

111 *P. rivulare* Brazil / Smith & Klein 7503 / 1956 / "Stinging hairs . . ."

112 *P. rudgei* Venezuela / B. Maguire 33257 / 1952 / "especially around ant nests" S. AM. / Martyn 89 / 1929 / "On Ant Hills . . ."

113 *P. urvilleanum* Argentina / A. L. Cabrera 989 / 1929 / " 'Jobancillo' "

114 *Paspalum millegrana* Jamaica / W. Harris 11620 / 1913 / " 'Sour Grass' "

115 *Pharus latifolius* Cuba / A. Luna 195 / 1920 / " 'Fruta de perro' "

116 *P. glaber* Argentina / J. E. Montes 749 / 45 / " 'Capihi-arroz' o 'Arróz de perro' "

117 *Piptochaetium fimbriatum* U.S.: Tex. / L. C. Hinckley 439 / 35 / "Pinyon Rice Grass"

118 *Schizachne purpurascens* U.S.: N.Y. / Torr. s.n. / ca. 1845 / "Purple Wild-oat"

119 *Schmidtia kalahariensis* S.W. Africa / R. Seydel 471 / 1955 / "Sogenanntes Sauergras, von säuerlichem Geruch"

120 *Setaria geniculata* Mexico / J. Swallen 2226 / 1957 / " 'sacate amarga' "

121 *S. sphacelata* Honduras / A. Molina R. 10178 / 1961 / " 'Pasto Gusanillo' "

122 *S. lilloi* Argentina / S. Venturi 1035 / 1920 / "Yerba Buena"

123 *S. vulpiseta* Brazil / Smith, Klein & Schnorrenberger 11751 / 1957 / " 'capim veado' "

124 *S. verticillata* S.W. Africa / R. Seydel 755 / 56 / "Die Samen von den Eingeborenen durch Reiben zwischen den Haenden von den Borsten befreit, liefern gekocht eine mehlige Feldkost"

125 *S. palmifolia* New Guinea / Watson 13 / 54 / "AGARABI USE: sections of the stem of the young plant are steamed in a bamboo tube and then eaten" / "AGARABI NAME: mai"
New Guinea / Watson 34 / 54 / "Also used to rub back & forth in nostrils at initiation of girls & boys. Is very cutting" / "masaPitnɇ?"

126 *Sorghum arundinaceum* Brazil / G. & L. T. Eiten 4891 / 1962 / "n.v. 'cana de macaco' "

127 *S. verticilliflorum* Australia: Queensland / L. E. Walter M2065 / 1960 / "Suspected of causing deaths of 8 cows, no symtons noted"

128 *Spartina alterniflora* U.S.: Mass. / L. B. Smith 1317 / 1945 / "Odor strong and rancid"

129 *Sporobolus indicus* Cuba / A. Luna 619 / 1920 / " 'Pitilla' "

130 *Stipa neesiana* U.S.: N.Y. / Torr. s.n. / ca. 1845 / "Tuscarora Rice. Water Oats"

131 *S. robusta* U.S.: N.M. / J. A. Gillmore s.n. / 1895 / " 'Sleepy Grass' "

132 *S. brachychaeta* Argentina / F. Kurtz 6676 / 1889 / " 'Paja cebadilla' "

133 *Streptochaeta spicata* Brazil / F. C. Hoehne 17226 / 1915 / " 'Arroz do matto' "

134 *Trachypogon plumosus* Brazil / Black & Lobato 50-9660 / 1950 / " 'Arroz bravo' "

135 *Trichachne insularis* El Salvador / S. Calderon 866 / 1922 / " 'Zacate amargo' "
Jamaica / W. Harris 11475 / 1912 / " 'Sour grass' "

136 *Tricholaena rosea* El Salvador / P. Standley 19663 / 1922 / " 'Ilusión' "

.37   *T. repens* Tanganyika / R. E. S. Tanner 1162 / 1952 / "Uses: in making banana beer" / "Kisukuma: Masomasaji"

.38   *Tripsacum pilosum* Mexico / H. Cutler 4019 / 1940 / " 'Maicillo' "

.39   *Zizaniopsis miliacea* U.S.: Tex. / V. L. Cory 50853 / 1945 / " 'southern wildrice' "

CYPERACEAE

.40   *Bulbostylis capillaris* Paraguay / W. Archer 4780 / 1936 / "Used as mouthwash for toothache" / " 'partillo chico' "

.41   *Carex hoodii* U.S.: Utah / S. Tillett 411 / 1954 / "Roots with pungent, spicy odor"

.42   *Carex* indet. Argentina / J. Montes 14689 / 1955 / "Raices olor a limon" / " 'Pipi' 'Pasto limon' "

.43   *Cyperus flavus* U.S.: Ariz. / Kearney & Peebles 12059 / 1935 / "Roots aromatic"

.44   *C. pringlei* U.S.: Ariz. / Kearney & Peebles 14944 / 1940 / "roots aromatic"

.45   *C. articulatus* El Salvador / P. Standley 21454 / 1922 / "Remedy for toothache" / " 'Santul' "
Venezuela / F. Tamayo 4029 / 1955 / "Las personas de la localidad suponen pueda ser ésta la planta productora de la borrachera, por lo que se recollectó muestra para análisis químico" / " 'funcia' "

.46   *C. luzulae* Guatemala / J. Steyermark 49198 / 1942 / "Reputed that one can cook the root and serve as a drink to be used with 'atol de maiz' or with 'atol' from sapote seeds" / " 'suchipaite' "

.47   *C. surinamensis* Mexico / Y. Mexia 8924 / 1937 / " 'Cebollin' "

.48   *C. confertus* Jamaica / G. Proctor 27506 / 1966 / "Rhizome aromatic"

.49   *C. corymbosus* Brazil / P. Baksta 21149 / 1968 / "Capim cultivado. O rizoma raspado pasto em infusão com agua e bebido contra dor de barriga" / "Ta dëxka Pa-punise-ko (Capim para dor de barriga)"
Venezuela / J. Steyermark 10743 / 1973 / "plant used as poison for arrow darts; take the rhizome and make a powder from it, then place on tip of arrow for killing"
Brazil / P. Baksta 21148 / 1968 / "Capim cultivado em terreno arenoso. O cha do rizoma e tido pelos indios tukanos como poderoso anti-concepcional e esterilizante. Tomam ea lua nova ou na lua cheia" / "Ta-Dëxka pona manise-ko (Capim para não ter filho"

.50   *C. friburgensis* Paraguay / Lourteig 1958 / 1967 / "Picada propiedad"

151   *C. prolixus* Surinam / Stahl 149 / 1941 / "A Cyperus cultivated by the Creoles and used as medicine. brought by the slaves from Africa" / "ningre blondre adios"
Surinam / Stahl 140 / 1941 / "sold on the market in Paramacaibo for medical purposes (the rhisoma) . . . of American *not* African origin"

152   *C. sphacelatus* Fr. Guiana / Black, Vincent & D'Ange 54-17456 / 1954 / "Erva perfumada"

153   *C. subnodosus* Peru / G. Klug 619 / 1929 / "Medicinal plant" / " 'Piri-piri' "

154   *C. compressus* Tanganyika / R. E. S. Tanner 1949 / 1955 / "aromatic"

155   *C. cristatus* S.W. Africa / R. Seydel 539 / 1955 / "Angenehmer, kalmusartiger Geruch"

156   *C. flabelliformis* Tanganyika / R. E. S. Tanner 2299 / 1955 / "Roots chewed for stomachache in children"
Tanganyika / R. E. S. Tanner 3140 / 1956 / "Sambaa: Zila"

157   *C. polystachos* Ghana / J. K. Morton A 923 / 54 / "Roots aromatic"

158   *C. pustulatus* Ghana / J. Morton s.n. / 1953 / "Sweet Scent"

159   *C. rotundus* Angola / J. Teixeira 1343 / 1956 / "Vivaz, tuberosa, de bolbilhos-comestiveis" / " 'Dumba' (Lunyaneca)"
Mariana Is. / F. Fosberg 24946 / 1946 / "tubers chewed for toothache" / " 'humatac' "

160   *C. tenax* Tanganyika / R. E. S. Tanner 2779 / 1956 / "roots aromatic"

161   *C. usitatus* Bechuanaland / H. Wild 5073 / 1960 / "Damaras use oil from buds as a scented pomade"

162   *C. bulbosus* Maldive Is. / F. Fosberg 36912 / 1956 / "roots and tubers very fragrant, sometimes used as incense" / " 'gok kulanduru' "

163   *Cyperus* indet. Samoa / E. Christophersen 2394 / 1931 / "Bulb for perfume" / " 'mumuta' "

164   *Eleocharis dulcis* Rhodesia / E. Robinson 1695 / 56 / "Plant locally called 'Masungu' and used for making salt"

165   *Evandra aristata* Australia: Queensland / C. White 5403 / 1927 / " 'Wild Oats' or 'Kangaroo Grass' "

166   *Fimbristylis cymosa* Marshall Is. / F. Fosberg 41414 / 1960 / "root fragrant with peppery fragrance"

167   *Kyllinga alba* Ghana / J. K. Morton A 3292 / 58 / "roots aromatic"

168   *K. erecta* subsp. *africana* Nigeria / J. Lowe 2069 / 70 / "Rhizome smells of thyme"

169   *K. squamulata* N. Rhodesia / E. Robinson 2148 / 57 / "Plant Citrus-scented when fresh"

170   **K. monocephala** China: Kwangsi / To & Ts'ang 12083, 12707 / 1924 / "Sheung hon ts'o, (Double cold-or possibly Catch cold-herb); Ch'un nai ts'o, (Spring milk herb); Chu sung ts'o, (Boehmeria lofty herb)"

171   **Mariscus jamaicensis** Dominican Republic / E. Valeur 631 / 1931 / " 'Caña Amarga' "

172   **M. soyauxii** Ghana / J. Morton A604 / 1954 / "Roots aromatic"

173   **Rhynchospora cyperoides** Dominican Republic / R. A. & E. S. Howard 9882 / 1946 / "called wild quinine locally and used for curing fever" Puerto Rico / W. Stimson 1685 / 1965 / "The seeds are boiled and used for toothache by the local people"

174   **R. caracasana** Brazil / D. B. Pickel 914 / 1925 / " 'Capim santo' "

175   **Scirpus corymbosus** S. Africa / A. Mogg 14843 / 1927 / "vlei poisoning"

176   **Scleria hirtella** Venezuela / J. Steyermark 55536 / 1944 / "Roots sweet" / " 'granca' "
Venezuela / J. Steyermark 52712 / 1945 / "Boil sweet fragrant rhizome in water and drink for colds; called 'malojillo' in Cumanacoa and sold for treating stomach aches, diarrhea, and vomiting; cook or crush up in crude condition" / " 'resfriade sabanero' "

177   **S. purdiei** Venezuela / J. Steyermark 58540 / 1944 / "Rhizome placed in luke warm water used in treating colds" / " 'atucamiri' "

178   **Scleria** indet. Venezuela / Steyermark & Nilsson 47 / 1960 / "If one has nose-bleed and headache, take angles of stem which cut and place up nose until bleeding stops (practised by Indians Taurepan-Aekuna)"

179   **Scleria** indet. Brazil / A. Silva 31 / 1944 / "The sharp edges of the leaves cut the skin of animals and man" / " 'tiririca' "

180   **S. glabra** N. Rhodesia / E. A. Robinson 1739 / 56 / "Plant strongly Citrus-scented"

181   **S. suaveolens** Rhodesia / E. Robinson 1458 / 56 / "Fresh plant smell when bruised of Citrus"

ARECACEAE

182   **Acrocomia** indet. Fr. Guiana / W. Broadway 289 / 1921 / "Fruit edible, sold in the market"

183   **Acrocomia** indet. Fr. Guiana / W. Broadway 934 / 1921 / "Fruits command ready sale in the markets. Seeds eaten raw"

184   **Arenga engleri** Hainan / S. Lau 1861 / 1933 / "the trunk can be molded into powder called Ye Fun which is edible"

185   **Astrocaryum** indet. Peru / Killip & Smith 28698 / 1929 / "Fruit edible"

186   *Bactris balanoidea* Costa Rica / Burger & Liesner 6681 / 1969 / "fruits with edible (sour) outer portion"

187   *Bactris* indet. Panama / S. Hayes 942 / 1860 / "Small Palm with edible fruit"

188   *B. maraja* Venezuela / J. Steyermark 87406 / 1960 / "fruit edible, black"

189   *Bactris* indet. Venezuela / Steyermark, Bunting & Blanco 101415 / 1968 / "Fruit edible" / " 'macanillo' "

190   *Bactris* indet. Venezuela / Aristeguieta & Lizot 7378 / 1970 / "Fruto comestible. Indios Guaicas (Yanomanö)"

191   *Balaka seemannii* Fiji Is. / O. Degener 15174 / 1941 / "young fruit edible" / " 'Mbalaka' (Serua)"

192   *Balaka* indet. Fiji Is. / O. Degener 14764 / 1941 / "fruit bright orange-red, succulent; kernel edible" / " 'Mbalaka' (Sabatu)"

193   *Brahea promineus* Mexico / H. Gentry 5897 / 1940 / "Palma de Coca"

194   *Caryota ochlandra* Hainan / Fung 20241 / 1932 / "cultivated . . . the tender hearts of the terminal buds eaten as a vegetable"

195   *C. cumingii* Philippine Is. / A. Elmer 9757 / 1908 / "millions of black ants make their homes amongst the bracts" / " 'Batican' "

196   *Catoblastus* indet. Ecuador / W. Camp E-1390 / 1944 / "The 'cabbage' small but tasty"

197   *Chamaedorea fusca* Guatemala / J. Steyermark 38123 / 1940 / "inside of leaf base edible in raw state"

198   *C. schiedeana* El Salvador / P. Standley 20165 / 1922 / "Pith eaten sweet" / " 'Cuiliote varón' "

199   *C. wendlandiana* El Salvador / P. Standley 19388 / 1922 / "Young spathes cooked and eaten" / " 'Paraya' "

200   *Chambeyronia* sp. aff. *C. microcarpa* New Hebrides / S. Kajewski 428 / 1928 / "The fruit of this species is eaten"

201   *Chelyocarpus chuco* Brazil / Prance, Rodrigues, Ramos & Farias 8717 / 1963 / " 'Carnaubinha' "

202   *Colpothrinax wrightii* Cuba / A. Curtiss 364 / 1904 / " . . . it is eaten green by birds, and also boys"

203   *Copernicia cerifera* Brazil / H. Cutler 12425 / no year / "To clean the blood, take as tea"

204   *Desmoncus prostratus* Brasil / C. Lindman 21256 / ca. 1900 / "pulpa eduli" / " 'Urumbámba' "

205   *Desmoncus* indet. Venezuela / Wurdack & Adderley 43486 / 1959 / "inner cortex sometimes used to make 'cebucanas' (casabe and manioco squeezers)" / " 'Voladora' "

206 *Geonoma oxycarpa* Haiti / E. L. Ekman 3781 / 1925 / " 'Coco macaque' "

207 *Goniocladus* indet. Fiji / O. Degener 14793 / 1941 / "fruit said to be greenish yellow when ripe; heart bud and kernel edible" / " 'Tanga' in Sabatu dialect"

208 *Guilielma gasipaes* Ecuador / E. Heinrichs 411 / 1933 / "Fruchtfleisch(rot) und junge Blätter essbar"

209 *Hyphaene ventricosa* S.W. Africa / R. Rodin 2617 / 1947 / "Alcoholic beverage made from sap" / "Ovambo names: 'Oshivale' (young tree) 'Omulunga' (large tree)"

210 *Jessenia weberbaueri* Peru / Killip & Smith 28845 / 1929 / "fruit used for oil" / " 'Hunguravi' "

211 *Livistona saribus* Malay Peninsula / M. Henderson 23665 / 1930 / "fruit . . . eaten by Sakai"

212 *Livistona* indet. Australia: Queensland / C. Trist 17 / 1931-32 / " 'Cabbage Palm' "

213 *Phoenix humilis* var. *pedunculata* S. India / E. Erlanson 5502 / 1934 / "eaten by bears"

214 *Phytelephas microcarpa* Peru / Killip & Smith 29333 / 1929 / "fruit produces a valuable oil" / " 'Yarina'; 'tagua' "

215 *Pinanga philippinensis* Philippine Is. / A. D. E. Elmer 8376 / 1907 / "The Igorotes cut the young inflorescence out of the leaf sheaths for food"

216 *Plectocomia* indet. Philippine Is. / Aguilar & Espiritu 31020 / 1928 / "young seed said to be poisonous" / "Marua"

217 *Socratea* indet. Ecuador / Grubb, Lloyd, Pennington & Whitmore 1667 / 1960 / "Young apices eaten raw or boiled in water" / "Pambil (Quechua)"

218 *Veitchia joannis* Fiji / O. Degener 15102 / 1941 / "young leaves & bud eaten raw like salad" / "Niusawa (Serua)"

219 *V. vitiensis* Fiji / A. Smith 8808 / 1953 / "heart edible as a salad" / " 'Sakiki' "

220 *Vitiphoenix smithii* Fiji / O. Degener 14765 / 1941 / "Fijians eat infl. cooked and kernal raw" / " 'Kaivatu' (Sabatu)"

221 *Arecacea* indet. New Guinea / Terr. Papua & New Guinea Dept. For. MDD-105 / 1968 / "A wild betel nut, eaten like a betel, one of many wild varieties" / "BONWALI - betel nut"

ARACEAE

22 *Alocasia macrorrhiza* Philippine Is. / A. Elmer 8818 / 1907 / "The roots are sometimes used to heal or sooth cuts or bruises" / " 'Absaba' "

223   *Anthurium crassinervium* Guatemala / P. Standley 24053 / 1922 / " 'Piña de montaña' "

224   *A. huamaliense* Ecuador / W. von Hagen 135 / 35 / "When young Jivaro eats tubers" / "Ippa"

225   *A. huegeli* Ecuador / J. Steyermark 53764 / 1943 / "Snakes are reputed to eat the fruit of this plant"

226   *A. nymphaeifolium* Venezuela / J. Steyermark 55806 / 1944 / "leaves used in preparing arepa (corn meal dough) — putting dough on leaf and moving from one leaf to another" / " 'corazon' "

227   *A. pentaphyllum* Venezuela / J. Steyermark 61035 / 1945 / "after the fruit matures birds and small mammals eat the fruit" / " 'mamurillo' "

228   *A. scolopendrinum* Venezuela / J. Steyermark 62812 / 1945 / "araguatas (monkeys) eat fruit" / " 'piña de mono' "

229   *Anthurium* indet. Ecuador / J. Steyermark 54794 / 1943 / "This is one of the sweetest-smelling plants in Ecuador" / " 'hoja de lucma' "

230   *Caladium esculenta* Ecuador / J. Steyermark 54893 / 1943 / "Cultivated for edible corm" / " 'papachina' "

231   *C. schomburgkii* Venezuela / J. Steyermark 76073 / 1953 / "Used as a good luck plant for fishing" / " 'murange' "

232   *Dieffenbachia paludicola* Brit. Guiana / S. S. & C. L. Tillett 45316 / 1960 / "old inflorescence full of maggots"

233   *Dracontium longipes* Brazil / Prance, Maas, Woolcott, Monteiro & Ramos 16400 / 1971 / "the root corm used as a cure for sting-ray bites by Deni Indians" / " 'Mapidzu' "

234   *Dracontium* indet. Peru / G. Klug 254 / 1929 / "For snake bites" / " 'Zergin' "

235   *Homalomena* indet. New Guinea: Papua / C. Carr 11660 / 35 / "The stem is cut up and used as ointment in conjunction with coconut oil" / "Native name IVA IVA"

236   *Montrichardia* indet. Panama / R. Williams 660 / 1908 / "seeds roasted & eaten"

237   *M. linifera* Venezuela / J. Steyermark 60938 / 1944 / "leaves used for applying on back pains, or side pains; fruit edible" / " 'rabano' "

238   *Philodendron anisotomum* El Salvador / P. C. Standley 19199 / 1922 / " 'Anona conte, Anona de conde' "

239   *P. scandens* Costa Rica / Burger & Liesner 6953 / 1969 / "surfaces of spadix whitish and smelling like a cheese spread"
Peru / Killip & Smith 25403 / 1929 / "Narcotic, used to produce sleep"

240   *P. scandens* subsp. *oxycardium* Mexico / Aguirre & Reko 334 / 48 / "fruit edible" / " 'chapiz' "

41 *P. lacerum* Cuba / Herb. Collegii Pharmaciae Neo-Eboracensis s.n. / 1840 / "Used in pains (as fomentation with oil) and externally in dropsy" / "magusey macho"

42 *P. acutatum* Surinam / Kramer & Kekking 2068 / 1960 / "Base of spadix emanating an anise-like odour" / " 'saparilla' "

43 *P. deflexum* Venezuela / J. Steyermark 74785 / 1953 / ". . . with ball of stinging ants at base"

44 *P. guttiferum* Venezuela / J. Steyermark 56383 / 1944 / "broken leaves or stem emitting odor of rose-geranium"

45 *P. krauseanum* Brit. Guiana / S. S. & C. L. Tillett 45539 / 1960 / "maggott infested"

46 *P. myrmecophilum* Venezuela / J. Steyermark 87400 / 1960 / "Spathe spicy-fragrant"

47 *P. pedatum* Brit. Guiana / For. Dept. of Brit. Guiana WB 140 / 48 / "Perennial herb used against snake bites"

48 *P. roraimae* Venezuela / J. Steyermark 59825 / 1944 / "inner part of shoot used by Indians to put on boils or parts infected by worms to kill the worms" / " 'cupita' "
Venezuela / J. Steyermark 75210 / 1953 / "cut petiole with fragrant spicy odor"

49 *Pothos ovatifolius* Philippine Is. / A. D. E. Elmer 9618 / 1908 / "leaf petioles . . . making excellent protective homes for black ants . . . spikes . . . with an unpleasant sour dough odor" / " 'Palaguiwon' "

50 *Schizocasia brancaefolia* New Guinea / P. van Royen 3481 / 1954 / "Wood white with itsching white milky juice"

51 *Syngonium vellozianum* Surinam / Jonker-Verhoef & Jonke 4782 / 1953 / "white, skin irritating latex"

52 *Taccarum caudatum* Boliva / H. Rusby s.n. / 1921 / "Arrow-poison"

53 *Xanthosoma helleborifolia* W.I.: Guadeloupe / P. Duss 3295 / 1893 / "Vivace por ses tubercules formes endedam il qui ne se mangent" / "Malanga poison Malanga cochon Malanga batau"

54 *X. jacquinii* W.I.: Guadeloupe / P. Duss 3296 / 1897 / "Toutes les parties de la plante, mais surtout en flours exhalent une odeur nausé abonde" / "Malanga brulant"
W.I.: Guadeloupe / P. Duss 3692 / 1897 / "les feuilles seules servent comme plante alimentaire" / "Colalon"

55 *X. brevispathaceum* Ecuador / W. von Hagen 258 / 35 / "CULTIVATED 'EAT' " / "Ambi-sangu-(Sanku) Jiv."

56 *Xanthosoma* indet. Brazil / W. Archer 8250 / 1942 / "Leaves boiled as food, rhizome also eaten" / " 'taióba' "

257   *Xanthosoma* indet. Hawaii / O. Degener 20102 / 1948 / "Very 'itchy'; fed
to pigs"

FLAGELLARIACEAE

258   *Flagellaria guineensis* Tanganyika / R. E. S. Tanner 3736 / 1957 / "strongly
aromatic" / "Sambaa: Mbegembege   Swaheli: Mhenzilani"

RESTIONACEAE

259   *Hypolaena fastigiata* Australia: New S. Wales / J. Boorman 202 / 1909 /
"mageren Boden bedeckende Pflanze, welche vom Vieh oft bis zu den
Wurzeln abgeweidet wird"

XYRIDACEAE

260   *Xyris elliottii* U.S.: Fla. / P. P. Sheehan s.n. / 1912 / "Medicinal plants of
the Seminole Indians. Used for colds and pulmonary disorders. The plant
is boiled, and the brew, while still warm, is rubbed on the patients breast" /
"Yellow-eyed grass"

ERIOCAULACEAE

261   *Paepalanthus bifidus* Brazil / A. Silva 210 / 1944 / " 'capim mortinha' "
262   *Syngonanthus glandulosus* Venezuela / J. Steyermark 58621 / 1944 / "Boil
entire plant for toothache" / " 'guanak' "

RAPATEACEAE

263   *Cephalostemon angustatus* Brazil / Philcox, Fereira & Bertoldo 3605 / 1967
/ "v.n. 'Capim-Cebola' "

BROMELIACEAE

264   *Aechmea huebneri* Brasil / Pires, Rodrigues & Irvine 50568 / 1961 / "Small
wasp inhabited plant"
265   *Ananas ananassoides* Venezuela / G. Agostini 223 / 1964 / " 'Piña' "
266   *Billbergia* sp. aff. *B. brachysiphon* Brazil / Prance, Maas, Kubizki et al.
12039 / 1971 / "Plant much visited by humming birds"
267   *Bromelia sylvestris* Mexico / H. Gentry 7072 / 1944 / "Ripe fr. yellow,
eaten raw or cooked by natives" / "Ahuama"

68   *B. wercklei* Honduras / S. Glassman 2068 / 48 / "Fruit used in soft beverages" / " 'pinuela' "
El Salvador / S. Calderon 1543 / 1923 / " 'Motate, Piña de cerca' "

69   *Bromelia* indet. Mexico / Rose, Standley & Russell 13482 / 1910 / "Bought in market"

70   *B. chrysantha* Venezuela / Aristeguieta & Zabala 7023 / 1969 / "Frutos comestibles. Las escamitas basales de las hojas se utilizan para desinfectar las heridas" / " 'Chigüichigüi' "
Colombia / H. Smith 2344 / "The pulp of the berry is eaten and has a pleasant flavor"

71   *Deinacanthon urbanianum* Argentina / Schreiter 50128 / 1922 / "flor hedionda olor a perro muerto"

72   *Dyckia brasiliana* Brasil / Irwin, Souza & Reis dos Santos 10268 / 1965 / "Rooted in termite nest"

73   *Greigia sodiroana* Ecuador / W. Camp E-4704 / 1945 / "Individual fruits quite sweet and Giler and Prieto say plant is more abundant in northern provinces and fruit is sold in Otovalo market under name of 'Piñuela' "

74   *Guzmania lingulata* W.I.: Guadeloupe / P. Duss 3403 / 1893 / "Ananas sauvage"

75   *Pitcairnia pulverulenta* Brazil / Maas, Kubitzki, Steward et al. P13140 / 1971 / "Visited by hummingbirds"

76   *P. pungens* Ecuador / W. Camp E-2172 / 1945 / "root is ground and infusion made which is said to be good for kidneys and liver" / " 'Quinde-Sungana' "
Ecuador / W. Camp E-1940 / 1945 / "Roots, ground and cooked, used as diuretic"

77   *Puya gummifera* Ecuador / W. Camp E-5198 / 1945 / "favorite food for bear; stems also fed to pigs and 'cuys' (guinea pigs)" / " 'Achupalla' "

78   *P. medica* Peru / E. Cerrate 1333 / 1952 / "El agua donde se ha hervido el tallo lo usan para curar casos de pneumonia" / " 'Achupalla-blanca' "

79   *P. raimondii* Peru / Iltis & Ugent 1288 / 1963 / "Birds nest in between leaves, apparently get caught by recurved marginal leaf spines and die there"

80   *Tillandsia usneoides* U.S.: La. / J. Riddell 54 / 1840 / "medicinal" / "Spanish moss"

81   *T. bulbosa* Panama / T. Croat 6098 / 1968 / "ant infested"

82   *T. fasciculata* Cuba / Herb. Collegii Pharmaciae Neo-Eboracensis s.n. / 1841 / "eaten by the cattle" / "curujey de lámpara"

83   *T. paraensis* Venezuela / Maguire & Wurdack 34906 / 1953 / "ant-infested"

COMMELINACEAE

284 *Campelia zanonia* Costa Rica / Ir. P. Grijpma 7 / 1969 / "stems fleshy, containing a brown, sticky, ill-scented juice"

285 *Commelina angustifolia* U.S.: Fla. / P. P. Sheehan s.n. / 1919 / "Medicinal Plants of the Seminole Indians. Used to soothe irritations. The mucilaginous sap from the inflorescence is applied to the feet to cure the so-called 'ground-itch' " / "Dew-flower"

286 *C. diffusa* Argentina / J. Montes 14649 / 1955 / "Uso. néctar de flor - medic. pop." / " 'Santa Lucia' "
Society Is. / E. Quayle 3 / 1921 / "Used for stock food"

287 *C. erecta* Argentina / J. Montes 27542 / 58 / "El néctar de las flores es usada para curar afecciones de la vista" / " 'Santa Lucia' "

288 *C. bracteosa* Tanganyika / R. E. S. Tanner 3162 / 1956 / "Sambaa: For sore eyes, sap squeezed into eyes. Used in magic medicines for stopping bad luck" / "Sambaa: Nkongo"

289 *C. nudiflora* China: Kwangtung / W. T. Tsang 21510 / 1932 / ". . . edible for pigs" / "Sai Yeung Chuk Ko Tso"

290 *Tradescantia pulchella* Mexico / G. Hinton 1745 / 32 / "for bruises" / "Yerba del pollo"

291 *Commelinacea* indet. Colombia / Killip & Smith 19669 / 1927 / "Remedy for fevers, etc." / " 'Calaguala' "

PONTEDERIACEAE

292 *Monochoria vaginalis* Hainan / S. Lau 557 / 1932 / "Edible" / "Tsin to kwoh"
Ceylon / F. Fosberg s.n. / 1969 / " 'plant eaten, flowers considered especially good' "

293 *M. vaginalis* var. *plantaginea* Hainan / Lei 409 / 1933 / "Soft leaves are edible" / "Aap Sit Tso"

294 *Pontederia rotundifolia* Colombia / Hashimoto, Ishikawa & Garcia-Barriga 18685 / 1965 / "Se usa para quitar parálisis fasciales" / "N.v. 'Amarón borrachero' "

JUNCACEAE

295 *Luzula multiflora* var. *congesta* Belgium / E. Marchal 10 / 1869 / "Bonne plante fourragére, aimée des bestiaux" / "Luzule multiflore, var. agglomérée"

LILIACEAE

96   *Albuca* indet. S.W. Africa / R. Seydel 245 / 1954 / "Von Vieh und Wild gern und ohne Schaden gefressen"

97   *Allium drummondii* U.S.: Tex. / C. L. & A. A. Lundell 8342 / 1940 / " 'Wild onion' "

98   *A. textile* U.S.: Utah / K. Diem 9 / 1956 / "Prairie Onion"
U.S.: Minn. / L. Moyer s.n. / 12 / "Pomme de Terre"

99   *A. validum* U.S.: Nev. / W. A. Archer 6605 / 1938 / "Strong onion odor"

00   *A. glandulosum* Mexico / G. Hinton 4487 / 33 / "edible" / "Ajo Cimarón"

01   *A. arenarium* Germany / Herb. of Columbia Univ. s.n. / 1836 / Kuraz-wiebel"

02   *A. bakeri* China: Kweichow / Steward, Chiao & Cheo 324 / 1931 / "Uses: Vegetable" / "Chiu Tsai"

03   *Allium* indet. China: Kwansi / W. Tsang 24563 / 1934 / "fr. edible" / "Kau Tsoi"

04   *A. rubellum* India W. Koelz 1874 / 1931 / "with onion flavor"

05   *Aloë kirkii* Tanganyika / R. E. S. Tanner 3024 / 1956 / "Zigua: for toothache, sap mixed with tobacco and put on tooth" / "Zigua: Kikoli"

06   *A. secundiflora* Tanganyika / R. E. S. Tanner 3796 / 1957 / "Masai—roots for making beer . . . unpleasantly aromatic" / "Masai—Endadaiyoku"

07   *Asparagus* indet. S.W. Africa / R. Seydel 180 / 54 / "Nach Dinter sollen die jungen Triebe einzelner Arten essbar sein" / "Wilder Spargel"

08   *Asparagus* indet. N. Rhodesia / W. Bainbridge 594 / 61 / "on a small ant-hill"

09   *Asphodelus fistulosus* S. Australia / J. Anway 353 / 1965 / " 'Wild Onion' "

10   *Bulbine stenophylla* S.W. Africa / R. Seydel 4213 / 65 / "unangenehmer Geruch"

11   *Bulbinella hookeri* New Zealand / E. E. Collin 128 / 1934–35 / " 'Maori onion' "

12   *Fritillaria recurva* U.S.: Calif. / Frémont 346 / 1845 / " 'The rice-root plant' "

13   *F. przewalski* China: Kansi / J. Rock 12461 / 1925 / "used as medicine" / "Penue"

14   *Hemerocallis flava* Burma / F. Kingdon Ward 17543 / 1948 / "Said to be a wild plant, but is usually not far from villages, and the leaves are eaten locally"

15   *Ophiopogon microcarpa* Tanganyika / R. E. S. Tanner 3394 / 1956 / "Bondei: For stomach ache in children, roots boiled and the water drunk"

16   *Ornithogalum* indet. Brit. Guiana / J. Olgivie s.n. / 1913 / "Fish-poison plant of the Wapisiana tribe. Seeds only used" / " 'purowan' "

317   *O. calcicolum* S. W. Africa / R. Seydel 3264 / 63 / "giftig u. gefaehrlich, besonders fuer Schafe" / "gelbbluehender 'Kap-slangkop' "

318   *O. thyrsoides* S. Africa / R. Baylis 4016 / 1967 / "Poisonous to stock"

319   *Reineckea carnea* China: Kweichow / Y. Tsiang 7480 / 1930 / "Purchased from market. A medical plant"

320   *Schoenocaulon officinale* Mexico / G. Hinton 13226 / 38 / "Extremely unpleasant odor"
       Venezuela / A. Lawrence 888 / 1935 / " 'Said to have been shipped to Europe in large quantities during world war, for manufacturing poison gas from dried seeds' "

321   *Smilacina liliacea* U.S.: Mont. / Dr. V. Havard s.n. / 89 / "Fruit bright red, edible"

322   *Streptopus amplexifolius* Alaska / Mr. & Mrs. S. Oliver 57 / 1946 / "Locally called 'watermelon' "

323   *Toxicoscordion paniculatum* U.S.: Nev. / J. Henrichs 40 / 1937 / "Fls. . . . bitter odor"

324   *T. venenosus* U.S.: Ore. / A. Wood 851 / 1866 / "Poison Camass"

325   *Trillium albidum* U.S.: Calif. / C. Wolf 9149 / 1937 / "Large fleshy fruits are eaten by animals"

326   *T. luteum* U.S.: Ky. / J. & R. Freeman 472 / 1966 / "Plants strongly lemon-scented"

327   *T. maculatum* U.S.: S. C. / E. Marquand s.n. / 1931 / "Birth-root"

328   *T. recurvatum* U.S.: Ill. / Princeton Univ. Herb. s.n. / 85 / "Three leaved Nightshade"

329   *Tulbaghia violacea* U.S.: Fla. / Ray & Stoner 10996 / 1961 / "with strong odor of garlic / "society garlic"

330   *Urginea altissima* Tanganyika / R. E. S. Tanner 1086 / 1952 / "Sukuma: for men and cattle who have swellings or boils and for cattle that refuse to suckle calves; roots pounded up green and put on swellings or udders as a paste. Very astringent" / "Kisukuma: Kipayapayanda"

331   *U. sanguinea* S. W. Africa / R. Seydel 4084 / 64 / "Sehr giftig" / "echter slangkop"

332   *Liliacea* indet. Bahama Is. / W. Corbin C 124 / 1973 / "Called 'Poison Lily' . . . flowers . . . have an odor as strong as jasmine"

SMILACACEAE

333   *Smilax domingensis* Mexico / Hernandez, Ramos & Cedillo 491 / 1968 / " 'popo medicinal' "
       Brit. Honduras / P. Gentle 6626 / 1948 / " 'Wild Sarsaparilla' "

34   *S. beyrichii* Bahama Is. / Small & Carter 8897 / 1910 / "berries black and edible"

35   *S. poeppigii* Venezuela / J. Steyermark 61240 / 1945 / "Cook root for fevers; stem used for making guarares" / " 'bejuco de corona' "

36   *S. schomburgkiana* Brit. Guiana / A. Persaud 175 / 1924 / "Wild Yam"

37   *Smilax* indet. Brit. Guiana / For. Dept. Brit. Guiana F 3452 / 52 / "tubers have strong tonic properties" / "Cockshun"

38   *Smilax* indet. Brazil / M. da Silva 98 / 1942 / "Tea made from fleshy roots taken for rheumatism" / " 'japecanga' "

39   *Smilax* indet. Brazil / R. Fróes 1800 / 1932 / "Roots used as treatment for syphilis" / " 'Japecauga' "

40   *S. ferox* Nepal / Kanai, Hara & Ohba 722054 / 1972 / "Young fruits . . . sour and astringent"

41   *S. laevis* China: Kwangtung / W. Tsang 20766 / 1932 / "fruit, red, edible" / "Chim Mei Ip Ma"
Thailand / Singhasthit & Smitinand 92 / 1948 / "Flowers edible"

42   *S. riparia* China: Kwangtung / W. Tsang 21338 / 1932 / "fruit . . . edible" / "Ma Mei Kit T'ang"

43   *S. thomsoniana* Hong Kong / Y. Taam 2098 / 1941 / "fr. red, edible"

44   *Smilax* indet. China: Kwangsi / W. Tsang 32830 / 1934 / "fr. red; edible" / "Yuen Ip Ma Kap Tang"

45   *Smilax* indet. China: Kwangsi / W. Tsang 24019 / 1934 / "fr. red, edible" / "Chim Ma Kap Tang"

46   *Smilax* indet. China: Kwangsi / W. Tsang 34176 / 1934 / "fr. yellow, edible" / "Tai Ma Kap Tang"

47   *S. leucophylla* Philippine Is. / A. D. E. Elmer 12613 / 1911 / "the natives eat the green fruits and are said to be delicious when mature" / " 'Banang' "

AMARYLLIDACEAE

48   *Curculigo recurvata* Philippine Is. / A. D. E. Elmer 8862 / 1907 / "The leaves are frequently used by the Igorotes as cups for drinking water; the two ends are stoved in and held in the boat shaped receptacle" / " 'Tokan' "

49   *Hymenocallis* indet. Brasil / O. White s.n. / 1921 / "Bulbs used in medicine"

50   *Hypoxis decumbens* Argentina / J. Montes 14755 / 1955 / " 'Cebollin' "

51   *Stenomesson aurantiacum* Ecuador / J. Steyermark 53671 / 1943 / " 'cebolleta' "

AGAVACEAE

352 **Beaucarnea stricta** Mexico / K. & E. Roe 1925 / 1965 / "Alcohol"

353 **Cordyline fruticosa** Caroline Is. / F. Fosberg 32088 / 1950 / "leaves wrapped around cassava for food / " 'Nsis' "
New Britain / J. Waterhouse 27026 / 1934 / "Leaves used for covering food. Planted on graves, etc." / "Taqete, Tavea, Luqa"
Samoa / D. Garber 554 / 1921 / "Berries chewed; ls. formerly for lava lavas" / " 'ti' "

354 **Dasylirion serratifolium** Mexico / Y. Mexia 2626 / 1929 / "Alcoholic distilled drink can be made from it" / " 'Sotól' "

355 **Yucca constricta** U.S.: N.M. / Dr. Gregg s.n. / 47 / "root used for amole (a substitute for soap)" / "Palmilla"

VELLOZIACEAE

356 **Vellozia alexandrinae** Venezuela / T. Lasser 1774 / 1946 / "viscosas y aromáticas"
Venezuela / J. Steyermark 60326 / 1944 / " 'comida de venado' "

357 **Xerophyta pauciramosa** Tanganyika / Michelmore 1152 / 1935 / "Stems used for tooth brushes and for brushing dirt out of grain or flour" / "Chifuti (Chifipa), Chipiagiyu (Kiswahili)"

358 **X. simulans** Uganda / A. Thomas 4046 / 1941 / "almond scented"

TACCACEAE

359 **Tacca leontopetaloides** Caroline Is. / M. Evans 499 / 1965 / "Root eaten" / " 'MOGMOG' "
Mariana Is. / M. Evans 713 / 1965 / "Dried and ground for starch. Crushed and mixed with sugar and water for heart trouble" / " 'GABGAB' "
Fiji / A. Smith 6742 / 1947 / "edible" / " 'Yambia' "
Caroline Is. / D. Anderson 1026 / 1949 / "Used as a starch" / " 'Mokmok' or 'moka mok' "
Caroline Is. / F. Fosberg 32128 / 1950 / "tubers eaten" / " 'Chubochub' "
Caroline Is. / F. Fosberg 25550 / 1946 / "tuber made into starch called 'melkin' (or American) flour, used for making bread. This learned from Chamorros; plant ignored previously" / " 'chobchob' "
Caroline Is. / Fosberg & Wong 25445 / 1946 / "some say tuber is eaten, others say not" / " 'mogmog' "
Marshall Is. / F. Fosberg 27000 / 1946 / "tubers grated to extract starch which is much used for food. Harvested about October" / " 'mokmok' "
Marshall Is. / F. Fosberg 34398 / 1952 / "tubers eaten, dug in December and January when plant commonly dies" / " 'mokemok' "

Marshall Is. / F. Fosberg 33938 / 1951 / "starch from tubers eaten very commonly, mixed with pandanus pulp, or with coconut to make puddings" / " 'mokemok' "

DIOSCOREACEAE

60 *Dioscorea composita* Mexico / Y. Mexia 9226 / 1938 / "eaten by toucans"

61 *D. pilosiuscula* W.I.: Guadeloupe / P. Duss 3809 / 1896 / "racine edible" / " Ignam a eau"

62 *D. stegelmanniana* Brazil / R. Fróes 1860 / 1932 / "Reported as a poison" / " 'Cara brava' "

63 *Dioscorea* indet. Colombia / Schultes & Cabrera 16287 / 1952 / "Tanimuka = *nee-koo-ree-ká* ('peach-palm vine'); Makuna = *ho-tá-mee-see-ma;* Yukuna = *wan-hña-goó* ('wild yam'); Mirana = *kö-ně-ó-pě* ('wild paint plant')"

64 *Dioscorea* indet. Argentina / J. Montes 14809 / 1955 / "medicinal" / "N.V. 'Isipho milhombres' "

65 *Dioscorea* indet. Venezuela / Aristeguieta & Lizot 7383 / 1970 / "Raíz larga y comestible, pero es amarga; la comen sólo en casos de mucha escasez. Indios Guaicas (Yanomanö)"

66 *D. linearicordata* China: Kwangtung / W. T. Tsang 20815 / 1932 / "tubers edible" / "Tsok Ko Shi"

67 *D. persimilis* China: Kwangtung / Lingnan (To & Ts'ang) 12798 / 1924 / "Chuk ko shu, (Bamboo pole potato)"
Hainan / McClure & Tsang 839 / 1929 / "Tubers eaten by Lois (fide Wai Tak)"

68 *Dioscorea* indet. New Guinea: Papua / C. Carr 13875 / 35 / "The tuber is edible" / "Native name TAITUKAVA"

69 *D. cumingii* Philippine Is. / A. Elmer 10395 / 1908 / "With a distinct sweet oily odor"
Philippine Is. / A. Elmer 8692 / 1907 / "parts of the large roots are cooked for food / " 'Kasi' "

70 *D. hispida* Philippine Is. / A. Elmer 14307 / 1912 / "The tubers are good food" / " 'Bagay' in Manobo"

IRIDACEAE

71 *Eleutherine plicata* W.I.: Martinique / P. Duss 1950 / 1882 / "tubercules employés contre les rhumatismes" / "Chalotte savanne"

72 *E. bulbosa* Peru / J. Schunke 971 / 1966 / "Antifertility. The Shipibas indians drink it as an infusion blended with Dichromena ciliata (Cyperaceae)" / " 'Yahuar Piripiri' "

Brit. Guiana / Altson 518 / 1926 / "Bulb used as a hunting 'bina' (charm) by Patamonas, crushed and rubbed into cuts on legs" / " 'Mari-va' (Patamona dialect)"

373   *Homeria* indet. S.W. Africa / R. Seydel 3640 / 1963 / "Vielleicht giftig"

374   *Iris hexagona* U.S.: Fla. / P. Sheehan s.n. / 1919 / "Medicinal plant of the Seminole Indians. Used as a remedy for shock following alligator-bite" / "Blue-iris"

375   *I. dichotoma* China: Kwangtung / W. Tsang 20612 / 1932 / "for medicine" / "Yea Tang Tsat"

376   *Lapeyrousia bainesii* S.W. Africa / R. Seydel 419 / 63 / ". . . Wurzelknolle, essbar"

377   *Romulea rosea* W. Australia / J. Anway 240 / 1965 / " 'Onion Grass' "

378   *Sisyrinchium scabrum* Mexico / J. Beaman 2692 / 1959 / "Poisonous to burrows"

379   *Trimezia* indet. Venezuela / Trujillo 5009 / 1960 / "N.V. 'Cebolleta' "

380   *Tritonia crocosmifolia* W.I.: Guadeloupe / P. Duss 4211 / 1905 / "il pousse à poison"

381   *Iridacea* indet. Puerto Rico / J. Stevenson 2152 / 1914 / " 'Cebollito' "

382   *Iridacea* indet. Brazil / G. & L. T. Eiten 9797 / 1969 / "n.v. 'batata de perdiz'. (The perdiz bird digs up & eats corm) Also 'batata de tatu' (Tatu is armadillo)"

HELICONIACEAE

383   *Heliconia collinsiana* var. *collinsiana* El Salvador / P. Standley 19330 / 1921 / " 'Platanillo' "

384   *H. hirsuta* var. *villosa* Venezuela / J. Steyermark 60916 / "Rhizomes eaten and cultivated / " 'puwa' "

385   *Heliconia* indet. Venezuela / J. Steyermark 61303 / 1945 / "The mashed up root cooked in water and then put in aceite de castilla put on snake bite of the bushmaster (guaima) and also drink the same 3 times daily, cures the bite of this samke; my workman Celestino Antuare was bitten by a guaima and cured himself with this and with another remedy called 'reina toro'; wasps were seen more frequently resting on the upper shoot of this plant than on any other plant" / " 'bijan' "

ZINGIBERACEAE

386   *Aframomum* indet. Nyasaland / L. Brass 17777 / 1946 / "fruit red, palatable, eaten by the natives"

87 *Alpinia galanga* China: Kwangsi / W. Tsang 24124 / 1934 / "fr. red, edible" / "Shan Keung Tze"

88 *Alpinia* indet. China: Kwangtung / F. McClure 461328 / 1921 / "Leaves and stem boiled to make tonic tocrsash for sores and boils / "Wong Keung, Mo T'o"

89 *A. oxyphylla* Hainan / H. Fung 20131 / 1932 / "a highly esteemed ingredient in many Chinese drugs and medicinal teas" / "Yiu Tsz Cha"
Hainan / C. Lei 643 / 1933 / "The hard taste in the fruit may be removed by treating with salt. The soup made from boiling the fruits may help digestion" / "Walk Di"

90 *A. fraseriana* Brit. N. Borneo / B. Evangelista 943 / 1929 / "edible" / "Talidus"

91 *Amomum* indet. Papua / L. Brass 3604 / 1933 / "Pulp edible & very palatable; something like a sour apple apple in flavour" / "GORAKH"

92 *Curcuma* indet. India / W. Koelz 3224 / 1931 / "yellow root used to color food" / " 'Haldi' "

93 *Hedychium coronarium* Costa Rica / Holm & Iltis 45 / 1949 / "Very spicy-fragrant"

94 *Hornstedtia* indet. Borneo / J. & M. S. Clemens 22163 / 1929 / "Via Dyak who use as substitute for vinegar"

95 *Kolowratia elegans* Philippine Is. / A. D. E. Elmer 7866 / 1907 / "seeds . . . eaten by natives"
Philippine Is. / Ramos & Convocar 83372 / 1931 / "The fruit of this plant is edible"

96 *Renealmia aromatica* Panama / T. Croat 15128 / 1971 / "cut parts with sweet aroma"

97 *R. occidentalis* var. *occidentalis* Brit. Honduras / P. Gentle 4184 / 1942 / " 'wild ginger' "

98 *R. alpinia* W.I.: Montserrat / J. Schafer 594 / 1907 / " 'wild ginger' "

99 *Zingiber officinale* Brit. Guiana / A. C. Smith 2835 / 1937 / "entire plant boiled and used externally as bath to reduce fever by Waiwais"

00 *Zingiber* indet. China: Kweichow / Steward, Chiao & Cheo 611 / 1931 / "Fl. scales edible" / "Yang Ho"

01 *Zingiberacea* indet. Nicaragua / Proctor, Jones & Facey 27052 / 1966 / "Root used for fever and snakebite"

COSTACEAE

02 *Costus scaber* W.I.: Grenada / R. Howard 10555 / 1950 / "juice used for sore eyes"

CANNACEAE

403   *Canna edulis* Panama / J. Duke 13628 / 1967 / "Fruits used for treating the 'obradera' of children" / " 'achinachinata' Choco; 'bandera' Spanish"

MARANTACEAE

404   *Calathea* indet. Surinam / G. Stahel 38 / 1939 / "Main carbohydrate food of the wild Wawa-Indians (without agriculture!) on the upper Oelemani River"

405   *Maranta* indet. Venezuela / Aristeguieta & Lizot 7356 / 1970 / "Utilizan la raiz para embrujar y matar mujeres. Indios Guaicas (Yanomanö)"

406   *Thalia geniculata* Dominican Republic / Mareano & Jiménez 4957 / 1965 / "N.V. = PLATANILLO"

ORCHIDACEAE

407   *Aceras anthropophora* Algeria / Alston & Simpson 37633 / 1937 / "Leaves smell musky"

408   *Catasetum barbatum* Brit. Guiana / A. Pinkus 208 / 1939 / "flowers with peppermint odor"

409   *Caularthron bilamellatum* Panama / T. Croat 7762 / 1969 / "pseudobulbs with ants"

410   *Dendrobium longicornu* China: Kwangsi / W. T. Tsang 24348 / 1934 / "Used for medicine" / "Sai Shek Kam Chai Ts'o or Wong Ts'o"

411   *Disa zombica* Nyasaland / L. Brass 16114 / 1946 / "tubers eaten by the natives"

412   *Epidendron erubescens* Mexico / W. Camp 2611 / 1937 / ". . . plant seems to have a special religious significance for the Zapotecans"

413   *E. imatophyllum* Panama / T. Croat 7775 / 1969 / "ant infested"
Brazil / Argent & Richards 6641 / 68 / "Roots growing amongst ants nest"

414   *E. bahamensis* Bahama Is. / R. A. & E. S. Howard 10207 / 1948 / " 'Wild onion' "

415   *E. brachyphyllum* Ecuador / W. H. Camp E-2027 / 45 / "Infusion of fls. used as diuretic" / " 'Flor de Cristo' "

416   *E. ibaguense* var. *schomburgkii* Surinam / B. Maguire 24695 / 1944 / "in ant nest"

417   *E. nocturnum* Brit. Guiana / S. S. & C. L. Tillett 45337 / 1960 / "strong, sweet, heavy fragrance"

418   *Gymnadenia orchidis* China: Kansu / J. Rock 13120 / 1925 / "odor of vanilla"

19 *Habenaria straminea* Canada: Newfoundland / Fernald, Long & Fogg 1543 / 1929 / "Vanilla-scented"

20 *Malaxis parthonii* var. *denticulata* Panama / R. Seibert 170 / 1935 / "Faint odor of tobacco"

21 *Nervilia aragoana* Caroline Is. / R. Kanehira 1127 / 1930 / "used for medicine"

22 *Pleïone bulbocoides* China: Yunnan / J. Rock 24700 / 1932 / "roots edible"

23 *Spiranthes cernua* Canada: Nova Scotia / Fernald, Long & Linder 20918 / 1920 / "odor disgustingly pungent"

SAURURACEAE

24 *Anemopsis californica* Mexico / C. V. Hartman 5 / 1894 / "Roots used for fever" / "Herba de la mansa"

PIPERACEAE

25 *Peperomia maculosa* Dominican Republic / Gastony, Jones & Norris 170 / 1967 / "Juice strongly and pungently aromatic"

26 *P. pellucida* Dominican Republic / J. Jiménez 5845 / 1970 / "Found mostly in private gardens for medical purposes" / "Siempre fresca"

27 *P. rotundifolia* Cuba / Bro. Leon 12666 / 1925 / "the leaves crushed smells anise" / " 'anis de sierra' "
Surinam / Lanjouw & Lindeman 1880 / 1949 / "pikien fauroe-sopo (Sur. = little bird's soap)"

28 *P. acuminata* Venezuela / J. Steyermark 55246 / 1944 / "Leaves used as seasoning, either dried or boiled" / " 'culantro de montana' "

29 *P. macrostachya* Surinam / J. Lindeman 4783 / 1953 / "growing on ants' nest"

30 *P. peltigera* Ecuador / W. Camp E-5025 / 1945 / "expressed sap of lvs used to alleviate fevers, inflammations, and bad coughs. In asking about the derivation of the name, Prieto smiled and said: 'Well, the leaf is somewhat in the form of a burro's foot-print, but when anybody has a very bad cough—like 'tos ferina' (whooping cough) we say that they 'sound like a burro' and so the plant good for such coughs we call 'Pata con panda' because 'panda' is the Quiche name for a burro' "

31 *P. tafelbergensis* Venezuela / T. Lasser 1407 / 1946 / "Usada contra catarros crónicos"

32 *Peperomia* indet. Ecuador / C. Fuller 123 / 53 / "For heart pain, leaves and stems mashed, juice extracted cold. Dose: 1 Oz." / "Churu quihua; Quichua"

433   *P. guamana* Mariana Is. / F. Fosberg 35516 / 1954 / "medicinal" / " 'pot puput' "

434   *P. ponapensis* Marshall Is. / F. Fosberg 26734 / 1946 / "used medicinally" / " 'rebij rege' "

435   *Piper auritum* Nicaragua / Proctor, Jones & Facey 27282 / 1966 / "Used for snake bites" / " 'Santa Maria' "

436   *P. crassinervium* Costa Rica / W. Burger 4160 / 1967 / "with celery-like odor when crushed"

437   *P. patulum* El Salvador / N. Fassett 28922 / 1951 / "Said to repel wood-ticks" El Salvador / P. Standley 22046 / 1922 / " 'Chile' "

438   *P. tricuspe* Panama / Kirkbride & Duke 1125 / 1968 / "Used by the Choco for headaches & pains in the ribs. Cultivated"

439   *P. urophyllum* Costa Rica / Burger & Guillermo Matta 4241 / 1967 / "Spicy aromatic when crushed"
Peru / J. Schunke V. 1176 / 1966 / "Las hojas usan en infusión para provocar insomnia" / " 'Huayusu hembra' "

440   *P. villiramulum* Panama / R. Seibert 549 / 1935 / "Real pepper aroma"

441   *P. aduncum* Puerto Rico / W. Stimson 1615 / 1965 / "According to a native the fruit is eaten when black and ripe and is very sweet" / " 'higuillo' "
Peru / P. Hutchison 1449 / 1957 / "Leaves used to brew a tea to treat coughs"

442   *P. amalgo* Puerto Rico / W. Stimson 3355 / 1966 / "According to native this plant is medicinal and is good 'para baños y para guarapos.' Also, when one is cut, one chews the leaves of this plant and then puts the resulting chewed mass on the cut. It stops the bleeding" / " 'higuillo oloroso' "

443   *P. dilatum* W.I.: St. Vincent / H. H. & G. W. Smith 125 / 1890 / "All the Artanthes are called 'Roman Candle' or 'Mastumec' (Mal d'estomac corrupted) - & the fruit is occasionally used as a substitute for Pepper"

444   *P. ossanum* Cuba / Fre. Leon 11727 / 1924 / " 'Canilla de muerto' "

445   *P. tuberculatum* W.I.: Tobago / L. Andrews 3-42 / 1963 / "Soap bush"
Colombia / H. H. Smith 383 / 1898–99 / "The washerwomen sometimes use the leaves of this plant to scent clothes. The growing plant has a strong odor of pepper"
Venezuela / W. Archer 3118a / 1935 / "Fish poison"

446   *P. andreanum* Ecuador / W. Camp E-1982 / 1945 / "fruiting spike ('mazorca') rubbed onto teeth as dentifrice" / " 'Rabo de Raton' "

447   *P. arboreum* var. *hirtellum* Brazil / Ratter & Ramos / 514 / 1967 / "Pimenta longa, Pimenta de Macaco"

48 *P. callosum* Peru / J. Schunke V. 4047 / 1970 / "Los nativos utilizan en infusión para provocar insomnio y en maceración para clamar los dolores reumáticos" / " 'Huayusa' "
Ecuador / Cazalet & Pennington 7613 / 62 / "Lvs . . . , smelling slightly of mint"
Peru / Killip & Smith 23617 / 1929 / " 'Anis' or 'anisillo' "
Peru / Killip & Smith 25399 / 1929 / "leaves boiled to make a tea"

49 *P. lenticellosum* Ecuador / J. Steyermark 52913 / 1943 / "Leaves used as spice; roots used for menstruation purposes" / " 'huabiduca' "
Ecuador / W. Camp 1740 / 45 / "Plant-parts spicy-aromatic, with something of odor of Benzoin"

50 *P. marginatum* Surinam / Lanjouw & Lindeman 1165 / 1948 / "with strong anise-scent" / "anijsie-wiwirie (Sur.) (anise weed)"

51 *P. oblongatum* Peru / Y. Mexia 8173 / 1936 / " 'Matico' "

52 *Piper* indet. Brazil / Prance, Maas et al. 13940 / 1971 / "Leaves used as a cure for coughs & sore throats. Leaves are made into a tea which is drunk"

53 *Piper* indet. Brazil / Prance, Maas et al. 13932 / 1971 / "Roots used as cure for tooth-ache. The roots are chewed and apparently act as a mild anaesthetic. They produce a tingling sensation in tongue and mouth" / " 'Washi' "

54 *Piper* indet. Venezuela / J. Steyermark 60986 / 1945 / "The root crushed for killing fish as a barbasco and also for toothache; the roots when bitten and chewed cause the tongue and lips to sense a paralyzed biting feeling; . . . leaves also have biting effect" / " 'raiz de candela' or 'raiz de muela' "

55 *P. guineense* Nigeria / P. Richards 3041 / 1935 / "used by natives in childbirth and for stomach-ache" / "Vern. name IYERI"

56 *P. hancei* China: Kwangtung / Lingnan (To & Ts'ang) 12184, 12557 & (McClure) 13738 / 1924 / "Ka ku (False betel pepper); P'o she (Embracing snake); Ka lau (False betel pepper)"

57 *P. laetispicum* Hainan / F. McClure 20122 / 1932 / "Leaves and fruits aromatic. The Lois say the leaves are used locally to chew with betel nut"

58 *P. sarmentosum* Hainan / C. Lei 112 / 1932 / "The juice made by putting the root of Pak Ma Pou Gun in salt may heal sore throat" / "Wa Lou"

59 *P. arborescens* Philippine Is. / P. Esguerra 31365 / 1931 / "used for chewing"
Philippine Is. / A. Elmer 9463 / 1908 / "with a biting pepery flavor" / " 'Malabouyoc' "

60 *P. erectum* Melanesia: Bougainville / J. Waterhouse 62 / 33 / "Used in native medicine" / "Pumpupuri"

61 *P. flavifructum* New Guinea: Papua / L. Brass 3926 / 1933 / "pulp sweet; seed pungent"

462　　**P. korthalsii** Philippine Is. / A. Elmer 8760 / 1907 / "the leaves are cut up and placed in water with rice—causing it to ferment or turn sour. This beer or wine is made in quantities and is called 'Tape' " / " 'Dawodniaso' " Philippine Is. / A. Elmer 8550 / 1907 / "leaves put in solution for making rice beer = 'Tapai' " / " 'Dawod' "

463　　**P. umbellatum** var. **subpeltatum** Philippine Is. / A. D. E. Elmer 9669 / 1908 / "with an unsavory green peppery taste" / " 'Bayag-bayag' "

464　　**Piper** indet. New Guinea: Papua / C. Carr 11639 / 35 / "Bark, leaves and fruit eaten by the natives" / "Native names are the vine POPO, the leaf RAUPA and the fruit VAGA"

465　　**Pothomorphe peltata** Brit. Guiana / W. A. Archer 2528a / 1934 / "Fish poison. Cultivated by the Warrau Indians. When used as a fish poison the leaves are pounded with the leaves of Clidabium, and the water is made frothy. The leaves are also used as poultice for bruises and swellings. Oil applied first" / " 'Duburi banato' (Arawak) means Sting ray leaf. 'Cow hoof' "
　　　　Peru / J. Schunke V. 1452 / 1967 / "Hierba medicinal para infecciones de heridos / " 'Santa Maria Sacha' "

466　　**P. umbellata** Venezuela / J. Steyermark 61261 / 1945 / "Fresh leaf placed on head good for headache" / " 'santa maria' "

467　　**Sarcorhachis sydowii** Ecuador / Cazalet & Pennington 5098 / 61 / "Fls. smelling of lemon"

468　　**Piperacea** indet. Argentina / J. Montes 127 / 44 / "es medicinal" / " 'Pariparota chica' 'P. macho' "

LACISTEMATACEAE

469　　**Lacistema aggregatum** Brit. Honduras / P. Gentle 3233 / 1940 / " 'Bastard coffee' 'Mountain water wood' "

CHLORANTHACEAE

470　　**Chloranthus glabra** China: Kwangtung / W. T. Tsang 20963 / 1932 / "medicine" / "Tsau Long Ch'a"
　　　　China: Kwangtung / W. T. Tsang 20052 / 1932 / "Medicinal value" / "Chow Long Tzar"

471　　**C. officinalis** Philippine Is. / A. Zwickey 161 / 1938 / "fls. used in making perfume" / " 'Makadadag' "

472　　**Hedyosmum glabratum** Venezuela / J. Steyermark 55435 / 1944 / "leaves coriaceous, used in making chimin (tobacco)" / " 'caribita' "

473　　**H. racemosum** Ecuador / W. Steere 8005 / 1943 / "much used in medicine because of very aromatic leaves" / " 'Grancillo' or 'Guayusa' "

474 *Hedyosmum* indet. Peru / R. Scolnik 1441 / 1948 / "Para el estómago y los riñones" / "N.V. supinune"

475 *Hedyosmum* indet. Peru / R. Scolnik 1088 / 1948 / "antireumático: en potaciones" / "N.V. aitacopa"

476 *Hedyosmum* indet. Bolivia / R. F. Steinbach 591 / 1966 / "Flores y hojas olor penetrante a perfume, gusto a canela. Infusión usada como bebida estimulante"

SALICACEAE

477 *Populus balsamifera* U.S.: Ore. / M. W. Gorman 977 / 1899 / " . . . & is used as a med in coughs & colds" / "Tree = Thoh. Gum on buds = Dä-la"

MYRICACEAE

478 *Myrica pensylvanica* Canada: Prince Edward I. / Erskine & Smith 2006 / 1953 / " 'bay' scented leaves"
U.S.: N.Y. / E. H. Day s.n. / 76 / " 'Bayberry'—'Wax-Myrtle' "

479 *M. pubescens* var. *pubescens* Ecuador / W. Camp E-2643 / 1945 / "one of the ingredients of the infusion with which women are bathed the first five days after child-birth" / " 'Laurel' "
Bolivia / J. Steinbach 5042 / 1930 / "La fruta exuda una grasa saponifera" / "Jabon vegetal"
Colombia / E. Little 7296 / 1944 / "Oil extracted from seed" / " 'olivo' "

480 *M. rubra* China: Yunnan / A. Henry 9015 / no date / "fruit small, but edible" / " 'yang-ma' "
China: N. Luchen / Ching & Chun 5938 / 1928 / "widely used by Miu and natives when dried or fresh"

481 *M. adenophora* Hainan / C. Lei 388 / 1933 / "fruit deep red edible" / "Yeung Mui Muk"

482 *M. esculenta* Burma / F. Kingdon Ward 65 / 1938 / "Fruit eaten"

JUGLANDACEAE

483 *Carya glabra* U.S.: Fla. / Cooley, Monachino & Eaton 5702 / 1958 / "Bruised leaves strongly fragrant"

484 *C. illinoensis* U.S.: Tex. / V. Cory 52353 / 1946 / " 'pecan' "

485 *C. lecontei* U.S.: La. / D. Stone 431 / 1958 / "more or less sweet meats"

486 *C. leiodermis* U.S.: La. / D. S. & H. B. Correll 9717 / 1938 / "young kernels not bitter to taste"

487 *C. myristicaeformis* U.S.: Ark. / G. Letterman s.n. / 1913 / "Nutmeg Hickory"

488 *C. ovalis* U.S.: Tenn. / B. W. Ellertsen 8503 / 1939 / "Oval Pignut Hickory"
U.S.: Mass. / D. E. Stone 1488 / 1963 / "odoriferous"

489 *C. texana* U.S.: Fla. / W. A. Murrill s.n. / 48 / "kernel bitter; cult."

490 *C. tomentosa* U.S.: N.Y. / A. Granger s.n. / 1899 / "Pig Nut"
U.S.: La. / C. A. Brown 1551 / 1927 / " 'sweet meat' "

491 *Juglans cinerea* U.S.: Penn. / W. Canby s.n. / ca. 1895 / "Bark of root a tonic laxative"

492 *J. mollis* Mexico / F. Ventura A. 1395 / 1970 / "comestible"

493 *J. neotropica* Peru / J. Wurdack 1535 / 1962 / "Fruit green, edible when ripe" / " 'Nogale' "
Ecuador / Grubb, Lloyd, Pennington & Whitmore 1007 / 1960 / "Fruits eaten"

494 *Juglans* indet. S. AM. / R. Espinosa 2425 / 48 / "Arbol muy propagada en toda la hoya de Ibarra, donde se lo aprecia por las nueces" / "Nogal"

BETULACEAE

495 *Betula glandulosa* U.S.: Calif. / L. C. Wheeler 3858 / 1935 / "Very palatable to cattle"

496 *Corylus americana* U.S.: N.Y. / Herb. Columbia Univ. / ca. 1845 / "Wild Hazilnut or Filbert"

FAGACEAE

497 *Castanopsis carlesii* China: Kwangtung / W. Tsang 21313 / 1932 / "fruit, . . . edible" / "Taai Mei Yuen Shue"

498 *C. concinna* China: Kwangtung / W. Tsang 21069 / 1932 / "edible" / "Hung Yuen Shue"

499 *C. fordii* China: Kwangtung / Lingnan (To and Ts'ang) 12767 / 1924 / "Used in mushroom culture" / "Tau fung shu (Bean wind tree)"

500 *C. javanica* Philippine Is. / A. D. E. Elmer 10579 / 1909 / "The nuts of this species are large and are eaten by the Bogobos. When they have no rice or by rice crop failure they gather these as substitute" / " 'Cahau-te-Ulayan' "

501 *Lithocarpus cooperta* Philippine Is. / A. D. E. Elmer 10480 / 1909 / "wood tasteless but with a distinct sourish smell-like sour dough" / " 'Ulayan' "

502 *L. woodii* Philippines / P. Esguerra 31370 / 1931 / "Fruit edible" / "Teledeg or Palayen"
Philippines / A. Elmer 8733 / 1907 / "fruits fattening wild hogs" /
" 'Tidgtog' "

503 *L. wrayi* Sumatra / W. N. & C. M. Bangham 1097 / 1932 / "Fruits edible"

504 *Quercus brachystachys* Guatemala / J. A. Steyermark 50606 / 1942 /
"Leaves used for wrapping up tamales" / " 'col' "

505 *Q. glaucoides* Mexico / G. Hinton 9198 / 36 / "Toasted acorn edible" /
"Tocuz"

506 *Q. reticulata* Mexico / G. Hinton 13704 / 39 / "Fruit edible"

507 *Q. cornea* China: Kwangtung / W. T. Tsang 21687 / 1932 / "fruit . . . edible" / "Sai Ip Poi Kwo Shue"
Hainan / C. I. Lei 900 / 1933 / "fruit . . . edible"

508 *Q. litseoides* China: Kwangtung / Levine 6304 / 1921 / "Cooked with fackras a medicine" / "Lui Kung Kwoh"

509 *Q. fenzeliana* Hainan / Chun & Tso 43897 / 1932 / "edible when matured"

510 *Q. silvicolarum* Hainan / H. Fung 20101 / 1932 / "leaves edible"

ULMACEAE

511 *Celtis crassifolia* U.S. & CAN. / C. W. Short s.n. / no date / "fruit eatable"

512 *C. pallida* Mexico / Dr. Gregg s.n. / 47 / "fruit yellow—edible"
Mexico / Dr. Gregg s.n. / 47 / "leaves used for pain in head, stomach, etc." / "Acébuche"

513 *Celtis* indet. Argentina / E. Sesmero 343 / 1945 / "frutos pequeños—comestibles" / " 'Tala' "

514 *Celtis* indet. Venezuela / J. Steyermark 90021 / 1961 / "fruit . . . edible"

515 *C. sinensis* China: Kwangtung / S. Lau 612 / 1932 / "edible" / "Wong huk tsz"
China: Kwangtung / W. Tsang 21331 / 1932 / "fruit, black, edible" / "Ye Shong Shue"

516 *Gironniera subaequalis* Hainan / C. Lei 380 / 1933 / "fruit green poisonous" / "Wong Kui"

517 *Trema micrantha* Costa Rica / H. Stork 4125 / 1937 / "leaves used as a poultice for stiff necks or limbs" / " 'Capaslan' "
Panama / N. Bristan 1040(3) / 1967 / "fruits . . . much eaten by birds"

518 *T. orientalis* Tanganyika / R. E. S. Tanner 3207 / 1956 / "Sambaa: For recovery from witch craft, roots cooked with stomach of goat and eaten" / "Sambaa: Mshinga"

519   *Ulmus rubra* U.S.: Del. / W. M. Canby s.n. / ca. 1875 / "A demulcent"

MORACEAE

520   *Artocarpus altilis* Ecuador / Little & Dixon 21159 / 1965 / " 'pepa de pan'
       'pan del norte' "
       Guyana / Omawale & Persaud 76 / 1970 / "Breadfruit"
       New Hebrides / S. F. Kajewski 394 / 1928 / "Breadfruit 'Ne-marl' "
       Papua / C. Carr 11633 / 35 / "Fruit green, eaten by the natives" / "Native
       name UNU"

521   *A. incisa* Peru / Killip & Smith 25238 / 1929 / "fruta de pan, cultivated"

522   *A. heterophyllus* China: Kwangtung / F. McClure 00524 / 1921 / "Fruits
       eaten" / "Poh Loh"

523   *A. styracifolius* China: Kwangsi / R. Ching 7654 / 1928 / "red when ripe
       edible"

524   *Artocarpus* indet. China: Yunnan / A. Henry 13015 / no year / "fruit edi-
       ble but small" / " 'Lou-tzi-yen-tai-kuo' "

525   *A. cumingianus* Philippine Is. / A. D. E. Elmer 12609 / 1911 / "fruits . . .
       soft and edible when fully ripe. The skin meat and seed membrane are
       very soft and juicy and have a strong vinegar-like taste" / " 'Bariwas' "

526   *A. dasyphylla* Brit. N. Borneo / G. Pascual 829 / 1928 / "Fruit eadible"

527   *Brosimum bernadettae* Panama / T. Croat 10306 / 1970 / "frts . . . being
       eaten with outer shell thrown to ground; milky sap"

528   *B. guianense* Brit. Honduras / P. Gentle 3400 / 1940 / " 'Wild breadnut' "
       Brit. Honduras / P. Gentle 9241 / 1956 / " 'Bastard Breadnut' "

529   *B. lactescens* Brit. Honduras / P. Gentle 3307 / 1940 / " 'Breadnut' "

530   *B. alicastrum* Ecuador / C. O. Janse 275 / 1966 / "Latex . . . sabor dulce"

531   *B. gaudichaudii* Brazil / Ratter, de Castro & Ramos 383 / 1967 / "The
       roots are used locally to produce a medicine used in the treatment of der-
       matitis" / "Inharezinhe do Cerrado"

532   *B. parinarioides* subsp. *parinarioides* Brasil / Pires, Rodrigues & Irvine
       51401 / 1961 / "latex abundant, white, pleasing taste—said to have curative
       value for lungs"
       Brit. Guiana / A. Smith 2901 / 1938 / "milky latex, sometimes used to
       adulterate balata"

533   *B. utile* Venezuela / A. Bernardi s.n. / 1956 / "frutos comestibles" / "N.V.
       Vacuno"

534   *Cecropia peltata* Cuba / no collector s.n. / no year / "The juice of the tops
       and the tops themselves used externally in pains" / "Cuban name: yagruma
       hembra"

535   *C. lyratiloba* Brazil / P. Carauta 4295 / 1964 / "Em mutualismo com formigas vermelhas" / " 'Embaúba' "

536   *C. orinocensis* Venezuela / Steyermark & Bunting 102717 / 1970 / "Tree with ants"

537   *Chlorophora tinctoria* Brazil / Y. Mexia 4475 / 1930 / "somewhat milky juice. Said to be used for toothache" / " 'Tajuba' "

538   *Clarisia ilicifolia* Brazil / G. T. Prance et al. 10622 / 1971 / "Fruit much sought after by Indians for cooking & eating" / " 'Hulina' "

539   *Coussapoa glaberrima* Venezuela / Maguire, Steyermark & Maguire 53517 / 1962 / "used in extraction of teeth" / " 'Sararay'–'Waremba-yed' "
Venezuela / J. Steyermark 90736 / 1962 / "Fruit reputed to soften teeth from gums for extraction" / " 'cani' "

540   *C. schunkei* Peru / J. Schunke 130 / 1935 / "fruit brown, edible" / " 'Ubilla' "

541   *Ficus lapathifolia* Mexico / C. L. Lundell 1242 / 1932 / "fruit eaten" / " 'Alamo', 'Higo' "

542   *F. microchlamys* Mexico / G. Hinton 3254 / 1933 / "Fruit edible" / " 'Ceiba' "

543   *F. aurea* W.I.: Little Cayman / W. Kings 87 / 38 / "Edible"

544   *F. citrifolia* Dominican Republic / W. Abbott 2713 / 1923 / "Edible to some extent"

545   *F. membranacea* Cuba / Britton, Earle & Wilson 4723 / 1910 / "fruit . . . when ripe, taste of figs"
Cuba / Fre. Leon 11918 / 1924 / " 'jagüey de vaca' "

546   *F. urbaniana* Cuba / J. Acuña 14618 / 1947 / "Introducido como anthelmintica"

547   *F. killipii* Peru / A. Weberbauer 7538 / 1926 / "drunk by women to prevent sterility" / " 'Aceite Maria' "

548   *F. maxima* Brazil / Prance, Silva & Pires 59120 / 1964 / "Used locally as a vermifuge"

549   *F. monckii* Argentina / J. E. Montes 14753 / 1955 / "Frutos comestibles" / " 'Higuera brava' "

550   *Ficus* indet. Venezuela / J. J. Buza 405 / 1959 / "Madera contiene latex blanco, se le extras aceite medicinal" / "Chuaro"

551   *Ficus* indet. Peru / C. M. Belshaw 3444 / 1937 / "Milky sap used for rheumatism" / " 'Renaco' "

552   *Ficus* indet. Colombia / H. García y Barriga 18052 / 1964 / "El latex se usa para sacar los parásitos en menguante"

553   *Ficus* indet. Colombia / E. L. & R. R. Little 9644 / 1945 / "White latex, used as purgative" / " 'higuerón' "

554   *Ficus* indet. Brit. Guiana / Brit. Guiana For. Dept. 26000 / "edible" / "Kumakmballi"

555   *Ficus* indet. Venezuela / F. Tamayo 4006 / 1955 / "Fruto . . . es comido por los cochinos, pajaros y palomas. Se usa como estantillos vivos porque prende por estacas guesas. La corteza del tallo, que es fibrosa y resistente, se usa marrada a la muñeca para curar las falseaduras de esta parte del cuerpo" / " 'matapalo' "

556   *Ficus* indet. Colombia / H. García y Barriga 17584 / 1962 / "Medicinal para la tos ferina"

557   *F. gnaphalocarpa* S.W. Africa / R. Seydel 2043 / 1959 / "essbaren Fruechten"

558   *F. mallotocarpa* Liberia / P. Konneh 162 / 1951 / "Used by the natives for treatment of Gonorrhea"

559   *F. zambesiaca* Rhodesia / W. R. Bainbridge 823 / 1963 / "Fr. eaten by numerous birds & animals"

560   *F. cavaleriei* China: Kwangsi / W. T. Tsang 23850 / 1934 / "Fr. black, edible" / "Ngau Nin Tze Shue"

561   *F. clavata* China: Szechuan / W. P. Fang 2416 / 1928 / "Fruit . . . edible"

562   *F. congesta* China: Kwangsi / W. T. Tsang 24077 / 1934 / "fr. edible" / "Ngau Nin Tze Muk"

563   *F. formosana* China: Kwangtung / W. Tsang 20838 / 1932 / "Fr. blackish, edible" / "Ngan U Shuh"
China: Kwangtung / Lingnan (To & Ts'ang) 12129 / 1924 / "Ngau nai t'o, (Cow milk peach)"

564   *F. pyrifolia* China: Kwangtung / W. T. Tsang 21386 / 1932 / "fruit . . . edible" / "Taai Ip Ngau Ue Shue"

565   *F. pyriformis* China: Kwangsi / W. T. Tsang 24450 / 1934 / "fr . . . edible" / "Tai Yeung Ngau Nin Shue"
China: Kwangtung / Fung Hom 00044 / 1931 / "Fruit very dry but with a palatable flavor"

566   *F. sarmentosa* var. *thunbergii* Ryukyu Is. / Walker & Tawada 7176 / 1951 / "Frts. pubescent, edible"

567   *F. stenophylla* China: Kwangtung / W. Tsang 21247 / 1932 / "fruit black, edible" / "Cheung Ip Ngau Nye Shue"

568   *F. variolosa* China: Kwangsi / W. Tsang 24049 / 1934 / "fr. black, edible" / "Ngau Nin Shue"
China: Kwangtung / Lingnan (To & Ts'ang) 12605 / 1924 / "Tai ngau nai shu, (Large cow milk tree)"

569 *Ficus* indet. China: Kwangtung / W. T. Tsang 21039 / 1932 / "fruit, black, edible" / "Au Ue Tse"

570 *Ficus* indet. China: Kwangsi / W. T. Tsang 24544 / 1934 / "fr . . . edible" / "Ngau Nin Shue"

571 *Ficus* indet. China: Kweichow / Y. Tsiang 6701 / 1930 / "Fruit . . . edible"

572 *F. hispida* Hainan / H. Fung 20296 / 1932 / "Fruits crushed and mixed with salt water can cure diarrhoea" / "Ngau Yue Shue (Milk Tree)" Hainan / C. Lei 90 / 1932 / "fruit edible. Fruit boiled with water cures diarrhea" / "Ngou Yui Shu"

573 *F. barnesii* Philippine Is. / A. D. E. Elmer 13868 / 1912 / "figs . . . when in the right mature state are eaten by the natives, and really have a good figy flavor" / " 'Basicong' "

574 *F. benguetensis* Philippine Is. / A. D. E. Elmer 8609 / 1907 / "the wild cats feed on its fruits" / " 'Tabul' "

575 *F. caudatifolia* Philippine Is. / J. Fontanoza 71 / 1929 / "eaten by birds" / "Dalakit"

576 *F. copiosa* New Britain / J. Waterhouse 247 / 1934 / "Fruit is eaten and also young shoots as spinach" / "Kaqua"

577 *F. cumingii* Philippine Is. / R. Williams 942 / 1904 / "Fruit red, edible"

578 *F. fastigiata* Philippine Is. / A. D. E. Elmer 8577 / 1907 / "Fruits are eaten"

579 *F. granatum* New Hebrides Is. / S. Kajewski by J. Wilson 956 / 1929 / "Natives eat it" / "Nating"

580 *F. hylophila* New Guinea: Papua / L. J. Brass 5792 / 1934 / "receptacles yellow-green; sweet & palatable"

581 *F. integrifolia* Philippine Is. / A. D. E. Elmer 89492 / 1907 / "its woods carved into 'Colden' lids — Coldens are iron pots the Igorotes use to cook their chow in" / " 'Tubang' "

582 *F. leptorhyncha* Sumatra / W. N. & C. N. Bangham 773 / 1932 / "Fruits . . . edible"

583 *F. manilensis* Philippine Is. / A. D. E. Elmer 7171 / 1906 / "figs . . . sweet and edible"

584 *F. minahassae* Philippine Is. / A. L. Zwickey 20 / 1938 / "fruit edible; water in trunk available for drinking" / " 'Hagimit' 'Gimit' "

585 *F. mindanaensis* Philippine Is. / R. Williams 2708 / 1905 / "Fruit dark red, edible"

586 *F. papuana* New Guinea: Papua / C. E. Carr 11640 / 1935 / "Ripe fruit eaten by the natives" / "VEIYA"

587   *F. prolixa* Mariana Is. / F. Fosberg 25004 / 1946 / "fruit . . . used mixed with vinegar for internal pains or to cure cuts" / " 'nunu' "

588   *F. prolixa* var. *carolinensis* Caroline Is. / D. Anderson 772 / 1949 / "bark pounded and used as pack for headache" / " 'Au' Fruit called 'won' "

589   *F. pupinervis* Philippine Is. / F. Paraiso 249 / 1929 / "Sap cures herpes" / "Ray-ya-Ray ya Dialect Ilocano"

590   *F. ramentacea* Caroline Is. / F. R. Fosberg 32425 / 1950 / "bark chewed like betel nut" / " 'Kubul' "

591   *F. tinctoria* var. *tinctoria* New Britain / J. Waterhouse 243 / 1934 / "Fruit is edible and young shoots are cooked and like spinach" / "Tagataga"

592   *F. tinctoria* var. *neo-ebudarum* Caroline Is. / F. R. Fosberg 24570 / 1946 / "The figs are cooked and eaten" / " 'awan' or 'auan' "
Caroline Is. / F. R. Fosberg 25467 / 1946 / "fruit eaten" / " 'haual' "
Mariana Is. / F. R. Fosberg 24977 / 1946 / "fruits eaten, mashed and made into cakes during war" / " 'hodda' "
Marshall Is. / D. Anderson 3602 / 1950 / "fruit cooked (boiled)" / " 'tobro' "

593   *F. vitiensis* Fiji / A. Smith 7502 / 1953 / "fruit edible" / " 'Lolo' "
Fiji / O. Degener 15224 / 1941 / "Fig eaten by Fijians" / " 'Komba' in Serua"

594   *F. wassa* New Britain / J. Waterhouse 269 / 1934 / "leaf & fruit edible"

595   *Ficus* indet. Borneo / J. & M. S. Clemens 29585 / 1932 / "Edible, quite sweet and testey"

596   *Helicostylis coriacea* Brazil / R. L. Fróes 22563 / 1947 / "Juice considered deadly poison"

597   *H. tomentosa* Surinam / G. Stahel (Indian coll.) s.n. / 1967 / "Sacred tree of Caraib and Arowak Indians"
Brazil / Prance, Steward, Ramos & Monteiro 11240 / 1971 / "Fruit eaten by Indians" / " 'Xubaco' (Uaicá-Mucajai)"

598   *Malaisia scandens* Hainan / C. I. Lei 668 / 1933 / "fruit red edible" / "Bo Pai Tz Tang"

599   *Morus microphylla* U.S.: Tex. / R. McVaugh 7231 / 1945 / "fruit red, tart, with more flavor than cultivated mulberry"

600   *M. rubra* Bermuda / Brown, Britton & Bisset 2066 / 1914 / "fruit black, delicious"

601   *M. alba* Formosa / A. Henry 134 / ca. 1890 / "all use as medicine" / " 'Sane' "

602   *Myrianthus holstii* Nyasaland / L. J. Brass 17615 / 1946 / "fruit . . . eaten by natives"

503 *Naucleopsis mello-barretoi* Brazil / G. T. Prance et al. 15563 / 1971 / "Sap used by Makú for arrow poison, bark slashed, sap extracted & painted on arrow tips with feather"

504 *N. ulei* Peru / A. G. Ruíz 186 / 1964 / "frutos . . . comestibles por los nativos aislados. Usado por leña y carbón" / " 'Pinsha caspi' "

505 *Olmedia asperula* Brit. Guiana / D. Fanshawe F148 / 2757 / 1938 / "tree has a scent of almond essence" / "Vern. Name—Ituri-ishi-lukudo"

506 *Perebea mollis* subsp. *mollis* Peru / J. Schunke 319 / 1935 / "fruit red, edible" / " 'Chimicua' "

507 *P. xanthochyma* Ecuador / Cazalet & Pennington 7785 / 62 / "Frt a seeded fleshy berry (edible)"

508 *Pourouma apiculata* Brazil / G. T. Prance et al. 9952 / 1969 / "fruit much sought after and eaten by Indians"

509 *P. chocoana* Ecuador / Little & Dixon 21070 / 1965 / " 'uva' "
Ecuador / Játiva & Epling 2037 / 1966 / "Aqueous liquid texture and flavor like coco milk" / " 'Raices sancudas' "

510 *P. cuatrecasasii* Colombia / J. Cuatrecasas 7299 / 1939 / "N.v. 'uvo silvestre' "

511 *P. minor* Peru / J. Schunke V. 2139 / 1967 / "Los frutos son comestibles" / " 'Uvilla' "

512 *P. simplicifolia* Bolivia / O. Buchtien 2122 / 1909 / " 'Uva menuda' "

513 *P. steyermarkii* Venezuela / J. Steyermark 60665 / 1944 / "fruit . . . edible" / " 'caibari-cai-yek' "

514 *P. substrigosa* Peru / Y. Mexia 6201 / 1931 / "inhabited by small ants" / " 'Uvilla' "

515 *P. subtriloba* Bolivia / M. Cardenas 1990 / 1921 / "Fruit edible"

516 *Pourouma* indet. Venezuela / C. A. Blanco C. 1149 / 1971 / "Fruto morado, comestible agradable" / "Cucure"

517 *Pourouma* indet. Ecuador / P. J. Grubb et al. 1545 / 1960 / "Fruits . . . savoured by monkeys"

518 *Pourouma* indet. Brazil / J. J. Wurdack 2169 / 1962 / "fruit said to be edible" / " 'Shuvija' (Aguaruna) or 'Sacha unilla' "

519 *Pourouma* indet. Ecuador / P. J. Grubb et al. 1681 / 1951 / "Fruits delectable, tasting like grapes"

520 *Pourouma* indet. Brazil / M. Barbosa da Silva 130 / 1942 / "Seed imbedded in sweet pulp which is eaten by parrots and monkeys" / " 'imbauba rana' "

521 *Streblus asper* China: Kwangtung / Fung Hom 00510 / 1931 / "eaten by small boy" / "Koon Yam Muk"

622   *Trophis racemosa* Mexico / Y. Mexia 8844 / 1937 / "fruit said to be . . .
       edible" / " 'Ushi' "

623   *T. racemosa* var. *racemosa* Venezuela / L. Aristeguieta 5869 / 1965 / "la
       fruta después de cocinada, es comestible" / "Charo"

624   *Moracea* indet. Peru / J. Schunke Vigo 1067 / 1966 / "Tree medicinal. —
       The latex utilizan the Indians para fracturas de heusos"

URTICACEAE

625   *Boehmeria virgata* Fiji / O. Degener 15337 / 1941 / "Leaves used as remedy
       for boils" / " 'Ndrendre' "

626   *Dendrocnide ternatensis* New Guinea: Papua / L. J. Brass 5490 / 1933 /
       "stinging hairs on petioles"

627   *D. vitiensis* Fiji / O. Degener 15034 / 1941 / "for pain in chest take root
       and scrape its bark. Put this in coconut cloth with a little water. Then
       squeeze & drink liquid" / " 'Bolavatu' in Serua"

628   *Discocnide mexicana* El Salvador / P. Standley 22403 / 1922 / "stinging" /
       " 'Chichicaste, Pan caliente' "
       Mexico / G. Hinton 3676 / 33 / "The tuberous root will relieve the pain of
       the sting" / "Ortiga"

629   *Elatostema sessile* China: Kwangtung / W. Tsang 20742 / 1932 / "food for
       pigs" / "Pin Tsoi"

630   *Girardinia condensata* S. Africa / Stacy & Schlieben 8612 / 1961 / "stinging
       terrible"

631   *G. palmata* China: Kweichow / Steward, Chiao & Cheo 852 / 1931 /
       "Stinging herb" / "Ho Ma"

632   *Laportea aestuans* W.I.: Martinique / P. Duss 1389 / 1879 / "Les feiulles
       par infusion sont employées contre la toun" / "Ortie Crulanb"

633   *L. cuneata* Cuba / Ex Herb. Collegii Pharmaciae Neo-Eboracensis s.n. /
       1840 / "Decoction used in intermittent fevers — is a burning nettle" / "ortiga
       blanca"

634   *L. ruderalis* Caroline Is. / Fosberg & Evans 47142 / 1965 / "Used
       medicinally" / " 'hafalifal' "
       Marshall Is. / F. Fosberg 26716 / 1946 / "used for medicine, eaten by
       Japanese" / " 'nen kutakut' "

635   *Leucosyke corymbulosa* Fiji / Degener & Ordonez 13909 / 1940 / "Used by
       early settlers for tea"

636   *Myriocarpa stipitata* Peru / Y. Mexia 6252 / 1931 / "Leaves irritating to
       skin" / " 'Ishanga' "

537   *Parietaria debilis* Colombia / Killip & Smith 19666 / 1927 / "Remedy for tropical anemia" / " 'Baico' "

538   *Pilea microphylla* W.I.: La Guadeloupe et Dépendances / P. Duss 2192 / 1892 / "Est employé conre la fievre" / "petit teigne blanc"

539   *P. serpylacea* Ecuador / W. Camp 2489 / 1945 / "the juice of the fleshy leaves is used against fevers, and as an intestinal refresher" / " 'Prenadilla' "

540   *Pilea* indet. Colombia / García y Barriga, Hashimoto & Ishikawa 18605 / 1965 / "Se usa en Medicina popular como purgante"

541   *Pipturus argenteus* Papua / N. E. G. Crutwell 35 / 1947 / "receptacle swollen, edible"

542   *Urera caracasana* Panama / R. Seibert 608 / 1935 / "Contact with plant irritates skin"
Brazil / G. Prance et al. 15553 / 1971 / "Flowers rubbed over body as cure for scorpion bites by Makú Indians" / " 'Chiuk' "

543   *U. baccifera* Cuba / A. Curtis 608 / 1905 / "Very poisonous"

544   *Urera* indet. Tanganyika / R. E. S. Tanner 1178 / 1953 / "Sukuma tribe: Leaves rubbed on udders of animals refusing to suckle young. . . . great irritation upon touching" / "Kisukuma: Lubambi"

545   *Urtica subincisa* Mexico / G. Hinton 2844 / 32 / "For reumatism" / "ortiga"

546   *U. ballotaefolia* Ecuador / W. Camp 2464 / 1945 / "The root used as one of the components of a medicine much favored for pneumonia" / " 'Ortiga caballo chino' "

547   *U. magellanicum* Chile / W. Eyerdam 10523 / 1958 / "Nettle causing sharp stinging pain for many hours"
Peru / Hutchinson & Wright 4306 / 1964 / "Take as tea for 'molida' / " 'Shinuwa' 'Hortiga' "

548   *U. hyperborea* Kashmir / W. Koelz 2275 / 1931 / "Eaten as spinach. Delicious" / "Zatsut"

PROTEACEAE

549   *Finschia chloroxantha* New Hebrides / S. Kajewski 95 / 1928 / "Nuts eaten by natives" / " 'N'gye N'gye' "

550   *Hakea pugioniformis* Australia: New S. Wales / C. White 10247 / 1935 / "strongly honey scented"

551   *Lomatia hirsuta* Ecuador / B. & C. Maguire 44281 / 1959 / "used for liver difficulties" / " 'Rumpiache' "

552   *Lomatia* indet. Ecuador / W. Drew E-65 / 1944 / "Said to be medicinal: diuretic"

653 *Persoonia cornifolia* Australia: Queensland / M. S. Clemens s.n. / 1944 / "Fruits eaten by children"

654 *Roupala montana* Costa Rica / A. Jimenez 1276 / 1963 / " 'danto hediondo' "
Brazil / Richards, Argent & Ramos 552 / 1967 / "Vernacular name refers to the rather nitrogenous smell of the crushed tissues of some specimens (it means maned-wolf's urine)"

LORANTHACEAE

655 *Gaiadendron punctatum* Ecuador / W. B. Drew E-105 / 1944 / "A local medicinal . . . useful for constipation; its properties were not tested" / " 'voleta' "

656 *Loranthus exocarpi* Australia: Queensland / M. S. Clemens s.n. / 1946 / "Fruit eaten by aborigines"

657 *Peristethium polystachus* Panama / G. P. Cooper 205 / 1928 / " 'Mal ojo' "

658 *Psittacanthus calyculatus* Mexico / G. Hinton 827 / 32 / "The fruit is a favorite food of birds. The Indians are positive that berds deliberately sow the seed on trees"
Mexico / Y. Mexia 1087 / 1926 / " 'Malojo' "
Mexico / W. Camp 2551 / 1937 / "pulp of fruit pleasantly aromatic, its juice resembling raw latex"

659 *P. robustus* Brazil / Ratter, de Castro & Ramos 387 / 1967 / "Has local medicinal use against urinary diseases" / "ERVA DE PASSERINA"

660 *Psittacanthus* indet. Colombia / Schultes & Cabrera 16996 / 52 / "Makuna = *kan-san-wee-gaw* ('casa dela hormiga')"

661 *Loranthacea* indet. Peru / R. Williams 2538 / 1901 / " 'Verdolaga' "

VISCACEAE

662 *Phoradendron bolleanum* Mexico / Stanford, Retherford & Northcraft 149 / 1941 / "heavily grazed by goats"

663 *P. piperiodes* Venezuela / J. A. Steyermark 75508 / 1953 / "reputed that boiled plant is used for placing on wounds and cuts" / " 'ata-pik' "
Venezuela / J. J. Wurdack 302 / 1955 / "Fruit used to treat indigestion" / "Guate de pajaro"

664 *Viscum obscurum* S. Africa / R. Baylis 5124 / 1972 / "The rubbing on of berries reputed to cure warts"

665 *V. album* China: Hopei / J. C. Liu L.302 / no year / "Used in Chinese medicine"

SANTALACEAE

566   *Osyris lanceolata* W. Himalayas / W. Koelz 3140 / 1931 / "fruit orange, insipid, slightly nauseating"

567   *Santalum spicatum* W. Australia / C. T. White 5487 / 1927 / "Roots dug up, cut and used by aborigines as a source of water" / " 'Water Tree' "

568   *S. ellipticum* Hawaii / O. & I. Degener 28585 / 1963 / "rats or mice eat young kernels"

569   *S. insulare* Marquesas / Mumford & Adamson 421 / 29 / "perfume from the wood—grated & mixed with coconut oil"

OPILIACEAE

570   *Champereia manillana* Philippine Is. / M. C. Fenix 31099 / 1929 / "leaves edible" / "Panalayapin; Dialect Iloc"

OLACACEAE

571   *Cathedra caurensis* Venezuela / Aristeguieta & Zabala 7055 / 1969 / "El fruto es comestible por el cochino, peces" / " 'Yaguazil' o 'Asta blanca' "

572   *Heisteria macrophylla* Panama / Cooper & Slater 166 / 1927 / " 'Naranjillo colorado' "

573   *H. media* Brit. Honduras / P. Gentle 3516 / 1941 / "fls yellowish, fragrant" / " 'nance cimarron' "

574   *Liriosma adhaerens* Brazil / Prance & Silva 58572 / 1964 / "Used locally as a treatment for rheumatism"

575   *L. singularis* Brazil / J. Ratter et al. 1222 / 1968 / "Vernacular name: Mejo do Porco"

576   *Olax dissitifolia* Rhodesia / W. R. Bainbridge 429 / 1960 / "Eaten by people" / "Valley Tonga name MOYOWANSYA"

577   *Ptychopetalum olacoides* Brazil / R. de L. Froes 29608 / 1953 / "flõr branca, perfumuadissima"

578   *Scorodocarpus borneensis* Malaya / J. Sinclair 10558 / 1960 / "Twigs smelling like turnips but more foetid"

579   *Ximenia americana* Colombia / R. Romero-Castañeda 9632 / 1963 / "Frutos amarillos, que los moradores los consideran como vomitivo" / " 'Huevo de morrocoyo' "
Venezuela / L. Aristeguieta 2900 / '57 / "El fruto es considerado como venenoso" / "N.v.: Solimán"
Marshall Is. / F. Fosberg 34400 / 1952 / "weak bitter almond odor when crushed . . . pleasantly but strongly acid . . . 'fruit eaten' " / " 'kalikelik' "

680   *X. borneensis* N. Borneo / Patrick 43603 / 64 / "bark reddish, strong odour like onions . . . flower white fragrant, slightly onion like smell"

BALANOPHORACEAE

681   *Balanophora involucrata* China: Kwangsi / R. C. Ching 7020 / 1928 / "A valued tonic by natives"

ARISTOLOCHIACEAE

682   *Aristolochia wrightii* U.S.: Tex. / Herb. of Columbia College, New York s.n. / no year / " 'Used for ulcers' "

683   *A. grandiflora* Mexico / G. L. Fisher 35500 / 1935 / "A strong odor of putrid meat attracts the insects 100 or more feet from the flower" Honduras / Yuncker, Koepper & Wagner 8466 / 1938 / "carrion-like odor"

684   *A. schippi* Brit. Honduras / P. H. Gentle 2755 / 1939 / "Used as snake remedy" / " 'guaco' "

685   *A. anguicida* W.I.: Martinique / P. Duss 887 / 1880 / "Le suc des feuilles et des racines est employé contre les douleurs rhumatismales chroníques. C'est un antisyphílitique par excellence" / "Liane douce"

686   *A. trilobata* W.I.: Martinque / P. Duss 582 / 1881 / "Employé contre la morsure du serpent" / "Crífle caraibee"

687   *A. arcuata* Brazil / B. A. Krukoff's 6th Exped. to Braz. Amaz. 7542 / 1935 / "Component of Curare of Tecuna Indians"

688   *A. bicolor* Colombia / J. M. Idrobo 2215 / 1956 / "Medicinal, para cólicos" / "N.v. 'Tigre — huasca' "

689   *A. daemoniana* Brit. Guiana / For. Dept. of Brit. Guiana F1383 / 43 / "Rough, crumbly-barked rope with strong menthol scent" / "Boyarri"

690   *Aristolochia* indet. Surinam / Herb. of B. A. Krukoff 150 / 1941 / "much used by the Creoles as a febrifuge and extensively used by woman & in childhood" / "Loango — tété (Loango — vine)"

691   *Aristolochia* indet. Argentina / J. E. Montes 157 / 44 / "es medicinal" / " 'Siphó mil hombre S. Geuma' "

692   *A. contorta* China: Hopei / J. C. Liu L.2209 / 1929 / "used in Chinese medicine" / " 'Ma Tou Ling' (Horse bell; from the fruit)"

693   *Aristolochiacea* indet. Argentina / J. E. Montes 14809 / 1955 / "Medicinal" / "N.V. Isipho milhombres)"

POLYGONACEAE

694   *Coccoloba montana* Guatemala / E. Contreras 10548 / 1971 / " 'Uva' "

595   *C. reflexiflora* Brit. Honduras / P. Gentle 4150 / 1940 / " 'wild grape' "

596   *C. tuerckheimii* Honduras / C. & W. von Hagen 1058 / 1937 / " 'Almendra de monte' "

597   *C. diversifolia* Puerto Rico / E. Little, Jr. 13455 / 1950 / "Local name: uvilla"
Bahamas / Nash & Taylor 1452 / 1904 / "Berries eaten by children" / " 'Pigeon Plum' "

598   *C. krugii* W.I.: Brit. Virgin Is. / W. Fishlock 26 / 1918 / " 'Wild Grape' "

599   *C. plumieri* Jamaica / W. Harris 12023 / 1915 / "ripe fruits size of a grape, black, edible" / " 'Wild Grape' "

700   *C. venosa* W.I.: Antigua / J. Beard 351 / 1944 / " 'Wild grape' "
W.I.: Martinique / P. Duss 1744 / 1879 / "fruits blancs, mangeables" / "Raisiń coudre"

701   *C. zebra* Jamaica / D. Watt s.n. / 90 / "Wild Grape"

702   *C. caracasana* Venezuela / A. Bernardi 1077 / 1954 / "n.v.: 'Uvitos' "

703   *C. coronata* Venezuela / Aristeguieta, Blanco & Carillo 6845 / 68 / "frutos comestibles" / "N.v.: 'Uvero macho' "
Venezuela / W. Broadway 491 / 1923 / "fruit dark purple, edible"

704   *C. cruegeri* Venezuela / Curran & Haman 796 / 1917 / "Edible fruits"

705   *C. mollis* Surinam / J. Lanjouw 359 / 1933 / "Small tree full of ants"
Peru / G. Klug 1991 / 1931 / " 'Tangarana' (ant tree)"

706   *C. parimensis* Brit. Guiana / S. S. & C. L. Tillett 45828 / 1960 / "wood with foetid Eucalyptus—oil odor"

707   *C. sticticaulis* Brazil / E. Warming 126, 129/4 / 1866 / " 'Uva do Mato' "

708   *Emex australis* S.W. Africa / R. Seydel 4413 / 66 / "So lange noch jung u ohne Fruechte, gefressen"

709   *Eriogonum annuum* U.S.: Tex. / E. Whitehouse 10754 / 1945 / " 'wild buckwheat' "

710   *E. pyrolaefolium* U.S.: Idaho / Hitchcock & Muhlich 10542 / 1944 / "odor of strong urine"

711   *Fagopyrum esculentum* China: Kwangtung / W. Tsang 20651 / 1932 / "Use as food for pig" / "Sam Kwok Fung"

712   *Muehlenbeckia fruticulosum* Peru / Mr. & Mrs. F. E. Hinkley 48 / 1920 / "Use, medicinal" / "Local name 'Airampillo' "

713   *Oxygonum dregeanum* S. Africa / R. Bayliss 3013 / 1965 / "fls white—smell of honey"

714   *Polygonum hydropiper* U.S.: N.J. / H. Moldenke 11249 / 1939 / " 'Water-pepper' "

715 **P. phytolaccaefolium** Alaska / A. Hollick s.n. / 03 / "Used extensively as a pot herb. Boiled and eaten like spinach. Very palatable"
Alaska / Y. Mexia 2163 / 1928 / " 'Miner's greens', 'Wild rhubarb' "

716 **P. acuminatum** Brazil / F. Hoehne 554 / 1918 / " 'Herva de bicho' "

717 **P. ferruginea** Brasil / W. Archer 8368 / 1943 / "leaves eaten by the peixe boi — manatee" / " 'tabaco de peixe boi' "

718 **P. persicaria** Ecuador / R. Espinosa 16 / 1946 / "Se dice que tiene aplicaciones como insecticida"
Germany / Herb. Schleicher 2208 / 1837 / "Muekhenkraut"

719 **P. punctatum** Colombia / W. Archer 3312 / 1935 / "Dry plant has peppery odor" / " 'Barbasco' "
Argentina / J. Montes 27561 / 58 / "Usos: ahuyenta a pulgos y piojos" / " 'Caa-ité' 'Yerba del bicho' "
Argentina / J. Montes 14838 / 55 / "Uso: medicina popular" / " 'Caá-tay' 'Herba do Bicho' "
Paraguay / A. Woolson 761 / 56 / "Plant has bitter taste, becoming pungent on chewing" / "n.v. 'cáá-taf' "

720 **P. punctatum** var. **aquatile** Cuba / H. Fernando 197 / 1920 / " 'Hierba de sapo' "
Venezuela / W. Archer 2983 / 1935 / ". . . aromatic odor. Said to be very effective as a fish poison, but burns hands" / " 'Barbasco' "

721 **Polygonum** indet. Bolivia / R. Steinbach 28 / 1966 / "Hierba curativa (infusión de las hojas contra males hepáticos)" / " 'Wacchabarbero' "

722 **Polygonum** indet. Colombia / C. Rodriguez A. 36 / 1946 / "n.v. 'Barbasco' "

723 **P. caespitosum** China: Kwangtung / Lingnan (To Ts'ang) 12346, 12490 / 1924 / "Hung lat liu (Red peppery Polygonum)"

724 **P. senticosum** China: Kwangtung / To & Ts'ang 12466, 12784 / 1924 / "Used as medicine, for fever" / "Wo lim lak (Rice sickle thorn); Lo fu li (Tiger tongue)"

725 **P. chinense** Hainan / C. Lei 276 / 1932 / "fr. blue edible" / "For Tan Tang"

726 **P. hydropiper** var. **hispidum** Hainan / C. Lei 335 / 1933 / "flower red ill-smelling . . . This plant has strong hard tasted substance. Ground plant washed in water, the hard tasted substance may kill fish. It is used as medicine for killing fish" / "Tai Chung Lat Tso"

727 **P. orientale** Indo-China / G. Groff 00685 / 32 / "The bees are very fond of this — it would probably make a good honey plant" / "Shui Saat Liu"

728 **P. paniculatum** Ukhrul / F. Kingdon Ward 17890 / 1948 / "Flowers smell of honey"

29 *Rheum emodi* Asien / no collector s.n. / no year / "Wurzel dieser Art giebt ein trefflich tonisches, aber keineswegs purgirendes Mittel"

30 *R. nervosum* Mongolei / no collector s.n. / no year / "Eine der zahlreichen Arten dieser Gattung deren Wurzel als Radix Rhei in Handel kommen"

31 *R. rhaponticum* Asien / no collector s.n. / no year / "In ihrer Heimat wird diese Art nach Art der Rhabarber, bei uns aber nur von VeterinärArzten angewendet"

32 *R. webbianum* W. Himalayas / W. Koelz 2294 / 1931 / "delicious stalk"

33 *Rumex venosus* U.S.: Nev. / J. Sevy 37 / 1938 / "Sagebrush, poison horsebush"

34 *R. andinus* Ecuador / W. Camp 2052 / 1945 / " 'Cuchi-gulac' or Pig's lettuce"

35 *R. crispus* Bolivia / R. Steinbach 581 / 1966 / "para aplicación de cataplasmas"

36 *R. obtusifolius* Argentina / J. Montes 14706 / 1955 / "Uso: hojas tiernas comestibles"

37 *Rumex* indet. Argentina / Bertoni 1091 / 45 / "medicinal, algunos lo llaman tambien zarzaparrilla"

38 *Rumex* indet. Peru / Hutchison & Wright 4328 / 1964 / "Remedy for anemia" / " 'Aselga Blanca' "

39 *R. bequaerti* Nyasaland / L. J. Brass 17820 / 1946 / "leaves eaten by natives" / "Native name—Nsindajera (Chinyanja)"

40 *R. usambarensis* Tanganyika / R. E. S. Tanner 3853 / 1957 / "Masai—twigs chewed as a relish" / "Vernacular name: Masai—Engaisije"

41 *R. vesicarius* Teneriffa / no collector s.n. / ca. 1850 / "Wird wo er vorkommt, als Arznei und Speise verwendet"

42 *R. acetosa* Himalaya: Lahul / W. Koelz 351 / 1930 / "Used as sort of confection with chili & salt" / "milum loki Lauli & Kulu"

43 *Ruprechtia laxiflora* Argentina / D. Rodriguez 1124 / 1913 / " 'Duraznillo grande' "
Brasil / Lindeman & de Haas 3146 / 1966 / "visited by large numbers of flies"

44 *Symmeria paniculata* Brasil / W. A. Archer 8369 / 1943 / "Fruit used as bait to catch pacú and other fish" / " 'tucunará-arú' "
Venezuela / L. Williams 11983 / 1939 / "N.v.: Manteco rebalsero"

45 *Triplaris americana* subsp. *felipensis* Colombia / F. Fosberg 20172 / 1943 / "inhabited by large stinging ants"
Colombia / E. L. & R. R. Little 8271 / 1966 / "vara santa"

746   *T. peruviana* Peru / J. Schunke V. 819 / 1965 / "The indians drink an infusion of the bark in order to purify the blood" / " 'Tangarana' "

747   *Polygonacea* indet. Ryukyu Is. / S. Tawada 2248 / 1949 / "Used for fodder
and drugs"

CHENOPODIACEAE

748   *Atriplex acanthocarpa* MEX. & C. AM. / Dr. Gregg s.n. / 47 / "Animals
eat it" / "Quelito"

749   *A. hortensis* Cuba / H. van Hermann 789 / 1905 / "Cult. for salad greens"

750   *A. suberecta* S.W. Africa / R. Seydel 3585 / 63 / "Gelegentlich gefressen"

751   *Bassia birchii* Australia: Queensland / S. Blake 5352 / 1934 / "eaten by
camels only" / " 'Galvanized Burr' "

752   *Blitum capitatum* U.S.: Wis. / I. A. Lapham s.n. / no year / "Indian
strawberry"

753   *Chenopodium botrys* U.S.: Nev. / W. A. Archer 6840 / 1938 / "Plant
resinous and aromatic"

754   *C. capitatum* Canada: Manitoba / F. Fosberg 40073 / 1959 / "Fruiting
heads . . . insipidly sweetish"

755   *C. multifidum* U.S.: N.Y. / J. Carey s.n. / ca. 1850 / "Small-leaved
Wormseed"

756   *C. ambrosioides* Mexico / F. Müller s.n. / 1853 / "Wird vom Volk als
Gemüse gegessen" / "Epasotle"
Dominican Republic / J. de Js. Jimenez 5197 / 1967 / "Strong and
disagreably scented" / "C.N. = APAZOTE"
Brazil / O. Machado 75627 / 45 / "Herva de Santa Maria"
Ecuador / W. Camp 2463 / 1945 / "Young plants used to flavor soup; also
an infusion of leaves and roots used for colic" / " 'Payco' "
Paraguay / T. Morong 1543 / 1890 / "Our Guarani peons on the
Pilcomayo River attributed great medicinal virtue to the Roman Worm-
wood, which grows profusely along the banks. I frequently saw them
gathering the spikes and stripping the flowers and fruit into tin cups for the
purpose of steeping them into tea"
Brasil / W. A. Archer 8164 / 1943 / "Lv. used as cough remedy" / " 'men-
truz' "

757   *C. murale* Mexico / Dr. Gregg s.n. / 47 / "used for eruptions on the head"
/ "Quelito de perro"
W.I.: Curaçao / Curran & Haman 12 / 1917 / " 'Picante sjimaron' "
Ecuador / W. H. Camp 2441 / 1945 / "used as a wash in some skin infec-
tions" / " 'Palitaria' "

58 *C. hircinum* Ecuador / W. H. Camp 2445 / 1945 / "Fed to cattle; the seeds fed to game-cocks" / " 'Sacha quinia blanca' "

59 *C. quinoa* Colombia / J. M. Idrobo 2205 / 1956 / "Cultivada por un yerbatero (Silverio Roman) para mezclar con otras yerbas como remedio en varias enfermedades"

60 *Chenopodium* indet. Argentina / J. E. Montes 27533 / 58 / "medicinal, estomacal y para combatir los parásitos intestinales" / " 'Te jesuita' 'Caá-ré' "

61 *C. opulifolium* Tanganyika / R. E. S. Tanner 3335 / 1956 / "Zigua: For attracting person, plant burnt and the ash put in that persons food. For difficult childbirth, roots boiled and the water drunk by the woman"

62 *C. anidiophyllum* Australia: Queensland / S. Everist 1301 / 1935 / "Plant with a very strong odour of stale fish"

63 *Endolepis monilifera* Mexico / Dr. Gregg s.n. / 47 / "used for 'greens' " / "Quelito"

64 *Eurotia lanata* U.S.: Wyo. / A. Nelson 3607 / 1897 / "Accounted good sheep feed"

65 *Grayia spinosa* U.S.: Wyo. / A. Nelson 3698 / 1897 / "Seed and leaves form an appreciable part of the sheep feed of the region"

66 *Kochia americana* U.S./ Wyo. / A. Nelson 3743 / 1897 / "Eaten by sheep in winter"

67 *Suaeda torreyana* U.S.: Ariz. / W. Clute 54 / 1920 / "Abundant and so ill smelling as to be very offensive" / " 'Little Greasewood' "

68 *Chenopodiacea* indet. Bolivia / R. Steinbach 58 / 1966 / "Hierba . . . usada por los nativos contra males estomacales en infusion. Las hojas y la planta en general tienen un olor muy particular" / " 'Paecko' "

69 *Chenopodiacea* indet. Bolivia / R. Steinbach 45 / 1966 / "Hierba de hojas comestibles (crudas o cocidas)" / " 'Jatacko' "

70 *Chenopodiacea* indet. Argentina / F. Fortuna 3 / 1945 / " 'Quinoa' "

71 *Chenopodiacea* indet. Bolivia / R. Steinbach s.n. / 1966 / "comestible" / " 'Pachiyuyu' "

72 *Chenopodiacea* indet. Argentina / J. E. Montes 92 / 1944 / "es medicinal"

73 *Chenopodiacea* indet. Argentina / F. Ybarrola 2962 / 1945 / "usada en medicina casera"

74 *Chenopodiacea* indet. S. AM. / R. F. Steinbach 693 / 1967 / "usan la infusion para dolores estomacales"

75 *Chenopodiacea* indet. Tanganyika / R. E. S. Tanner 652 / 1952 / "Used to treat persistent finger sores"/ "Vernacular name: Kisukuma: Kabukibuki"

AMARANTHACEAE

776 *Achyranthes aspera* Tanganyika / R. E. S. Tanner 2693 / 1956 / "Zigua: For menstrual pain, roots boiled and drunk" / "Kizigua: Tara"
Tanganyika / R. E. S. Tanner 2972 / 1956 / "Zigua: Used in medicine for removing spirit possession" / "Zigua: Purule"
Tanganyika / R. E. S. Tanner 1190 / 1953 / "Plant ash used with tobacco to improve flavor" / "Kisukuma: Buhulula"
Tanganyika / R. E. S. Tanner 1974 / 1955 / "Roots boiled and water drunk, to relieve pain upon urination" / "Kibondei: Mbatimanga"

777 *A. bidentata* China: Kweichow / Stewart, Chiao & Cheo 133 / 1931 / "Herb used in medicine for kidney trouble" / "Niu Sheh"

778 *A. hackijoensis* Ryukyu Is. / S. Tawada 2239 / 1949 / "Used for food & drugs"

779 *A. mollicula* Ryukyu Is. / S. Tawada 2235 / 1949 / "Used for drugs"

780 *Aerva leucura* Tanganyika / R. E. S. Tanner 1504 / 1953 / "Sukuma tribe: Roots pounded, used in drink to combat poisoning secretly administered" / "Kisukuma: Lweja bafumu"

781 *Amaranthus hybridus* Mexico / G. Hinton 3015 / 32 / "Grain made into cake" / "Alegria"

782 *Amaranthus* indet. Argentina / J. Montes 14778 / 1955 / "Uso: medicina popular y forragera para cerdos" / " 'Caá-rurú-blanco' "

783 *A. dubius* Tanganyika / R. E. S. Tanner 2241 / 1955 / "Used as vegetable"

784 *Amaranthus* indet. Nyasaland / Newman & Whitmore 285 / 56 / "Leaves used in village as relish — cultivated"

785 *A. gangeticus* var. *angustior* Hainan / C. Lei 959 / 1933 / "fruit edible" / "Toy Tsoi"

786 *Amaranthus* indet. Hainan / C. Lei 684 / 1933 / "fruit edible"

787 *Calicorema capitata* S. W. Africa / R. Seydel 345 / 54 / "Sehr wertvoller Futterbusch fuer Wild u. Vieh"

788 *Celosia trigyna* Tanganyika / R. E. S. Tanner 5041 / 1960 / "Used for stomach pain" / "Kishubi: Ifungu"

789 *Centrostachys aquatica* Tanganyika / R. E. S. Tanner 1533 / 1953 / "Sukuma tribe: Roots pounded and cooked in porridge to treat syphilis in pregnant women" / "Kisukuma: Buhulula wa ngwisanga"

790 *Cyathula stringens* Tanganyika / R. E. S. Tanner 1281 / 1953 / "Sukuma tribe: Roots pounded and used as gargle to treat dry cough" / "Kisukuma: Yinza wa matogoro"

791 *Gomphrena globosa* El Salvador / S. Calderon 578 / 1922 / " 'Siempreviva' "

792 *Gomphrena* indet. Argentina / J. Montes 664 / 45 / "Uso: Medc. popular"

93   *Gomphrena* indet. Australia: Queensland / Mr. Hill s.n. / 1960 / "Suspected of causing paralysis in a 6 month old calf"

94   *Iresine calea* El Salvador / S. Calderon 143 / 1921 / " 'Siete pellejos, Comida de cabra' "

95   *I. paniculata* Mexico / G. Hinton s.n. / 39 / "Cure for Aslome bite" / "Aslome"

96   *I. celosia* Colombia / F. Fosberg 19724 / 1943 / "leaves rubbed and crushed in cold water for red irritated skin resulting from kidney trouble" / " 'contra fuego' "

97   *Iresine* indet. Argentina / J. Montes 766 / 45 / "Uso: Medicina popular" / " 'Caá-rurú huehi' "

98   *Iresine* indet. Argentina / Y. Peredo 386 / 1945 / "tiene olor desagradable"

99   *Pfaffia sericea* Argentina / J. Montes 544 / 1945 / "Sé usa en medicina popular. Sus raices, posee propiedades purgatorias" / " 'Batatilla' "

00   *Psilotrichum africanum* Tanganyika / R. E. S. Tanner 1368 / 1953 / "Leaves heated, steam used to treat eye trouble; roots pounded, taken in water for stomach pains; roots pounded, eaten for anklyostomiasis, rheumatism" / "Kisukuma: Nkumba mbizo"

NYCTAGINACEAE

01   *Abronia latifolia* Canada: Brit. Columbia / Calder, Savile & Taylor 23173 / 1957 / "sweet smelling . . . eaten by goats"

02   *Allionia* indet. Venezuela / F. Tamayo 3788 / 1950 / "Es pasto de todo ganado"

03   *Boerhaavia erecta* Mexico / G. B. Hinton et al. 9072 / 1936 / "Vernac. name: Pata Pollo"

04   *B. coccinea* Brazil / W. A. Archer 8388 / 1943 / "Root cooked as poultice for wounds. Root infusion taken as tea for liver complaints due to malaria. Formerly prepared locally and sold as liver remedy" / " 'pega-pinto' "
Paraguay / W. A. Archer 4776 / 1936 / "Sold in herb market extensively as a diuretic. Roots are used" / " 'caárurúpé' "
Tanganyika / R. E. S. Tanner 1283 / 1953 / "Leaves used to treat toothache — pounded, mixed with ground-nut oil, burnt in open lamp, fumes inhaled" / "Kisukuma: Kagigo"

05   *B. hirsuta* Argentina / J. E. Montes 515 / 1945 / "Forrajera para cerdos" / " 'Caá-rurú' "
Brazil / H. C. Cutler 12428 / no year / "Part of a collection of 42 remedies from one herb stall. For refreshing drink or diuretic. Take as desired. Twice this amount for 1 cent U.S." / "Pega-pinto"

806 *Boerhaavia* indet. Brazil / M. B. da Silva 43 / 1942 / "The fleshy root is sliced and dried in sun. The tea from dried root taken in morning while fasting. A cupful after bathing to purify the blood" / " 'solidonia', 'pega-pinto' "

807 *B. tetrandra* Marshall Is. / F. Fosberg 34163 / 1952 / "root said to be eaten" / " 'nat'to' "

808 *Commicarpus fruticosus* S. W. Africa / R. Seydel 451 / 1955 / "Vielleicht giftig?"

809 *C. plumbagineus* Tanganyika / R. E. S. Tanner 961 / 1952 / "Sukuma tribe: Roots used to treat stomach pains" / "Kisukuma: Namata"

810 *Commicarpus* indet. Tanganyika / R. E. S. Tanner 1031 / 1952 / "Sukuma tribe: pounded plant mixed with ghee and applied to swellings . . . very aromatic and sticky" / "Kisukuma: Ndingwandoni"

811 *Mirabilis jalapa* Peru / J. Schunke V. 1729 / 1967 / " 'Sacha pimienta' "

812 *Okenia hypogaea* Mexico / G. Hinton 6445 / 34 / "Uses: Medicinal" / "Vernac. name: Mata Cancer"

813 *Phaeoptilon spinosum* AFR. / R. Seydel 3177 / 1962 / "Die gefluegelten Fruechte, . . . bilden ein in der heissen trockenen Vorregenzeit hochgeschaetzter Futter fuer Wild und Vieh"

814 *Pisonia umbellifera* Fiji / A. C. Smith 7358 / 1953 / "fruit purple-tinged, very viscous, said to entangle pigeons" / " 'Ndainga' "

815 *Salpianthus purpurascens* Mexico / H. Gentry 6615 / 1941 / "Eaten as greens when tender" / "Fraile, quelite"

816 *Torrubia myrtiflora* Ecuador / Little & Dixon 21068 / 1965 / " 'jaboncillo' "

PHYTOLACCACEAE

817 *Agdestis clematidea* Mexico / F. Ventura A. 2627 / 1970 / "olor desagradable"
Mexico / Y. Mexia 8947 / 1937 / "Mephitic odor producing headache"

818 *Lophiocarpus polystachyus* S. W. Africa / R. Seydel 1244 / 57 / "Unangenehmer Geruch"

819 *Phytolacca icosandra* Mexico / G. Hinton 3860 / 33 / "Fruit macerated for washing clothes" / "Amolquelite"
Mexico / Y. Mexia 1661 / 1927 / "Leaves irritating to skin" / " 'Congeran' "

820 *Phytolacca* indet. Brazil / A. Silva 161 / 1944 / "Young shoots cooked as greens" / " 'carurú de soldado' "

821 *P. dodecandra* S. Rhodesia / S. M. Johnston 18443 / 1947 / "On termite mound. Unpleasant smell"

822 *Rivina humilis* Venezuela / T. Lasser 1388 / 1946 / "Las hojas hervidas son usadas como alimento"

23 *Trichostigma octandra* Haiti / E. C. & G. M. Leonard 15364 / 1929 / "fruit purplish black; astringent"

AIZOACEAE

24 *Glinus oppositifolius* Tanganyika / R. E. S. Tanner 1100 / 1952 / "Sukuma: for swollen fingers and toes—plant pounded up green, butter added and put on as paste" / "Kisukuma: Ndili va kuli"

25 *Hypertelis verrucosa* S. W. Africa / R. Seydel 4112 / 65 / "Von Kleinwild gefressen"

26 *Limeum suffruticosum* S. W. Africa / R. Seydel 573 / 1955 / "Wird vom Vieh gefressen"

27 *Mollugo verticillata* Argentina / Steinbach 2439 / 1916 / "pl. medicinal"

28 *Trianthema?* sp. nov. Tanganyika / R. E. S. Tanner 928 / 1952 / "Sukuma tribe: Plant burned and ash mixed with tobacco snuff 'to make strong' " / "Kisukuma: Kalandi"

29 *Aizoacea* indet. Peru / R. S. Williams 2538 / 1901 / " 'Verdolaga' "

PORTULACACEAE

30 *Portulaca lanceolata* U.S.: Tex. / V. L. Cory 51320 / 1945 / " 'purslane' "

31 *P. coronata* Honduras / Molina R. & Molina 25844 / 1970 / " 'Verdolaga' "

32 *P. pilosa* W.I.: Guadeloupe / P. Duss 2424 / 1892 / "Vulgo: Quinine"

33 *Portulaca* indet. Argentina / J. Montes 14804 / 1955 / "hojas tiernas comestibles" / " 'Verdolaga' "

34 *Portulaca* indet. Brazil / M. B. da Silva 171 / 1942 / "Eaten as greens" / " 'beldroega' "

35 *Portulaca* indet. Brazil / A. Silva 378 / 1946 / "plantas medicināis do mercado. Medicinal" / " 'amor crescido' "

36 *P. oleracea* Tanganyika / R. E. S. Tanner 2822 / 1956 / "Bondie: for asthma, pounded up and the liquid therefrom drunk" / "Kibondie: Msahau"

37 *Talinum aurantiacum* U.S.: Tex. / A. Schott s.n. / 52 / "The tuberous root having the taste of an irish potatoe is slightly bitter. It is, as I am told, with the Lipan Indians an antidote in fever & ague"

38 *T. paniculatum* Mexico / E. Matuda 17623 / 1948 / "used for cook"
Peru / C. M. Belshaw 3232 / 1937 / "always around cultivation; leaves edible"
Brazil / W. A. Archer 8389 / 1943 / ". . . sometimes cultivated. Stems & lvs. blanched for green salads, or cooked in soups" / " 'carurú' "

839   *T. caffrum* S. W. Africa / R. Seydel 497 / 55 / "Sehr geschaetztes futter fuer Wild und Vieh"

CARYOPHYLLACEAE

840   *Arenaria lanuginosa* Ecuador / W. H. Camp E-2523 / 1945 / ". . . used as a kidney remedy" / " 'Chinchi mani' "
Ecuador / W. H. Camp E-2698 / 1945 / ". . . used for the kidneys in connection with 'Drimaria', 'Yaco muyo', and 'Guarmi berros' " / " 'Chinchi mani' "

841   *Cucubalus baccifer* China: Kweichow / Steward, Chiao & Cheo 208 / 1931 / "Root said to be a tonic drug"

842   *Drymaria villosa* Guatemala / J. A. Steyermark 50528 / 1942 / "Reputed to be used in curing dysentery; boil plant and drink one glass with water of sugar, 3 times daily. (this was also stated to be dysentery remedy in San Juan Ixcoy)" / " 'millón' "

843   *D. cordata* Peru / J. Schunke 1987 / 1967 / "Las hojas utilizan para curar caracha" / " 'Challua Carilla' "
Ecuador / W. H. Camp E-2576 / 1945 / ". . . this one used for kidney and liver ailments" / " 'Drimaria del cerco' "

844   *Silene succulenta* Egypt / M. Drar 32/434 / 1930 / "Roots contain saponin. Employed by Bedouins for washing" / "Zazawa"

845   *S. venosa* Himalaya / W. Koelz 357 / 1930 / "Eat boiled when young (Charing Dashi)" / "Chaulzi"

846   *Spergula arvensis* U.S.: N.H. / E. G. Knight s.n. / 78 / " 'Corn Spurry' "
Canada: Québec / Boivin & Blain 412 / 1938 / " 'Grippe' "

847   *Stellaria longipes* Canada: Brit. Columbia / T. T. McCabe 7581 / 1940 / ". . . flourishing in great numbers about base of numerous large ant hills . . ."

848   *S. aquatica* China: Kwangtung / T. K. Peng 87, 2429 / 1921 / "Economic Uses: vegetable" / "Ngoh Cheung Tsoi"

849   *Vaccaria segetalis* U.S.: N.Y. / E. G. K. s.n. / 76 / " 'Cow Herb' "

850   *Caryophyllacea* indet. Argentina / A. G. Schulz 6.655 / 1947 / "Planta muy aromática"

851   *Caryophyllacea* indet. Ecuador / J. A. Steyermark 53264 / 1943 / "reputed to be good remedy for gonorrhea" / " 'chinchimani' "

NYMPHAEACEAE

852   *Castalia ampla* Mexico / J. N. Rovirosa 330 / 1890 / "*Pan-de-manteca* incolarum"

Surinam / Lanjouw & Lindeman 3104 / 1949 / "Vern. name: pankoekoewi-wirie (Sur., = pancake leaf)"

53 *C. elegans* Mexico / J. N. Rovirosa 674 / 1890 / "Nom. vernacul. *Pan-de-manteca*"

54 *Nymphaea amazonum* Argentina / T. M. Pedersen 4494 / 1957 / "smells strongly of acetone"

RANUNCULACEAE

55 *Aconitum heterophyllum* India / W. Koelz 21916 / 1948 / "native febrifuge"

56 *A. violaceum* W. Himalayas / W. Koelz 2912 / 1931 / "said to be very poisonous"

57 *Actaea alba* U.S.: N.J. / Herb. Torrey Club s.n. / 84 / " 'The root is of a dark brown color, and bears some resemblance to that of *Helleborus niger* for which it is said to be occasionally substituted. The taste is bitterish and somewhat acrid. In its operation on the system, the root is purgative and sometimes emetic, and is capable, in overdoses, of producing dangerous effects' " / "Red Baneberry, Christophskraut, Schwarzkraut"

58 *A. arguta* U.S.: Utah / K. L. Diem 16 / 1956 / "poisonous" / "Baneberry"

59 *A. rubra* Canada: Quebec / F. Fabius 74 / 46 / "Actée rouge (Poison de couleuvre)"
U.S.: N.Y. / T. F. Lucy 68 / 1899 / "Quite fragrant" / "Red Baneberry"
Canada: Newfoundland / R. H. Kimball 120 / 1919 / " 'Poison berry' "

60 *Anemone quinquefolia* U.S.: N.J. / Herb. Torrey Club s.n. / 84 / ". . . is said to act as a poison to cattle, producing bloody urine and convulsions. The anemones contain a peculiar crystallizable principle, named anemonin, convertible into anemonic acid by the action of alkalies" / "Windflower, Windröschen"

61 *A. alpina* Switzerland / Meisner s.n. / 1862 / " 'Gamswoad' (i.e. Gemsweide, Gemsfutter) bei den Bergleuten"

62 *Aquilegia canadensis* U.S.: N.Y. / J. F. Poggenburg s.n. / 84 / ". . . its sister, the European *Aquilegia vulgaris,* cultivated in our gardens, is thought to be poisonous. Like most of the Ranunculaceae, our Wild Columbine very likely possesses the same qualities" / "Wild Columbine, Adlerblume"

63 *A. formosa* Canada: Yukon Terr. / Porsild & Breitung 10504 / 1944 / "ill smelling"

64 *Cimicifuga racemosa* U.S.: N.Y. / Columbia College of Pharmacy Herb. s.n. / no year / "In Nord-Amerika ist die Wurzel als Radix Serpentariae nigrae officinell und gegen Schlangenbiss im Gebrauche"

865 *C. foetida* Himalaya: Lahul / W. Koelz 686 / 1930 / "Root used as spices. Root is fragrant" / "bodangir (China)"

866 *Clematis virginiana* U.S.: N.Y. / Torrey Bot. Club Herb. s.n. / 84 / ". . . occasionally used in domestic medicine . . . diuretic and sudorific, and will cure chronic rheumatism, indolent ulcers, and palsy" / "Virgin's Bower, Gemeine Waldvebe"

867 *C. dioica* El Salvador / P. C. Standley 19910 / 1922 / "Juice vesicant used for distemper in horses" / " 'Hierba del mendigo, Cabellos de ángel' " Mexico / T. MacDougall s.n. / 1969 / "Cure for 'pinto' " W.I.: Martinique / P. Duss 1761 / 1899 / "Liane aux serpents"

868 *C. haenkeana* Ecuador / W. H. Camp 2559 / 1945 / "The flowers are mashed in a vessel for 5 days and the mass then used to treat animals with mange — also used to kill lice on animals" / " 'Zehiguisa' "

869 *C. hirsuta* Kenya / Glover, Gwynne, Samuel & Tucker 2329 / 1961 / "Roots boiled by Masai and liquid mixed with milk and drunk as a cure for fever" / "Pisingwet (Kips.) Orkishushet (Masai)"

870 *C. leschenaultiana* China: Kwangtung / Lingnan (McClure) 13735 / 1926 / "Drug plant" / "Muk t'ung (Wood pith)"

871 *C. orientalis* var. *acutifolia* W. Himalayas / W. Koelz 2537 / 1931 / "said to be poisonous to touch" / "Beecho"

872 *Coptis teeta* Burma / F. Kingdon Ward 201 / 1939 / "febrifuge, diuretic, depressant stimulant, laxative, aphrodisiac, etc." / "Hwang lien (Chinese)"

873 *C. teeta* var. *chinensis* China: Szechuan / E. Faber 886 / no year / "Chinese drug plant" / " 'huang-lien' "

874 *Delphinium peregrinum* Kleinasien / Columbia College of Pharmacy Herb. s.n. / no year / ". . . wird in Neapel gegen Wechselfieber angewendet" / "Ausländischer Rittersporn"

875 *D. requienii* Corsica / Columbia College of Pharmacy Herb. s.n. / no year / "Liefert ein scharfes Pflanzengift"

876 *Helleborus foetidus* Germany / Columbia College of Pharmacy Herb. s.n. / no year / "Radix Hellebori foetidi" / "Stinckchristwurz"

877 *Hepatica americana* U.S.: N.J. / J. F. Poggenburg s.n. / 84 / "The leaves are mucilaginous and very slightly astringent; they were formerly much used in domestic practice in chronic coughs, but have fallen into entire neglect. Their infusion may be taken ad libitum. Name from a fancied resemblance to the liver in the shape of the leaves" / "Hepatica, Leberblümchen"

878 *H. nobilis* Germany / Meisner Herb. s.n. / 1826 / "Kraut u. Blume sind officinell" / "die Leberblume, das edle Leberkraut"

79 *Knowltonia brevistylis* S. Africa / J. Gerstner 6712 / 1948 / "Herb used medicinally"

80 *Ranunculus acris* U.S.: N.Y. / J. F. Poggenburg s.n. / 84 / "Crowfoot, when swallowed in the fresh state, produces heat and pain in the stomach, and, if the quantity be considerable, may excite fatal inflamation. The property for which it has attracted the attention of physicians is that of inflaming and vesicating the skin; and, before the introduction of the Spanish fly into use, it was much employed for this purpose" / "Tall Crowfoot or Buttercup, Hahnenfuss"

81 *R. hydrocharoides* var. *stolonifer* Guatemala / J. A. Steyermark 48309 / 1942 / "Reputed that if animals eat this plant, their liver becomes badly affected and they die" / " 'sandifuela' "

82 *R. biternatus* S. AM.: S. Georgia I. / R. C. Murphy 1919 / 1913 / "This species has a sharp spicy taste"

83 *R. bulbosus* Switzerland / Meisner Herb. s.n. / 1826 / "Sehr kaustisch — Besonders wird die Wurzel als blasen — ziehendes Mittel gebraucht" / "der knollige Hahnenfuss, Braunkraut"

84 *R. flammula* England / E. T. & H. N. Moldenke 9258 / 1936 / "juice acrid"

85 *R. ranunculoides* EUR. / Meisner Herb. s.n. / no year / "Chelidonium minus Officin." / "Milder Löffelkraut"

86 *R. multifidus* Kenya / Glover, Gwynne & Samuel 985 / 1961 / "Grazed by all domestic stock according to the Masai and Kipsigis (generally known to be poisonous)" / "Kipkoleit (Kipsigis) Olomba (Masai)"

87 *R. affinis* China: Hupeh / S. C. Sun 988 / 1933 / "Flower-incense"

88 *R. hirtellus* Himalaya: Lahul / W. Koelz 778 / 1930 / "Sharp, peppery taste" / "Tsemmi Mantok"

89 *R. laetus* Himalaya: Lahul / W. Koelz 619 / 1930 / "Flower used as a counterirritant" / "Dantuk bala"

90 *Ranunculacea* indet. Ecuador / J. A. Steyermark 52909 / 1943 / "herb . . . with stinging hairs" / " 'ortega' "

91 *Ranunculacea* indet. China: Szechuan / W. P. Fang 1045 / 1928 / "One kind of medicine"

LARDIZABALACEAE

92 *Akebia trifoliata* China: Kweichow / Y. Tsiang 5999 / 1930 / "Frt. green edible when ripe"

BERBERIDACEAE

893 *Berberis trianae* Colombia / J. M. Duque J. 2680 / 1946 / "N.v. 'Uña de gato' "

894 *Berberis* indet. Colombia / H. St. John 20848 / 1944 / " 'uvita' "

895 *Berberis* indet. Colombia / J. A. Ewan 16230 / 1944 / "lvs. only mildly tannic to taste"

896 *B. holstii* Malawi / A. J. Salubeni 388 / no year / "Said to be medicinal"

897 *B. petiolaris* Himalaya: Lahul / W. Koelz 690 / 1930 / "The root used to dye wool yellow" / "Kyarpe"

898 *B. pseudumbellata* Himalaya: Chamba / W. Koelz 1084 / 1930 / "Eaten by Kuluis. Make yellow color out of root" / "Gashambul"

899 *B. barandana* Philippine Is. / Quisumbing & Sulit 6311 / 1931 / "Fruit black, edible, used for dying"

900 *Epimedium acuminatum* China: Szechuan / W. P. Fang 802 / 1928 / "One kind of medicine"

901 *Mahonia nervosa* U.S.: Wash. / F. W. Johnson 1619 / 1913 / " 'Oregon Grape' "

902 *M. aquifolium* Mexico / H. S. Gentry 6248 / 1941 / "makes yellow dye for fabrics" / "Palo amarillo"

903 *M. longipes* Mexico / H. S. Gentry 6439 / 1941 / "used for dying wool yellow" / "Palo amarillo"

MENISPERMACEAE

904 *Abuta grandifolia* Brasil / R. de Lemos Fróes 20865 / 1945 / "usado para febres" / " 'Orelha de onça' "
Brazil / R. Souza 10426 / 1968 / "Fruit rather soft, with . . . juicy pulp which is edible" / "Pitombinha"
Brasil / Irwin & Soderstrom 6487 / 1964 / "Fruit . . . reputedly edible"
Brasil / Sidney & Onishi 1298, 519 / 1968 / "frutos amarelos, comestiveis e adocicados"
Brazil / W. R. Anderson 10512 / 1974 / "pulp around seed edible"

905 *A. grisebachii* Brazil / B. A. Krukoff's 6th Exped. to Brazil 7572 / 1935 / "Used in Curare"
Brazil / Prance, Dobzhansky & Ramos 20087 / 1973 / "Bush used as an arrow poison by Sanama Indians. Extract from boiling used pure, without other plant components" / " 'Macoli' (Sanama)"
Brazil / Prance, Fidalgo, Nelson & Ramos 21439 / 1974 / "Used as an arrow poison ingredient by Sanama Indians" / " 'Batewadodo' (Sanama)"

Colombia / R. Romero Castañeda 3837 / 1952 / "Frutos amarillo-par-
duscos, comestibles, acidulos"
Peru / Mathias & Taylor 3958 / 1959 / "very bitter" / " 'abuta negra' 'Tara-
pota-Chontasapa' "

06 *A. obovata* Venezuela / J. A. Steyermark 75344 / 1953 / "stems boiled in
water are used as a remedy for colds" / " 'huatacurán' "
Venezuela / Maguire, Wurdack & Keith 41700 / 1957 / "Sap used by
natives in treating 'pink-eye' " / " 'Bejuco amarillo' "

07 *A. rufescens* Brazil / Dr. Martius 46 / ca. 1830 / "Arbor venenum" / "Urari-
üva"

08 *A. solimoesensis* Peru / J. Schunke V. 5185 / 1971 / ". . . con sabor semia-
margo" / " 'Ampi amarillo' "
Brazil / N. T. Silva 845 / 68 / "Sample of stemwood for chemical studies"

09 *A. splendida* Brazil / Prance, Fidalgo, Nelson & Ramos 21440 / 1974 /
"Used as an arrow poison ingredient by Sanama Indians" / " 'Batewadodo-
peu' (Sanama)"
Peru / J. Schunke V. 23 / 1970 / ". . . la corteza con sabor amargo" /
"Abuta Negra"
Colombia / R. E. Schultes 3525 / 1942 / "probably same as 3522 but said
to be stronger for Curare" / "Nombre kofán: sa-pĕ-a"

10 *Abuta* indet. Peru / J. Schunke V. 7125 / 1974 / "El tallo . . . con sabor
muy amargo" / "Ampihuasca Negra"

11 *Anamirta cocculus* Philippine Is. / A. D. E. Elmer 11128 / 1909 / "The
flowers have a slight unpleasant odor" / " 'Array' "

12 *Anomospermum grandifolium* Ecuador / C. Fuller 113 / 53 / "Leaves only
used in preparing curare" / "Yana iluchi; Quichua"

13 *Antizoma calcarifera* S. W. Africa / R. Seydel 4263 / 65 / "Fischgift?"

14 *Arcangelisia flava* Philippine Is. / A. D. E. Elmer 12231 / 1910 / "The
natives use the leaves for head-pads when they have headache; the wood-
pulp for internal application to relieve stomach troubles" / " 'Alebodra' "

15 *Borismene japurense* Peru / Mathias & Taylor 5023 / 1960 / "Liana with
milky sap, very bitter"

16 *Caryomene* sp. nov. indet. Peru / J. Schunke V. 26 / 1970 / ". . . la corteza
. . . de sabor agrio dulce"

17 *Cissampelos pareira* Panama / Kirkbride & Duke 1370 / 1968 / "Used by
the Choco to see if the urine is OK or not. A primitive urinalysis, which
mixed with the urine, if a lot of foam results, indicates bad health"
Panama / J. A. Duke 14433(2) / 1967 / "Medicinal Plants of the Bayano
Cuna (Names provided by chiefs, witch doctors, or secretaries)"

Tanganyika / R. E. S. Tanner 1299 / 1953 / "Sukuma tribe: Roots pounded, taken orally in water to treat fever, taken dry for diarrhoea" / "Kisukuma: Nkuluwanti"

Tanganyika / R. E. S. Tanner 3327 / 1956 / "Bondei: For navel pains in children, leaves pounded up, soaked in water, and drunk" / "Bondei: Chegonde"

Tanganyika / R. E. S. Tanner 2938 / 1956 / "Zigua: For pains in the stomach, leaves pounded up and squeezed into water and drunk cold" / "Kizigua: Kigutwi cha buga"

918   *C. andromorpha* Venezuela / J. A. Steyermark 60765 / 1944 / "iguanas eat leaves of this as favorite plant" / " 'macumi' "
Brazil / Harley, De Castro & Ferreira 10742 / 1968 / " 'Taja Cobra' "

919   *C. fasciculata* Colombia / F. R. Fosberg 19997 / 1943 / "juice of stem said to be taken orally to cure snake-bite" / " 'añaguasca' "

920   *C. grandifolia* Venezuela / J. A. Steyermark 55963 / 1944 / " 'bejuco amargos' "

921   *C. laxiflora* Peru / G. Klug 2322 / 1931 / " 'Aypoyo' is employed as a remedy in diseases of the eye; the leaves are crushed, and of the juice from these, one or two drops are dropped into the affected eye"

922   *C. ovalifolia* Brazil / B. & C. K. Maguire 40210 / 1954 / "Roots said to be a remedy for venereal diseases"

923   *Cocculus diversifolius* Mexico / M. T. Edwards s.n. / 1937 / " 'Yerba mora' "

924   *C. hirsutus* Nyasaland / L. J. Brass 17562 / 1946 / "on termite mounds" / "Native name — Nangunega (Chinyanja)"

925   *Curarea candicans* Brazil / Fróes 67 / 1942 / "medicine for home made" / "Buta"
Brit. Guiana / Brit. Guiana For. Dept. 2689 / 1938 / "Leaves associated with bark collected for the Medical Research Council" / " 'Teteabo' (Arawak)"

926   *C. tecunarum* Brazil / Prance et al. 16453 / 1971 / "The stem crushed and placed in water & stirred. The mixture drunk as a contraceptive by Deni Indians" / " 'Bekú' "
Brasil / J. Barbosa Rodriguez s.n. / 1971 / "Com esta Menispermaceae, fazem os Indios Maulés das folhas, torrar o parica"
Brazil / B. A. Krukoff's 6th Exped. to Braz. Amaz. 7578 / 1935 / "Used in Curare"
Colombia / Fabriciano Diaz M. 36 / no year / "Corteza de sabor amargo . . . La corteza y las hojas mezcladas con otras plantas las utilizan los indios Huitotos para preparar veneno para las flechas. Luis Vargas, Indio-Huitoto" / "N.V. Taufe lleida (Huitoto)"

Colombia / Fabriciano Diaz M. 10 / no year / "Según el indio Luis Vargas los indios Huitotos la utilizan como veneno para las flechas, para esto mezclan la corteza y las hojas con otras plantas"
Ecuador / R. C. Gill 35 / 1938 / "extra potent curare" / "Yana Pala Arigo"
Ecuador / M. Olalla for R. C. Gill s.n. / 1946 / "Se machan el bejuco y se pone al cosimiento hasta que se reduzea a extracto" / "Zanispa pala Augo Quechua"

27 *C. toxicofera* Peru / J. Schunke V. 7125 / 1974 / "El tallo . . . con sabor amargo"
Peru / Y. Mexia 6321a / 1931 / "Said to be one of woody vines from which poison is made for darts of Indians"
Colombia / R. E. Schultes 5526 / 1943 / "Said to have been used formerly by Karijona Indians in arrow-poison"
Colombia / C. O. Grassl 10076 / 1943 / "Used as cure for fevers" / "n.v. 'abuta' "
Brazil / R. L. Fróes 26364 / 1950 / "fruto verde, comestivel"
Brazil / Prance et al. 13931 / 1971 / "Vicinity of Jamamadi Indian village . . . Bark used as ingredient of arrow and dart poison, mixed with Strychnos (13929), Annonaceae (13936) and Sapindaceae (13938)" / " 'Bicava' "

28 *Menispermum diversifolium* China: Kweichow / Steward, Chiao & Cheo 658 / 1931 / "Medicinal" / "Shan Wu Kwei"

29 *Orthomene schomburgkii* Brazil / G. T. Prance et al. P18634 / 1974 / " 'Pimenta de reino braba' "
Peru / F. Woytkowski 6302 / 1961 / "Natives say that no animal eats the fruit"
Peru / T. J. Zavortink 2285 / 1962 / "smell of chlorox in broken leaf"
Brazil / G. A. Black 47-1995 / 1947 / " 'Grão de guariba' "

30 *Pericampylus glaucus* China: Kwangtung / Lingnan (To & Ts'ang) 12096, 12654 / 1924 / "Im t'ang, (Salt vine)"
Philippine Is. / A. D. E. Elmer 8881 / 1907 / "the leaves are roasted or steamed and placed upon swellings and bruises—said to be a sure cure" / " 'Pisok' "

31 *Sciadotenia pachnococca* Brazil / Prance et al. 15558 / 1971 / "Root bark scraped for toothache by Makú Indians"

32 *S. peruviana* Peru / R. Kayap 97 / 1972 / "Not poisonous, beaten into pulp, put in tinaca and boiled well as blow gun dart poison" / "Local name: dúpam"

33 *S. toxifera* Peru / J. Schunke V. 4637 / 1971 / "Los indios lamistas utilizan para preparar curare; asi mismo toman en maceración con aguardiente para curarse de diabetes y la terciana. El tallo es aplastado con sabor amargo" / " 'Abuta Hembra' "
Peru / Mathias & Taylor 6110 / 1962 / "large liana used in curare mix" / " 'Abuta negra' "

Ecuador / R. C. Gill 36 / 1938 / "Use root scrap up only — curare" / "Yana Qlucha"
Colombia / R. E. Schultes 3866 / 1942 / "Root & stem used in Kofan curare" / "ya-pé-he"
Ecuador / C. C. Fuller 117 / 53 / "Used occasionally in curare prep." / "Soliman caspi; Quichua"

934   *Sciadotenia* sp. nov. indet. Peru / Mathias & Taylor 3933 / 1959 / "very bitter" / " 'ampihuasca' "

935   *Telitoxicum* sp. nov. indet. Brazil / B. A. Krukoff's 6th Exped. to Braz. Amaz. 7566 / 1936 / "Used in Curare"

936   *Tinomiscium* indet. Philippine Is. / Ramos & Edano 75340 / 1928 / "La fruta de esto hemos cogido del Dr. Guzzen por quimical investigacion"

WINTERACEAE

937   *Drimys granadensis* Costa Rica / /A. Mardríz V. AMV-25 / 66 / "Chile, Quiebra muelas"
Costa Rica / K. Lems 630712 / 1963 / "Bark aromatic"
Colombia / J. Triana s.n. / 1851–1857 / "Vulgo: *Palo de aji* á Bogota, *Quinon* á Pamplona, *Canelo* á Antioquia, *lupis* á Ocaño"

938   *D. brasiliensis* Brazil / P. Occhieni 708 / 1946 / " 'casca Danta' 'Para Tudo' "
Brazil / Reitz & Klein 3586 / 1956 / "Casca d'anta"

939   *D. brasiliensis* var. *roraimensis* Venezuela / J. A. Steyermark 58750 / 1944 / "Used for the bark, for headaches and fever" / " 'Hue-tu-yek"

940   *Drimys* indet. Ecuador / J. A. Steyermark 53339 / 1943 / "Bark is peppery and used for curing wounds or cuts on mules and other beasts of burden" / " 'cascasrilla picante' "

941   *Drimys* indet. Colombia / Fosberg & Villareal 20612 / 1943 / "bark extremely pungent to taste"

942   *D. lanceolata* AUS. / O. Kuntze 20113 / 1904 / " 'Pepper tree' "

ILLICIACEAE

943   *Illicium floridanum* U.S.: Miss. / Webster & Wilbur 3448 / 1950 / "highly aromatic"
U.S.: Fla. / Herb. Chap. s.n. / no year / " 'Anise tree' "

944   *I. pachyphyllum* China: Kwangsi / W. T. Tsang 24425 / 1934 / "fr. edible" / "Pat Kok Shue"

SCHISANDRACEAE

945   *Kadsura heteroclita* China: Kwangsi / R. C. Ching 7232 / 1928 / "all parts aromatic"

946   *K. longepedunculata* China: Kwangtung / W. T. Tsang 20826 / 1932 / "Fr. red, edible" / "Sai Ng'ang F'an Tün"

ANNONACEAE

947   *Alphonsea arborea* Philippine Is. / M. Oro 220 / 1929 / "Fruit . . . tasting sweet"

948   *Annona palmeri* Mexico / Y. Mexia 12219 / 1926 / " 'Platanillo' "

949   *A. pittieri* Brit. Honduras / P. H. Gentle 2931 / 1939 / " 'wild custard apple' "

950   *A. purpurea* El Salvador / P. C. Standley 20547 / 1922 / "fr. yellow, said to be unhealthy" / " 'Sincuyo, Sincuya' "

951   *Annona* indet. Mexico / Rose, Standley & Russell 14358 / 1910 / "Bought in market"

952   *Annona* indet. Mexico / Rose, Standley & Russell 13995 / 1910 / "Bought in market"

953   *A. jamaicensis* W.I.: Trinidad / W.E. Broadway 8704 / no year / "Fruit edible"
W.I.: Jamaica / W. Harris 12006 / 1915 / " 'Wild Sour Sop' "

954   *A. ambotay* Brazil / B. A. Krukoff 's 6th Exped. to Braz. Amaz. 7636 / 1936 / ". . . with very aromatic bark. Used in Curare of Tecuna Indians"

955   *A. dioica* Bolivia / H. H. Rusby 1352A / 1921 / "Fruit edible"

956   *A. excellens* Peru / J. Schunke V. 2493 / 1968 / " 'Sacha chirimoya' "

957   *A. hypoglauca* Colombia / E. L. & R. R. Little 9602 / 1945 / "Frt . . . fleshy, edible" / " 'anona' "

958   *A. scandens* Bolivia / H. H. Rusby 2079 / 1922 / "Edible fruit called 'Chirimolla' "

959   *A. sericea* Surinam / J. C. Lindeman 6133 / 1955 / "fruit green, edible" / "Nomen vernac.: zuurzak; oeroesoeroe (Car.)"

960   *Annona* indet. Brazil / W. A. Archer 8110 / 1943 / "Frt . . . Edible" / " 'araticú' "

961   *Annona* indet. Fr. Guiana / Fr. Guiana For. Service 7039 / 1955 / "Fruits comestibles—atteignent la grosseur d'une poire" / "Bouchi ATOUKOU (Idiome Poromako)"

962   *Annona* indet. Bolivia / H. H. Rusby 1608 / 1921 / ". . . large edible fruit, called 'Sinini' "

963 **Annona** indet. Brazil / A. Silva 54 / 1944 / "Flr. white, with fetid odor" / " 'biriba-rana' "

964 **A. chrysophylla** Tanganyika / R. E. S. Tanner 1185 / 1953 / "Fruits eaten . . . by Sukuma tribe: for snake bite when teeth of snake remain in flesh; for swellings or boils—roots pounded and used" / "Kisukuma: Mkonola"

965 **A. senegalensis** Tanganyika / R. E. S. Tanner 3961 / 1958 / "Zigua: for swollen stomach, roots and/or bark boiled and the water drunk" / "Zigua: Mtomojkwe"
Tanganyika / R. E. S. Tanner 3883 / 1957 / "Bondei: for pains in center of stomach and for palpitation; roots boiled and the water drunk" / "Swaheli: Mbokwe Bondei: Mtonkwe"
Tanganyika / R. E. S. Tanner 3926 / 1958 / "Bondei: for pain near navel in women, roots boiled and the water drunk" / "Bondei: Mtonkwe mzitu"

966 **Annona** indet. Samoa / E. Christophersen 345 / '29 / "Fruit is eaten, of root is made beer (fa'amafū)" / "Sasalapa"

967 **Artabotrys hongkongensis** China: Kwangtung / Y. Tsiang 2765 / 1929 / ". . . very aromatic"

968 **Asimina longifolia** var. **spatulata** U.S. & CAN. / Chapman s.n. / ca. 1850 / "A decoction of the roots of this plant *is said* to be the 'black drink' of the Indians, preparatory to their 'green corn dance'. If this be the plant it is diaphoretic and a *drastic purgative*"

969 **A. parviflora** U.S.: Tex. / V. L. Cory 52628 / 1947 / " 'papaw' "

970 **A. reticulata** U.S.: Fla. / R. Kral 2535 / 1956 / "similar to avocado in taste"

971 **A. speciosa** U.S.: Ga. / F. Harper 333 / 1931 / " 'Pawpaw"

972 **Cremastosperma cauliflorum** Ecuador / Cazalet & Pennington 7528 / 62 / "fls. with vanilla scent"

973 **Cymbopetalum penduliflorum** Brit. Honduras / P. H. Gentle 3290 / 1940 / " 'Mata boni', 'guanabano' "

974 **C. brasiliense** Brit. Guiana / A. C. Smith 2832 / 1937 / "Leaves mashed and used in water for bathing to reduce fevers by Waiwais"
Surinam / B. Maguire 23799 / 1944 / "bark and fruit used for ringworm"

975 **Dasymaschalon clusiflorum** Philippine Is. / A. D. E. Elmer 12803 / 1911 / "pedicels . . . juicy and of a pleasant sourish flavor"

976 **Duguetia panamensis** Panama / G. P. Cooper 418 / 1928 / "Wood fragrant"

977 **D. asterotricha** Brazil / Campbell, Nelson, Ramos & Insley P21255 / 1974 / "Bark scraped from stem, mixed with cold water and included in arrow poison with Strychnos 21257, Meliaceae 21259, 21269" / " 'Boa (Jamamadi)' "

978   *D. echinophora* Brazil / R. Fróes 1905 / 1932 / "bark used as a cure for fever" / " 'Ameijú' "

979   *D. flagellaris* Peru / J. Schunke V. 3840 / 1970 / "la corteza es aromática con sabor muy amarga y los indios toman en aguardiente de caña para combatir dolores reumáticos" / " 'Espintana Negra' "

980   *D. hadrantha* Peru / J. Schunke V. 3140 / 1969 / "la corteza es aromática"

981   *D. riberensis* Venezuela / Aristeguieta & Zabala 7075 / "Frutos comestibles" / " 'Anoncillo' "

982   *D. spixiana* Brazil / B. A. Krukoff's 6th Exped. to Braz. Amaz. 7659 / 1935 / "Used in Curare"

983   *D. stenantha* Brasil / Ducke 1796 / 1940 / " 'Mão de cabra' vel 'Pé de jabotí' "

984   *Fissistigma maclurei* Hainan / C. I. Lei 615 / 1933 / "fruit edible" / "Hak Tang Tz"

985   *F. oldhami* China: Kwangtung / W. T. Tsang 21414 / 1932 / "Fr. yellow, edible" / "Shan Lung Ngan T'ang"

986   *F. uonicum* China: Kwangsi / W. T. Tsang 23923 / 1934 / "fr. yellow, edible" / "Kok Lok Tsz Tang"

987   *Fusaea longifolia* Colombia / R. Romero Castañeda 4933 / 1954 / "N.v. 'Chirimoyo' "

988   *Goniothalamus amuyon* Philippine Is. / S. Reyes 29827 / 1925 / "It is said that Japanese used to cut this tree; also as they perhaps believed it to be second class Sandal wood. Wood is scented when newly cut. Locally, fruit is used for stomach trouble" / "Saguiat"

989   *Guatteria dumetorum* Panama / T. B. Croat 7738 / 1969 / "tree infested with ants"

990   *G. macrocarpa* Peru / J. Schunke V. 4294 / 1970 / "La corteza utilizan los nativos para curarse del rumatismo, tomando en aguardiente" / " 'Hicoja negra' "

991   *G. pleiocarpa* Peru / J. M. Schunke 153 / 1935 / "fruit yellow; edible" / " 'Anonia' "

992   *Guatteria* indet. Ecuador / W. von Hagen 286 / no year / "Bark used for rope-string to bind up rights in house construction. Also used medicinally for stomach worms" / "Yais Language Jiv."

993   *Guatteriopsis blepharophylla* Brazil / Harley & Souza 11118 / 1968 / "Fruit juicy purple, edible"

994   *G. sessiliflora* Peru / J. Schunke V. 2703 / 1968 / "La corteza es aromática y fibrosa" / " 'Auca Hicoja' "

995   *Mitrephora maingayi* Hainan / F. A. McClure 9593 / 1921 / ". . . very fragrant; fruits edible" / "name reported, Shan tsiu"

996 **Oxandra laurifolia** Haiti / E. C. & G. M. Leonard 12513 / 1923 / "fruit purplish black with sweetish resinous taste"
W.I.: Nevis / J. S. Beard 460 / 1944 / " 'Okra' "
W.I. / Herb. Collegii Pharmaciae neo-Eboracensis s.n. / no year / "bears a fruit of the size of one of the largest apples . . . eatable—meat white—with many seeds similar in appearance to those of the water melon. The leaves fragrant" / "Mamon"

997 **Oxymitra acuminata** Brit. N. Borneo / B. Agullana 838 / 1928 / "The fruit has a repulsive odor"

998 **Pachypodanthium simiarum** Port. Congo / J. Gossweiler 6981 / 1917 / "frts . . . devoured by birds & monkeys"

999 **Polyalthia mayumbensis** Port. Congo / J. Gossweiler 6845 / 1916 / "fl . . . pungently odiferous"

1000 **P. korinti** Ceylon / Wirawan, Cooray & Balakrishnan s.n. / 1969 / "fruit . . . with sweet taste"

1001 **P. suberosa** Siam / G. W. Groff 6095 / 1920 / "Fruits: edible"

1002 **P. nitidissima** Australia: Queensland / Smith & Webb 3126 / 1947 / "Leaves aromatic when crushed"

1003 **P. grandiflora** Philippine Is. / A. D. E. Elmer 12854 / 1911 / "the flowers harbor numerous black ants and more or less of white scale insects around its stalks"

1004 **Rollinia jimenezii** Brit. Honduras / W. A. Schipp 408 / 1929 / "Flesh white, very sour"

1005 **R. rensoniana** El Salvador / P. H. Allen 6843 / 1958 / "Fruit reported edible" / " 'Chulumuyu' "
El Salvador / Allen & van Severen 6920 / 1958 / "Fruits yellowish & edible when ripe" / "Local name reported as 'Churumuyu' "

1006 **R. williamsii** Bolivia / H. H. Rusby 841 / 1921 / "Fruit delicious"
Bolivia / R. S. Williams 538 / '01 / " 'Custard apple' "

1007 **Rollinia** indet. Bolivia / J. Steinbach 5064 / 1920 / "Su fragancia de cirhuelas se apérciba de lejos" / "Cuhuchi falso"

1008 **Sapranthus campechianus** Brit. Honduras / P. H. Gentle 3980 / 1942 / " 'bastard custard apple' "

1009 **S. longepedunculatus** Mexico / G. B. Hinton 10318 / 37 / "flower dark purple—stinks"

1010 **Unonopsis pittieri** Panama / Kirkbride & Duke 1237 / 1968 / "a favorite firewood in the mts. that is supposed to burn green, slightly aromatic; ants attracted to a gray aromatic substance in the axils of the inflorescences" / " 'coavi' (Choco), 'yayz' (Negro)"

011 *U. mathewsii* Peru / J. Schunke V. 1611 / 1967 / "se utilizan en maceracion del aguardiente para combatir al reumatismo" / " 'Hicoja negra' "

012 *Uvaria acuminata* Tanganyika / R. E. S. Tanner 2778 / 1956 / "fruits . . . with thick, white, sweet juice"
Tanganyika / R. E. S. Tanner 3374 / 1957 / "Sigua: For cough, leaves boiled and the water drunk" / "Zigua: Msofu Mbago"
Tanganyika / R. E. S. Tanner 3046 / 1956 / "Fruits eaten. Zigua: for sharp stomach pain; roots cooked with flesh of black chicken and eaten" / "Zigua: Msufu"

013 *U. lucida* Tanganyika / R. E. S. Tanner 3680 / 1957 / "Fruits edible" / "Bondei: Msofu"

014 *U. microcarpa* China: Kwangsi / W. T. Tsang 24537 / 1934 / "fr. yellowish red, edible" / "Shap Man Tsai Shan"
China: Kwangtung / F. A. McClure 00638 / 32 / "Leaves used to make yeast cakes (tsau peng)" / "T'ang Long"
China: Kwangtung / Siu Iu Nin 125 / 1928 / "Edible" / "Kai Kung Shue"

1015 *U. purpurea* China: Canton / F. A. McClure 8944 / 1922 / "fls. . . . very fragrant; leaves used to make cakes and rice whisky; also makes a fine ornamental" / "Sai t'ang chau kung, Tai t'ang chau kung"

1016 *Xylopia frutescens* Honduras / C. & W. von Hagen 1330 / 1938 / "wood used for harpoon sticks; oil from fruit used on hair" / " 'Sihna' "

1017 *X. glabra* Jamaica / Britton & Hollick 2792 / 1908 / "Crushed leaves spicy in odor"

1018 *X. aromatica* Nouvelle Grenade / J. Triana s.n. / 1851–1857 / "Vul: fruta del burro or Sembe"

1019 *X. chivantinensis* Brazil / G. C. G. Argent 6892 / 1968 / "bark . . . smell of melon. Staining mauve in contact with steel"

1020 *X. sericea* Venezuela / J. A. Steyermark 88433 / 1961 / " 'fruta de burro' "

1021 *X. surinamensis* Brasil / Pires, Rodrigues & Irvine 51447 / 1961 / "crushed fruit has medicinal scent"

1022 *Xylopia* indet. Brit. Guiana / D. H. Dans 166 / 1967 / "latex with resinous smell"

1023 *X. aethiopica* Liberia / G. P. Cooper 174 / 1929 / ". . . edible fruits"

1024 *Annonacea* indet. Bolivia / O. E. White 1847 / 1921 / "Edible fruit"

1025 *Annonacea* Brazil / R. Fróes 1774 / 1932 / "Fruit and wood yielding a poisonous substance. Timber used in building"

1026 *Annonacea* Brit. Guiana / D. B. Fanshawe F2103 / 44 / "ripe fr . . . edible with flavour of sour apricots" / "Vern. Name—Kakirio"

1027 *Annonacea* indet. Liberia / R. M. Warner 4 / 1942–43 / "Strong nerve medicine"

MYRISTICACEAE

1028   *Horsfieldia hellwigii* New Guinea / J. H. L. Waterhouse 318 / 1934 / "Sap is used as a bird-lime" / "Com. Name: Mu"

1029   *H. irya* Philippine Is. / M. A. Tamayo 6 / 1934 / 'The fruits are eaten by the people" / "Warun"

1030   *Iryanthera juruensis* Colombia / Schultes & Cabrera 13731 / 1951 / "Fruits for fish-bait" / "Puinave = shaw"

1031   *I. laevis* Peru / A. Acosta M. 2 / 1968 / " 'Copal Caspi Blanco' "

1032   *Myristica castaneaefolia* Fiji Is. / Degener & Ordonez 14134 / 1941 / "Native pigeons swallow whole seeds"

1033   *M. elliptica* Borneo / J. & M. S. Clemens 20118 / 1929 / "Visited by Tucons"

1034   *M. philippensis* Philippine Is. / Ahern's collector 99 / 1904 / ". . . the fruit being used somewhat for a condiment for the same purposes as the true nutmeg" / "T., Dugoan, Anis cahoy"

MONIMIACEAE

1035   *Daphnandra aromatica* Australia: Queensland / C. J. Trist 35 / 32 / " 'Canary Sassafras' "

1036   *Mollinedia ruae* Nicaragua / A. Molina R. 20447 / 1967 / " 'Anonillo' "

1037   *Siparuna colimensis* Mexico / Kerber 13490 / 1880 / "Scheinfrucht sehr beissend, sindt essbar" / " 'Limon Coáhuil' "

1038   *S. guianensis* Panama / T. B. Croat 15267 / 1971 / "sap with foul aroma"
Brazil / G. T. Prance et al. 13941 / 1971 / "Vicinity of Jamamadi Indian village . . . Leaves made into tea which is drunk as a cure for rheumatic pain"
Brit. Guiana / S. G. Harrison 1463 / 1958 / "Fruits . . . highly aromatic"
Brit. Guiana / Jenman 4093 / 1887 / "an infusion made by boiling the leaves used to produce sweating in fever" / " 'Moonerydany' or fever-bush"

1039   *S. nicaraguensis* Costa Rica / Burger & Stolze 4953 / 1968 / "Sharp, spicy odor when crushed"
Honduras / C. & W. von Hagen 1036 / 1937 / "fruit with Citrus odor" / " 'Limonocillo' "
Brit. Honduras / P. H. Gentle 3061 / 1939 / " 'bastard fig' "
Mexico / Y. Mexia 1528 / 1927 / "strongly aromatic" / " 'Mata-chinche' 'Manzanillo' "

040   *S. pauciflora* Costa Rica / Burger & Liesner 6851 / 1969 / "All parts spicy or sweet aromatic when crushed"
Panama / W. H. Lewis et al. 3066 / 1967 / ". . . lemon odor"

041   *S. caloneura* W. I.: Dominica / G. P. Cooper 56 / 1933 / "Leaf has a spicy odor and flowers are faintly fragrant" / " 'Petite Bous Marbre' (Little marble wood)"
W.I.: St. Vincent / H. H. & G. W. Smith 24 / 1890 / "the leaves when bruised, have a very strong and disagreeable odour"

042   *S. bifida* Peru / J. Schunke V. 2708 / 1968 / "la corteza tiene olor a limón'

043   *S. chiridota* Colombia / F. R. Fosberg 20167 / 1943 / ". . . herbage with lemon-grass odor"

044   *S. cuyabana* Brazil / A. Loefgren 13008 / 87 / " 'Limão bravo' "
Brazil / Prance & Silva 59024 / 1964 / ". . . leaves orange smelling"
Brazil / Y. Mexia 4418 / 1930 / "fruits have strong, pleasing odor of mint"

045   *S. decipiens* Brit. Guiana / Brit. Guiana For. Dept. F 3426 / 52 / ". . . flesh greenish, unpleasantly scented; juice leaves a temporary brown stain on the fingers" / "Muniridan"
Brit. Guiana / D. B. Fanshawe 2061 / 44 / "plant has an unpleasant smell" / "Muneridan"

046   *S. grandiflora* Colombia / F. R. Fosberg 20486 / 1943 / ". . . with a disagreeable odor . . . lvs. used for rheumatism" / " 'conjon de chucha' "

047   *S. magnifica* Peru / J. Schunke V. 4759 / 1971 / "Toda la planta tiene olor nauseabundo, al frotar con al mano" / " 'Picho Huayo' "

048   *S. riparia* Colombia / H. H. Smith 2508 / 1898–'99 / "The leaves have a very strong disagreeable odor. I attribute to this plant an eruption, similar to that of poison ivy which attacked my hands and those of the man, who was working with me; it lasted several days"

1049   *S. sarmentosa* Brazil / B. A. Krukoff's 7th Exped. to Braz. Amaz. 8512 / 1936 / "fr. taste of pepper"

1050   *Siparuna* indet. Venezuela / A. L. Bernardi 3311 / 1956 / "Hoja del diablo"

1051   *Siparuna* indet. Ecuador / J. A. Steyermark 54172 / 1943 / "Infusion of leaves drunk for stomach ache" / " 'limoncillo' "

1052   *Siparuna* indet. Brazil / M. Barbosa da Silva 16 / 1942 / "Leaves used in bath water for children to make their legs strong. Leaf aromatic" / " 'capitiú' "

1053   *Siparuna* indet. Ecuador / J. A. Steyermark 53555 / 1943 / "Related to class of Siparuna called 'guayusa' and used for preparing a tea"

1054   *Siparuna* indet. Ecuador / J. A. Steyermark 52620 / 1943 / "shrub with fragrance of certain species of *Piper*. Heated leaves applied to rheumatic areas of body" / " 'arna' "

1055 *Siparuna* indet. Venezuela / Aristeguieta & Lizot 7366 / 1970 / "Frutos comestibles. Indios Guaicas (Yanomanö)"

1056 *Siparuna* indet. Venezuela / A. L. Bernardi 6494 / 57 / "Medicinal, baños para reumatismo" / "N. V. Pata de danta"

1057 *Siparuna* indet. Colombia / J. A. Duke 9866(3) / 1967 / ". . . used medicinally by the La Nueva Negroes" / " 'limoncillo' "

1058 *Siparuna* indet. Venezuela / T. Lasser 1381 / 1946 / "Planta medicinal"

1059 *Siparuna* indet. Ecuador / J. A. Steyermark 52816 / 1943 / "Infusion made of the leaves for treating stomach-ache, and for giving women after childbirth; prepared in form of tea" / " 'guayusa' "

1060 *Monimiacea* indet. Peru / J. Schunke V. 1091 / 1966 / "Nombre vulgar: Limoncillo"

1061 *Monimiacea* indet. New Guinea / J. H. L. Waterhouse 284 / 1907-35? / ". . . edible fruit" / "Lapua"

LAURACEAE

1062 *Actinodaphne pilosa* China: Kwangtung / F. A. McClure 00570 / 1931 / "The shaved wood yields mucilage when soaked in water" / "Taai Loh Shaan, Taai Shaan Kau, Paau Fa"
China: Kwangtung / F. A. McClure 0051 / 1931 / "This mucilage is used (1) as size in making (a) P'ai Chi at Ma Haang (b) Kam Pok Chi (2) by women for hair dressing" / "Taai Shaan Kaau, Kaau Muk, P'aau Fa"

1063 *Aiouea* indet. Bolivia / O. E. White 1991 / 1921 / "Bark has strong odor of Coto Bark"

1064 *Alseodaphne gigaphylla* Java / Kostermans 9999 / 1955 / "Fls. smelling of nutmeg"

1065 *Beilschmiedia curviramea* Venezuela / L. M. Berti 522 / 1964 / "N.v.: Aguacatillo Moises"

1066 *B. mannii* Liberia / G. P. Cooper 314 / 1929 / "Be-ey, Zoe-kpor (Spicy Cedar)"

1067 *B. fordii* China: Kwangtung / Lingnan (To & Ts'ang) 12035 / 1924 / "Cheung shu (Camphor tree)"

1068 *Cassytha filiformis* Bahama Is. / S. R. Hill 2196 / 1974 / " 'Love vine' "
Caroline Is. / F. R. Fosberg 47300 / 1965 / "Used medicinally" / " 'erau' or 'tirh tige' "
Marshall Is. / F. R. Fosberg 34080 / 1952 / ". . . fruit . . . sweetish but with bad flavor" / " 'kaanin' "
Mariana Is. / M. Evans 726 / 1965 / "Boiled and mixed with other leaves and drunk for fever" / " 'BUDILIGAS' "

069   *Cinnamomum inunctum* China: Kwangtung / Y. Tsiang 2610 / 1929 / "Fruit green, greatly infested by insects"

070   *Cinnamomum* indet. India / T. R. Chand 6911 / 1953 / "root cinnamon-scented"

071   *Cinnamomum* indet. India / W. N. Koelz 30045 / 1952 / "Tree, leaves aromatic"

072   *C. carolinense* Caroline Is. / F. R. Fosberg 32378 / 1950 / ". . . pleasantly aromatic when broken. Brought in by informant Alkong to substantiate ethnobotanic information" / " 'Ooth' "

073   *C. doederleinii* Bonin Is. / F. R. Fosberg 31532 / 1950 / ". . . odor of Sassafrass when broken. 'Sassafrass' used for medicinal tea in old days"

074   *C. fitianum* Fiji Is. / A. C. Smith 614 / 1933 / "Bark gives an oil which is used on skin" / " 'Mbatho' "

075   *C. grandifolium* Brit. N. Borneo / Sator 813 / 1928 / "Trunk medicinal" / "Patosuk (Sungei)"

076   *C. massoia* var. *rotundatum* Dutch New Guinea / M. Moszkowski 476 / 1911 / "Bäumchen mit aromatisch riechender Rinde, die angeblich gegessen wird"

077   *C. mercadoi* Philippine Is. / A. D. E. Elmer 10473 / 1909 / "As soon as the tree was cut into a sweet aromatic odor was detected by me a few meters from the native axeman. As I inquired about the odor he pointed to the wood and bark and quick-handed me a piece of wood and placed a chunk of bark into his mouth; they use the bark as a spice or condiment in their cookings" / " 'Caningag' "

078   *C. mindanaense* Philippine Is. / A. L. Zwickey 384 / 1938 / "lvs. cooked with fish . . . bark has odor of cinnamon" / " 'Kumi' or 'Komi' (Lan.)" Philippine Is. / J. Fontanoza 75 / 1929 / "Leaves and bark of cinnamon flavor . . . medicinal and flavoring" / "Kanela (Viz.)"

079   *Cinnamomum* indet. Indonesia / Kostermans 18542 / 1961 / "Bark is used as a medicinal tea and sold in the market as 'kaju lawang' "

080   *Cinnamomum* indet. Indonesia / Kostermans 18749 / 1961 / "Bark sold as a medicine" / " 'kaju lantang' "

081   *Cinnamomum* indet. Brit. N. Borneo / J. & M. S. Clemens 51036 / 33 / "Strong sassafras odor"

082   *Cinnamomum* indet. Borneo / A. Kostermans 9283 / 1954 / "Wood . . . with a very agreeable spicy smell"

083   *Cinnamomum* indet. New Guinea / Aët (Exped. Lundquist) 425 / 1941 / "Bark yields a medicinal oil"

084   *Cryptocarya aschersoniana* Brasil / F. C. Hoehne s.n. / 1931 / "Nome vulgar: 'Batalha' ou 'Canela Batalha' "

1085   *C. moschata* Brasil / Reitz & Klein 2413 / 1956 / "Canela fogo"

1086   *Cryptocarya* indet. Indonesia / Kostermans & Anta 610 / 1949 / ". . . fr. green, pippery taste"

1087   *Endiandra muelleri* Australia: Queensland / C. J. Trist 18 / 1931–32 / " 'Rose Walnut' "

1088   *E. clavigera* Sarawak / Au & Chi S.24120 / 68 / "Fruit: Edible. Bark: Flaky with sticky milky exudation"

1089   *Endlicheria anomala* Peru / G. Klug 728 / 1929 / " 'Canela' "

1090   *Endlicheria* indet. Brazil / G. C. G. Argent in Richards 6916 / 1967 / "Wood highly aromatic" / "LOURO PRECIOSA"

1091   *Endlicheria* indet. Brasil / E. P. Heringer 1302/9108 / 63 / ". . . produtora de frutos comestives pelos passarinhos"

1092   *Licaria capitata* Honduras / C. & W. von Hagen 1032 / 1937 / " 'Aguacatillo' "

1093   *L. pittieri* Panama / Stern & Chambers 169 / 1957 / "wood with cedar-like odor" / " 'jigua negro' "

1094   *L. salicifolia* Puerto Rico / L. E. Gregory 216 / 1940 / " 'Canela' "

1095   *L. duartei* Brazil / W. Duarte de Barros 530 / 42 / ". . . madeira com forte perfume" / "Canela"

1096   *L. limbosa* Ecuador / E. L. Little 6470 / 1943 / "Bark . . . fragrant, like cinnamon" / " 'Canela' "

1097   *L. polyphylla* Surinam / J. C. Lindeman 5298 / 1954 / ". . . bark with sweet odour" / "pisie (Sur.)"

1098   *L. rigida* Surinam / J. C. Lindeman 5932 / 1954 / "bark with cinnamon-like fragrance" / "kaneelpisie (Sur. Dutch)"

1099   *L. simulans* Brazil / W. A. Archer 7608 / 1942 / "Bark has spicy odor. Used by local druggists in remedies for dysentery and similar troubles" / " 'pichuri-mirim' "
Brit. Guiana / For. Dept. Brit. Guiana F2545 / 45 / ". . . bark and wood strongly aromatic" / "Yekuru"

1100   *Licaria* indet. Bolivia / O. E. White 1049 / 1921 / "Yields a spurious Coto bark"

1101   *Licaria* indet. Bolivia / O. E. White 890 / 1921 / "Bolivian Cinnamon"

1102   *Licaria* indet. Brazil / A. C. Smith 2957 / 1938 / "tea made from bark"

1103   *Licaria* indet. Brit. Guiana / Maguire, Maguire & Wilson-Browne 45944 A / 1961 / ". . . infusion of fruit used against vomiting" / "vernacular name 'Kamakusa' "

1104   *Licaria* indet. Brazil / G. T. Prance et al. 15829 / 1971 / "leaves used for making tea"

105   *Lindera benzoin* U.S. & CAN. / Columbia College of Pharmacy Herb. s.n. / no year / "Die aromatische Rinde und Früchte werden in Nordamerika als Arzneimittel gebraucht. Der benzoeartige Geruch dieses Strauches hat früher zu der irrigen Meinung Veranlassung gegeben, die Benzoe komme von demselben" / "Gemeiner Benzoe-Lorbeer"
U.S.: N.Y. / Columbia College of Pharmacy Herb. s.n. / no year / "Wild Allspice. Fever-bush"

106   *L. communis* China: Kwangtung / Y. Tsiang 2742 / 1929 / "Fruit green (the natives press out oil for lighting lamps)"
China: Yunnan / A. Henry 9047C / no year / "tree—leaves used for incense. Fruit produces a good lighting oil" / " 'Huay' "

107   *L. subcaudatum* China: Kwangsi / Steward & Cheo 431 / 1933 / "Leaves used for incense" / "Shiang Kuei Tze"

108   *Lindera* indet. Burma / F. Kingdon Ward 21421 / 1953 / "Smells powerfully of lemon grass"

109   *Litsea glaucescens* var. *subsolitaria* Mexico / R. E. Schultes 637 / 1939 / "Leaves used as condiment" / " 'Laurel' "
Mexico / G. B. Hinton 3188 / 33 / "Sold in markets, used to flavour foods" / "Laurel"
Mexico / Smith & Tejeda 4485 / 1966 / "Voucher for plant samples used in USDA-Cancer Chemotherapy National Service Center anti-tumor screening program"
Mexico / A. J. Sharp 441173 / 1944 / ". . . smells like benzoin"
Guatemala / P. C. Standley 58756 / 1938 / "Sold commonly in markets. Used to flavor food" / " 'Laurel' "

110   *L. euosma* China: Yunnan / A. Henry 9185 / ca. 1890 / "edible" / " 'Leh mu-chiar-tzu' "
China: Kweichow / Y. Tsiang 5888 / 1930 / "Lvs. camphor-odor . . ."

111   *L. monopetala* China: Canton / F. A. McClure 9516 / 1921 / ". . . the Lois say, 'The juice of the tree causes violent itching if it comes into contact with the skin' " / "Ye po lo"

112   *L. cubeba* Indochina / Siu & Fung 709 / 32 / "The bark is used by Annamites to perfume water for washing their hands. The bark exudes a fragrance like that of Citronella" / "T'oi Chan"

113   *L. elongata* N.W. Himalaya / R. R. Stewart 11003 / 1930 / " 'Camelo Bark' "

114   *Litsea* indet. Nepal / Stainton, Sykes & Williams 6478 / 1954 / "Orange scent when crushed"

115   *L. reticulata* Australia: Queensland / Smith & Webb 3669 / 1948 / ". . . inner bark creamy-brown and aromatic" / "Local name—'Bolly Gum' "

1116   *L. garciae* Philippine Is. / H. Bautista 2 / 1929 / "Fruit edible when cooked
. . . pulp eaten — milky in color, taste like pili-nut — but eaten raw. Wood
for lumber and fruit for food" / "Pepe babae"

1117   *L. griseola* Philippine Is. / A. D. E. Elmer 10470 / 1909 / ". . . many small
brown ants inhabit the leaf and inflorescent bracts"

1118   *L. lancifolia* Philippine Is. / M. Tahir 796 / 1928 / "Medicinal" / "Tahuli
(Bisaya)"

1119   *Litsea* indet. Borneo / A. Kostermans 12675 / 1956 / "Bark with stinging
hairs (miang)"

1120   *Litsea* indet. New Guinea / D. Frodin NGF 26428 / 66 / ". . . sap
aromatic"

1121   *Litsea* indet. New Guinea / D. Frodin NGF 26560 / 66 / "Pleasant
aromatic odour"

1122   *Litsea* indet. Borneo / Kostermans 6991 / 52 / "Living bark . . . smelling of
prussic acid"

1123   *Litsea* indet. Borneo / A. Kostermans 13906 / 1957 / "Fr. sweet with
aromatic, pungent taste"

1124   *Litsea* indet. Borneo / A. Kostermans 8704 / 1953 / "Living bark . . . with
strong aromatic smell, of mint and KCN. The wood has a sweeter scent
than the bark"

1125   *Machilus thunbergii* China / A. Henry s.n. / ca. 1895 / "The source of the
wood shavings used by Chinese women as landoline for the hair"
Hong Kong / Y. W. Taam 1174 / 1940 / "flower yellow, fragrant"
Hong Kong / Y. W. Taam 1941 / 1941 / "flower pale yellow, ill-smelling"

1126   *Machilus* indet. Nepal / Stainton, Sykes & Williams 4907 / 1954 / "Flowers
greenish, very fragrant"

1127   *Nectandra coriacea* U.S.: Fla. / Brizicky & Stern 548 / 1956 / ". . . with
aromatic leaves"
Nicaragua / A. Molina R. 20454 / 1967 / " 'Aguacatillo' "
Panama / G. P. Cooper 551 / 1928 / "Rock sweetwood"
Mexico / J. V. Santos 3037 / 1944 / " 'Aguacatillo' "
Mexico / R. McVaugh 10477 / 1949 / " 'Aguacate del monte' "
W.I.: Grenadines / J. S. Beard 619 / 1945 / " 'Sweetwood' "
Virgin Is. / E. L. Little 16350 / 1954 / "Bark dark brown, spicy"
Jamaica / W. Harris 8746 / 1904 / " 'Cap-berry Sweetwood' "

1128   *N. brenesii* Costa Rica / H. E. Stork 4171 / 1937 / "Flowers with cinnamon
odor"

1129   *N. glabrescens* Panama / R. L. Dressler 3291 / 1967 / "flowers white, fragrant"
Honduras / C. & W. von Hagen 1135 / 1937 / " 'Aguacatillo' "

Mexico / T. MacDougall s.n. / 1965 / " 'Aguacatillo' "
Brit. Honduras / W. A. Schipp 1037 / 1932 / "Flowers . . . highly perfumed"
Dominican Republic / A. H. Liogier 11916 / 1968 / " 'AGUACATE LAUREL' "
Cuba / Bro. Leon 10702 / 1922 / " 'Aguacatillo' "
Colombia / O. Haught 2110 / 1936 / "Said to be poisonous" / "Local name 'Laurel' "
Venezuela / Bernardi 2025 / 1955 / ". . . habitado por hormigas feroces"
Venezuela / L. Aristeguieta 1728 / 1953 / "Tanto esta planta como muchas especies de *Lauraceas,* parece que son utilizadas, cuando están en fructificacion, como alimento por los Guácharos" / "N.v.: 'Laurel' "

130 *N. longicaudata* Brit. Honduras / C. L. Lundell 6401 / 1936 / " 'Aguacatiloo' "

131 *N. lundellii* Brit. Honduras / P. H. Gentle 2896 / 1939 / " 'wild pear' "

132 *N. martinicensis* Panama / Duke & Bristan 227 / no year / "unpleasant odor in tree as well as in the white flowers"
Mexico / G. B. Hinton 11473 / 37 / "Vernac. name: Aguacatillo"

133 *N. mollis* Mexico / E. Matuda 17247 / 1947 / "Local name: Pepe aguacate"
Guatemala / A. F. Skutch 1988 / 34 / "Fls. white, fragrant"
Panama / Stern & Chambers 111 / 1957 / " 'Aguacatillo' "
Brazil / Y. Mexia 5280 / 1930 / " 'Canella' "
Brazil / M. Barreto 1807 / 35 / "Canella cabelluda Canella babenta"
Colombia / Schultes & Cabrera 17596 / 1952 / ". . . aromatic wood" / "Puinave = pee-sheé"
Ecuador / Játiva & Epling 2039 / 1967 / " 'raices sancudas' "
Brazil / F. C. Hoehne 29388 / 1932 / " 'Canellão' "
Venezuela / Bernardi 1737 / 1954 / "Flores . . . perfumado" / "Laurel picapica"

134 *N. polita* Mexico / J. N. Rovirosa 475 / 1889 / "Laurel de chile"

135 *N. pichurim* Panama / Stern & Chambers 113 / 1957 / " 'Aguacate' "
Peru / J. Schunke V. 1661 / 1967 / "madera aromática" / " 'Moena blanca' "
Ecuador / Dodson & Thien 1172 / 1961 / ". . . ant infested"

136 *N. rubriflora* Mexico / E. J. Alexander 542 / 1945 / " 'Wild Aguacate' "
Brit. Honduras / W. A. Schipp 1164 / 1933 / "Flowers white, sweetly perfumed"

137 *N. sinuata* El Salvador / P. C. Standley 22968 / 1922 / " 'Aguacate de mico, Aguacamico' "
Honduras / A. & A. R. Molina 27896 / 1973 / " 'Aguacatillo' "
Mexico / E. Matuda 17294 / 1947 / "Local name: 'Tepe-aguacate' "

138 *N. surinamensis* Honduras / Yuncker, Dawson & Youse 6386 / 36 / "flowers very fragrant"

1139   *N. whitei* Panama / Stern, Chamber, Dwyer & Ebinger 1026 / 1959 / "wood and bark aromatic" / " 'signa' "

1140   *Nectranda* indet. Haiti / L. R. Holdridge 1407 / 1942 / "New leaves pungent"

1141   *N. baccans* Venezuela / Steyermark & Allen 8 / 1962 / ". . . broken twigs with unpleasant odor"
       Venezuela / Steyermark & Bunting 105318 / 1971 / "flowers fragrant"

1142   *N. berchemifolia* Colombia / A. E. Lawrence 471 / 1932 / "flowers white, very fragrant"

1143   *N. cissiflora* Brit. Guiana / For. Dept. of Brit. Guiana WB 384 / 48 / "bark and wood sweet scented"
       Brit. Guiana / S. S. & C. L. Tillett 45584 / 1960 / ". . . stamens green, fragrance of papaya fruit"
       Bolivia / H. H. Rusby 831 / 1921 / "Large tree, covered with stinging ants"
       Venezuela / A. L. Bernardi 6723 / 1957 / "Madera fuertemente aromatica, olor penetrante"

1144   *N. cymbarum* Venezuela / Wurdack & Adderley 43639 / 1959 / "wood extensively used in construction; oil expressed from wood for lamps and for topical application for rheumatism" / " 'Sasafras' "

1145   *N. globosa* Venezuela / F. J. Breteler 3950 / 1964 / "Bark very aromatic"
       Surinam / B. Maguire 2490 / 1944 / "flower white, fragrant"

1146   *N. grandiflora* Brazil / Smith, Reitz & Caldato 9607 / 1956 / " 'canela fedida' "
       Brazil / F. C. Hoehne 28112 / 1931 / " 'Canella amarella' "

1147   *N. kunthiana* Surinam / J. C. Lindeman 7466 / 1955 / "bark with strong disagreeable odour"

1148   *N. lanceolata* Brazil / F. C. Hoehne 28342 / 1931 / " 'Canella' "

1149   *N. longifolia* Ecuador / E. Heinrichs 760 / 1936 / "Aguacatillo"

1150   *N. matthewsii* Peru / J. Schunke V. 3246 / 1969 / "la madera es aromática"

1151   *N. nitidula* Brazil / F. C. Hoehne 28276 / 1931 / " 'Canella' "

1152   *N. puberula* Brazil / H. M. Curran 409 / 1918 / " 'Lorah Sassafras' "
       Brazil / T. Mexia 4918 / 1930 / " 'Canela Cherossa' "

1153   *N. saligna* Brazil / A. Loefgren 10514 / 87 / " 'Canella que fede' "

1154   *N. steyermarkiana* Venezuela / Steyermark & Allen 7 / 1962 / "bark & twigs have unpleasant odor"

1155   *N. turbacensis* Trinidad / Allen, Bhorai & Ramkissoon 337 / 1966 / ". . . bark with pronounced unpleasant odor when scraped"

1156   *N. warmingii* Bolivia / H. H. Rusby 2183 / 1921 / "This is 'Guapi C.', but not a Guarea"

57   *Nectandra* indet. Venezuela / Steyermark & Allen 6 / 1962 / "bark of twigs and leaves when bruised give off an unpleasant odor"

58   *Neolitsea phanerophlebia* China: Kwangtung / Lingnan (To & Ts'ang) 12326 / 1924 / "Tai ip cheung, (Large leaf Camphor)"

59   *N. cassia* Ceylon / Kostermans 23664 / 1969 / "Bark of young branchlets smells of Cinnamomum porrectum (aniseed like)"

60   *N. lancifolia* Ceylon / Kostermans & Wirawan 830 / 1969 / "Bark of twigs smelling of manggo"

61   *N. umbrosa* India / W. Koelz 20908 / 1948 / "seeds used for oil"
India / W. Koelz 22162 / 1948 / ". . . oil taken from seed for cooking here and in Garhwal"

62   *Ocotea brenesii* Costa Rica / A. F. Smith 2 / 1937 / "Flowers fragrant, clove odor"

63   *O. cernua* Honduras / C. & W. von Hagen 1275 / 1937 / " 'Aguacatillo' "
Panama / T. B. Croat 15559 / 1971 / ". . . with an aroma like fresh peaches"

64   *O. dendrodaphne* Panama / Allen, Stern & Chambers 170 / 1957 / "Fruit, leaves and wood aromatic" / " 'Encibe' "
Guatemala / A. F. Skutch 1980 / 1934 / ". . . occupied by small stinging ants"

65   *O. ira* Panama / Stern, Chambers et al. 581 / 1959 / ". . . highly aromatic"

66   *Ocotea* indet. Honduras / C. & W. von Hagen 1048 / 1937 / "Quetzal eats fruit" / " 'Aguacatillo' "

67   *O. leucoxylon* W.I.: Dominica / G. P. Cooper 41 / 1933 / "When cut the wood has a pleasant odor" / "Called 'Laurier marbre' or 'Sweetwood' by natives"
Cuba / Fre. Leon 12270 / 1924 / " 'Aguacatillo' "

68   *O. nemodaphne* Puerto Rico / E. L. Little 13027 / 1950 / "leaves . . . slightly fragrant to taste"
Cuba / E. L. Ekman 5962 / 1915 / " 'Canelillo' "

69   *O. wrightii* Puerto Rico / L. E. Gregory 159 / 1940 / " 'Canelon' "

70   *Ocotea* indet. W.I.: St. Kitts / J. S. Beard 311 / 1944 / " 'Pumpkin sweet-wood' "

71   *O. abbreviata* Surinam / B. Maguire 24018 / 1944 / ". . . bark used as spice"

72   *O. aciphylla* Brazil / R. Klein 62 / 1949 / "Canela amarela"
Surinam / Lanjouw & Lindeman 2279 / 1949 / "Wood very odorant" / "Vern. name: rozenhout (Sir.-D.)"

1173   *O. amazonica* Brazil / B. A. Krukoff's 4th Exped. to Braz. Amaz. 5425 / 1933 / "Wood has very strong & very unpleasant odor"

1174   *O. amplissima* Venezuela / F. J. Breteler 3406 / 63 / "Bark . . . slightly smelling like a mandarin. Twigs hollow, with ants. Bark of the twigs smelling like a mandarin. Cut fruit with strong terpentine odour. Pulp exuding a yellowish-green latex"

1175   *O. auriculata* Venezuela / Steyermark & Allen 10 / 1962 / "bark with a sharp, pungent fragrance"

1176   *O. babosa* Venezuela / Allen & Ruiz-Teran 26 / 1962 / ". . . mucilaginous when cut, unpleasant odor" / " 'Laurel babosa' "

1177   *O. caparrapi* Colombia / H. García y Barriga 07661 / 1939 / " 'ACEITE DE CAPARRAPI' o 'DE PALO' "

1178   *O. costulata* Brazil / Ducke 320 / 1936 / " 'louro camphora' "

1179   *O. finium* Venezuela / J. A. Steyermark 90842 / 1962 / "(bark fragrant)" Venezuela / Steyermark & Dunsterville 92944 / 1964 / "Birds eat the seed inside when ripe"

1180   *O. glomerata* Trinidad / Allen, Bhorai & Hamkssoon 352 / 1966 / ". . . odor unpleasant when bark scraped, pungent, acrid" / " 'Laurier Mattack' "

1181   *O. guianensis* Venezuela / L. Williams 14817 / 1942 / " 'Pichi' (Baniba), 'Picháda' (Kuripako)"

1182   *O. illustris* Brazil / R. Klein 45 / 1950 / "Canela sassafraz"

1183   *O. indecora* Brazil / G. M. Nunes 20910 / 1927 / " 'Canella preta' " Brazil / F. C. Hoehne 28425 / 1931 / " 'Sassafrazinho' " Venezuela / Bernardi s.n. / 1955 / "Laurel pimiento"

1184   *O. karsteniana* Venezuela / Bernardi & Lamprecht 6384 / 57 / "N.V.: Mamey" Ecuador / Little & Dixon 21244 / 1965 / " 'catangare' (estéril)"

1185   *O. kunthiana* Surinam / C. K. Allen 49 / 1962 / "bark, when cut has faint, more or less unpleasant odor, not pungent but rather like chlorophyll" / " 'Pisie, zwarte (harde, kl. bl.)' "

1186   *O. nitidula* Brasil / O. Vecchi s.n. / 1918 / " 'Sasafrazinho do campo' "

1187   *O. opifera* Peru / Y. Mexia 6329 / 1931 / ". . . strong, disagreeable odor" / " 'Moéna' "

1188   *O. pretiosa* Brazil / M. Barreto 1483 / 1935 / "Canella sassafraz—sandalo do Brasil" Brazil / Y. Mexia 5304 / 1930 / " 'Canella Preta' " Brasil / W. Hoehne 1166 / 1943 / "Arvore cultivada, medicinal" Brazil / Y. Mexia 4982 / 1930 / " 'Fruta do Pomba' "

1189   *O. proboscidea* Bolivia / M. Cardenas 1988 / 1921 / "Bark very aromatic"

0 *O. rubiginosa* Brasil / R. Klein 1534 / 1955 / "Canela pimenta"

1 *O. venenosa* Ecuador / H. V. Pinkley 538 / 66 / "Use: sehe'pa = curare, fruit used — not bark . . . bitter to taste" / "Kofán: gongïbe'kï sehe'pa"

2 *Ocotea* indet. Ecuador / Grubb, Lloyd et al. 1429 / 1960 / "Fruits . . . much liked by animals" / "Ajua (Quechua)"

3 *Ocotea* indet. Venezuela / J. J. Buza 7582 / 1959 / "n.v.: Sasafrás"

4 *Ocotea* indet. Brazil / Prance, Maas et al. 15561 / 1971 / ". . . Makú Indian village . . . Leaves used to make tea drunk for malaria fever" / " 'Yachmodóoupe' (Makú)"

5 *Ocotea* indet. Peru / T. R. Dudley 10860 / 1968 / "foliage & stems very pungent"

6 *Ocotea* indet. Brazil / B. A. Krukoff's 5th Exped. to Braz. Amaz. 6469 / 1934 / "Wood with very strong and pleasant odor & perhaps can be used in preparation of Rosewood essence" / " 'Louro rosa' "

7 *Persea americana* var. *drymifolia* Mexico / Dr. Gregg s.n. / 1847 / "Kernel used as remedy for colic — also for indelible ink" / "Aguacate"

8 *P. rigens* Panama / Stern & Chambers 56 / 1967 / ". . . wood highly aromatic" / " 'Sigua canela' "

9 *P. vesticula* Honduras / A. Molina R. 10250 / 1962 / " 'Aguacatillo' "

0 *Persea* indet. Costa Rica / P. H. Allen 5552 / 1950 / "Edible flesh . . . of a very mild, pleasant flavor. Monkeys eat the fruits "

1 *P. krugii* Dominican Republic / Bro. A. H. Liogier 12141 / 1968 / " 'ALMENDRITO' "
Dominican Republic / Bro. A. H. Liogier 11294 / 1968 / " 'CANELA DE TIERRA' "

2 *P. urbaniana* W.I.: Montserrat / J. A. Schafer 515 / 1907 / " 'Sweetwood' "
W.I.: Martinique / P. Duss 218 / 1882 / "Lourei avocat"

3 *Phoebe helicterifolia* Mexico / C. E. Smith 4398 / 1966 / "Voucher for plant samples used in USDA-Cancer Chemotherapy National Service Center anti-tumor screening program"

4 *P. johnstonii* Panama / Stern & Chambers 156 / 1957 / "Leaves aromatic" / " 'Aguacatillo' "

5 *P. mexicana* Honduras / C. & W. von Hagen 1227 / 1937 / " 'Aguacate negro' "
Mexico / E. Matuda 17463 / 1948 / "Local name: Pepe aguacate liso"
Honduras / A. Molina R. 8309 / 1957 / " 'Aguacatillo' "
Mexico / C. K. Allen 108 / 1962 / ". . . when crushed, odor of lemon"
Brit. Honduras / P. H. Gentle 2941 / 1939 / " 'wild pear' "
Costa Rica / O. Jimenez s.n. / 1960 / " 'Aguacatillo' "

6 *P. elongata* Cuba / Bros. Leon & Yau 9681 / 1921 / " 'Aguacatillo' "

W.I.: Guadeloupe / P. Duss 3899 / 1897 / " 'lourei canella' "

1207   *P. montana* Cuba / Columbia College of Pharmacy Herb. s.n. / 1841 / "Cuban name: aguacatillo"
Jamaica / W. Harris 8742 / 1904 / " 'Cap-berry Sweetwood' "
Jamaica / Proctor & Mullings 22007 / 1961 / "flowers greenish, unpleasantly scented but popular with bees"

1208   *P. cinnamomifolia* Peru / A. Gutierrez R. 100 / 1963 / " 'Canela moena' "

1209   *P. stenophylla* Brazil / F. C. Hoehne 28317 / 1931 / " 'Canella de folha estreita' "

1210   *Phoebe* indet. Colombia / G. Gutierrez V. 446 / 1943 / "N.V. 'Aguacatillo' "

1211   *Phoebe* indet. Trinidad / C. K. Allen 216 / 1965 / " 'Laurier Spicy' 'Laurier Canelle' "

1212   *P. declinata* Sumatra / Kostermans 12342 / 1957 / "Exocarp juicy . . . slightly acid with strong aromatic taste" / "modang seluang"

1213   *Pleurothyrium williamsii* Brazil / R. Fróes 53 / 1941 / "Wood causes injury to skin. Vouchers for gum samples in connection with Chicle substitute project" / "Stauba"

1214   *Urbanodendron verrucosum* Brazil / Gomes 1220 / 61 / "madeira muito aromática"

1215   *Lauracea* indet. Mexico / R. McVaugh 10243 / 1949 / "odor strong & pleasant when crushed"

1216   *Lauracea* indet. Guatemala / S. S. White 5414 / 1945 / "Bark aromatic, with odor resembling sassafras" / " 'pimiento' "

1217   *Lauracea* indet. Costa Rica / L. R. Holdridge 6591 / 1971 / "Called 'Quizarra manteca' "

1218   *Lauracea* indet. Panama / T. B. Croat 8150 / 1969 / "Both wood and fls. with sweet odor"

1219   *Lauracea* indet. Dominican Republic / Bro. A. H. Liogier 13113 / 1968 / "wood and foliage aromatic" / " 'Sasafras' "

1220   *Lauracea* indet. Dominican Republic / Bro. A. H. Liogier 12089 / 1968 / " 'AGUACATE LAUREL' "

1221   *Lauracea* indet. Venezuela / C. K. Allen 21 / 1962 / "odor of branchlets somewhat unpleasant"

1222   *Lauracea* indet. Venezuela / Steyermark & Allen 9 / 1962 / "brts (twigs) when broken give out pungent, somewhat spicy fragrance that is very pleasing"

1223   *Lauracea* indet. Venezuela / L. Ruiz-Terán 440 / 1961 / "N.v.: Aguacatillo"

4   *Lauracea* indet. S. AM. / R. Espinosa 1662 / 47 / "Frutos aromaticos, usado como medicinal en el Sur del Ecuador y el Norte del Perú" / "N.V.: Ischpingo"

5   *Lauracea* indet. Brazil / C. K. Allen 76 / 1962 / "bark reddish, pungent sweet when cut" / " 'Louro' "

6   *Lauracea* indet. Colombia / Schultes, Raffauf & Soejarto 24419 / 1966 / "Alkaloid strongly positive"

7   *Lauracea* indet. Colombia / Schultes, Raffauf & Soejarto 24421 / 1966 / "Alkaloid negative"

8   *Lauracea* indet. Venezuela / J. de Bruijn 1552 / 1967 / "Flowers with an unpleasant sweet odour"

9   *Lauracea* indet. Venezuela / Allen & Ruiz-Terán 27 / 1962 / "slight odor of 'mapurito' when cut bark & brts" / " 'Laurel negro' "

0   *Lauracea* indet. Peru / T. B. Croat 18198 / 1972 / "plant with many large ants"

1   *Lauracea* indet. Venezuela / Steyermark & Gibson 95794 / 1966 / " 'sassafras' "

2   *Lauracea* indet. India / T. R. Chand 2773 / 1950 / ". . . leaves widly in India as spice" / " 'Jezpatta' Hindi"

3   *Lauracea* indet. India / F. Kingdon Ward 17054 / 1948 / "the flowers and young leaves when crunched up are very aromatic"

HERNANDIACEAE

4   *Gyrocarpus americanus* Honduras / P. C. Standley 20741 / 1922 / " 'Bálsamo' "
Fiji Is. / O. Degener 15329 / 1941 / "Extract of bark in cold water drunk if body aches" / " 'Wiriwiri' "

5   *G. jatrophifolius* Mexico / Y. Mexia 928 / 1926 / "Disagreeable odor" / " 'Palo amargo' "

6   *Hernandia jamaicensis* Jamaica / G. R. Proctor 22604 / 1962 / " 'Popnut' "

7   *H. peltata* Caroline Is. / F. R. Fosberg 25782 / 1946 / ". . . used medicinally by women during childbirth" / " 'dogo' "

8   *H. sonora* Mariana Is. / M. Evans 724 / 1965 / "Bark boiled and given to mothers at childbirth" / " 'BANALU' "

PAPAVERACEAE

9   *Argemone mexicana* Mexico / R. McVaugh 10110 / 1949 / " 'chicalote' "
El Salvador / S. Calderon 189 / 1921 / " 'Cardo santo' "

Mexico / Herb. of A. Wood s.n. / no year / "Das Kraut dieser Pflanze ist als Herba Cardui flavi officinell"

Cuba / Herb. Collegii Pharmaciae Neo-Eboracensis s.n. / 1840 / "Contains a yellow juice like 'chelidonium'. Root used in decoction against dropsy — is diuretic" / "Carlos santo"

Bermuda Is. / W. M. Rankin s.n. / 1897 / " 'Stinging Thistle' "

1240 *A. subfusiformis* subsp. *subfusiformis* S. AM. / A. Luna 382 / 1920 / " 'Cardo Santo' "

1241 *Argemone* indet. Bolivia / F. Steinbach 127 / 1966 / "(hojas arrugadas) pétalos en infución es medicinal para afecciones bronquiales"

1242 *Bocconia arborea* El Salvador / P. C. Standley 20001 / 1922 / " 'Sangre de toro, Sangre de chucho' "

1243 *B. frutescens* Mexico / Torrey Herb. 202 / no year / "They eat col of the leaf"
Colombia / F. R. Fosberg 20485 / 1943 / ". . . mature fruits mashed with water and used for mange-cure"
Colombia / J. M. Duque Jaramillo 2618 / 1946 / ". . . con látex rojizo para curar parásitos de la piel" / "N.v. 'Trompeto' Cund., 'Curador' Ant."

1244 *Bocconia* indet. Argentina / F. Rojas 11274 / 1944 / " 'Arbol hillo' 'Suncho amargo' "

1245 *Meconopsis speciosa* China: Tibet / J. F. Rock 22430 / 1932 / "Root used as medicine"

1246 *Papaveracea* indet. Brazil / G. & L. T. Eiten 2401 / 1960 / "Plant with rank odor & lemon yellow sap"

FUMARIACEAE

1247 *Corydalis flavula* U.S. & CAN. / Dr. Bernhardt s.n. / 1888 / "colic-weed"
U.S.: N.Y. / E. S. Burgess s.n. / no year / "Colic-root"

1248 *Corydalis* indet. India / W. Koelz 21506 / 1948 / ". . . very fragrant"

1249 *Fumaria declinatis* Germany / N. Basil s.n. / 1828 / ". . . officinalem magis accedit"

BRASSICACEAE

1250 *Brassica kaber* U.S. & CAN. / F. Fabius 185 / 1946 / "Wild mustard"

1251 *B. integrifolia* Liberia / L. C. Okeke 18 / 1950 / "Wild lettuce"

1252 *Cardamine bonariensis* Ecuador / W. H. Camp E-2626 / 1945 / "Used with the 'Drimarias' (caryophyllaceae) to cure ailments of the liver and kidneys" / " 'Guarmi — berros' . . . 'Female' water-cress"

253 *C. ovata* Ecuador / W. H. Camp E-2170 / 1945 / "An infusion of this plant used as a diuretic"

254 *Cardamine* indet. Ecuador / J. A. Steyermark 53314 / 1943 / "Used for aching liver and kidney; boil the plant in water and place the herbs of this in the water and use with 'yakamuya' and 'verdo lago' as an infusion" / " 'verros' "

255 *Cardamine* indet. New Guinea / L. J. Brass 30519 / 1959 / "leaves pungently flavored"

256 *Caulanthus inflatus* U.S.: Calif. / M. Vail s.n. / 1941 / " 'Squaw Cabbage!' "

257 *Cibotarium divaricatum* Mexico / Stanford, Retherford & Northcraft s.n. / 1941 / "heavily grazed by goats"

258 *Coronopus didymus* U.S.: N.Y. / E. S. Burgess s.n. / 1888 / "Wart-cress"

259 *C. pinnatifida* U.S.: Fla. / H. Moldenke 1112 / 1930 / "Plant very strong-scented"

260 *Descurainia sophia* U.S.: W. Va. / W. Frye 6228 / 1944 / "Introduced in chicken feed"

261 *Diplotaxis muralis* Ecuador / F. Prieto E-2549 / 1945 / " 'Mostaza' " or " 'Sacha berros' "—" 'false watercress' "

262 *Erysimum cheiranthoides* U.S.: Vt. / H. J. Banker 130 / 47 / "Worm-seed Mustard"

263 *Lepidium campestre* U.S.: Mass. / F. C. MacKeever N217 / 1959 / "Cow Cress"

264 *L. lasiocarpum* var. *orbiculare* U.S.: Tex. / V. L. Cory 53446 / 1947 / " 'peppergrass' "

265 *L. montanum* subsp. *angustifolium* U.S.: Tex. / V. L. Cory 51959 / 1946 / " 'peppergrass' "

266 *L. nitidum* var. *nitidum* U.S.: Calif. / no collector s.n. / 1927 / "Pep-pergrass"

267 *L. perfoliatum* U.S.: Wash. / R. W. Haegele 13149 / '26 / "Peppergrass"

268 *L. virginicum* var. *pubescens* Mexico / H. S. Gentry 5847 / 1940 / "Infused for tea" / "Lipasote"

269 *Lepidium* indet. Argentina / J. E. Montes 14672 / 55 / "*Uso:* Cuando tierne es comestible-raíz medic. popular" / " 'False mastuerso' "

270 *L. capitatum* var. *typicum* W. Himalayas / W. Koelz 2437 / 1931 / "Fls. . . . with mustard odor"

271 *Malcolmia africana* U.S.: Nev. / A. H. Holmgren 997 / 1941 / "Grazed by cattle"

272 *Pegaeophyton scapiflorum* W. Himalayas / W. Koelz 2152 / 1931 / "mustard odor"

1273 *Rorippa palustris* U.S.: N.Y. / E. G. Knight s.n. / 1876 / "Marsh Cress"

1274 *R. sylvestris* U.S.: Mass. / F. C. MacKeever N833 / 1964 / "Creeping Yellow Cress"

1275 *Sisymbrium altissimum* U.S.: Mass. / F. C. MacKeever N546 / 1961 / "Tumble-Mustard"

1276 *Streptanthus cordatus* U.S.: Nev. / N. H. & P. K. Holmgren 5235 / 1971 / "Specimens grazed"

1277 *S. longirostris* U.S.: Nev. / A. H. Holmgren 1065 / 1941 / "Grazed by cattle"

1278 *S. maculatus* U.S.: Ark. / Dr. Leavenworth s.n. / no year / " 'Arkansas cabbage' "

1279 *Brassicacea* indet. Argentina / J. Semper 601 / 1945 / "para ensalada" / "N.V. 'berro' "

CAPPARACEAE

1280 *Cadaba kirkii* N. Rhodesia / W. R. Bainbridge 447 / 1961 / "Used in the old days for making salt" / "Valley Tonga name KATYOLOTYOLO"

1281 *Cadaba* indet. Tanganyika / R. E. S. Tanner 3175 / 1956 / "Zigua: For low back ache, leaves burnt and the ash rubbed into small cuts over painful area" / "Zigua: Mnyuluka Mpala"

1282 *Capparis cynophallophora* Bahamas / Nash & Taylor 1148 / 1904 / " 'Wild Orange' "

1283 *C. frondosa* Cuba / F. Leon 11858 / 1924 / " 'Palo de cochino' "

1284 *C. angulata* Ecuador / J. A. Steyermark 54842 / 1943 / " 'sapote de perro' or 'sapote gomoso' "

1285 *C. eustachiana* Colombia / R. Romero-Castañeda 9379 / 1962 / " 'mal enterrado' "

1286 *C. nemorosa* Colombia / O. Haught 3934 / 1944 / "eaten by cattle" / " 'toco' "

1287 *C. salicifolia* Paraguay / T. M. Pedersen 4189 / 1956 / "fruit said to be edible"

1288 *C. sola* Peru / F. Woytkowski 7191 / 1962 / "roots and fruit very bitter; medicinal" / " 'Nina-caspi' "
Brazil / B. A. Krukoff 7667 / 1935 / "Used in Curare"

1289 *C. elaeagnoides* Tanganyika / R. E. S. Tanner 986 / 1952 / "flowers white, slightly aromatic. Sukuma tribe: very astringent, pounded roots used as poultice to alleviate testicular pain, boils" / Vernacular name: "Kisukuma: Lubisu"

90   *C. kirkii* Tanganyika / R. E. S. Tanner 1091 / 1952 / "flowers cream-white, aromatic" / Vernacular name: "Kisukuma: Msangwasangwa"

91   *C. stuhlmanii* Tanganyika / R. E. S. Tanner 2513 / 1955 / "flowers white, aromatic"

92   *C. tomentosa* Tanganyika / R. E. S. Tanner 1052 / 1952 / "flowers pink and yellow, strongly aromatic. Sukuma: for painful eyes, roots pounded up, water added and the juice squeezed into the eyes" / Vernacular name: "Kisukuma: luwisu"
Tanganyika / R. E. S. Tanner 3189 / 1956 / "Zigua: For swellings, roots pounded up and mixed with water and applied as a paste" / "Zigua: Tungula ngosa"

93   *C. koi* China: Kwangtung / Buswell & Levene 6420 / 1921 / "Edible very good flavor" / "Wat tau Kai"

94   *C. sepiaria* Hainan / S. K. Lau 4 / 1932 / "The root can be used to heal boils" / "Kung So Fa, Tsing P'ei Tsz, 'So Flower' "

95   *Cleome gynandra* Bahama Is. / A. E. Wight 141 / 1905 / "Slight peppery odor, and taste of pods bitter"
Tanganyika / R. E. S. Tanner 586 / 1952 / "Leaves used as vegetable; leaves put in nostrils for cold; leaves rubbed into small cuts in temple for headache" / "Kisukuma: Ngagani"
Tanganyika / R. E. S. Tanner 974 / 1952 / "Sukuma tribe: . . . (reported to cure in 5 minutes) . . . flowers white, aromatic" / "Kisukuma: Migagani"

96   *C. rutidosperma* W.I.: Virgin Is. / W. C. Fishlock 159 / 1918 / " 'Wild Mustard' "

97   *C. spinosa* Haiti / R. J. Seibert 1755 / 1942 / "Rare — plant with a pungent odor"

98   *C. spinosa* var. *longicarpa* Brazil / R. Fróes 1794 / 1932 / "seeds used for sinapisms and leaves used for fever" / " 'Mucambe' "

99   *C. hirta* Tanganyika / R. E. S. Tanner 559 / 1952 / "flower magenta, aromatic. Leaves used as vegetable" / "Kisukuma: Kakunguni"
Tanganyika / R. E. S. Tanner 966 / 1952 / "flowers pale mauve, aromatic. Used as a vegetable; much liked by goats" / "Kisukumu: Lusaga; Kakununhi"

00   *C. monophylla* AFR. / R. Tanner 4521 / '59 / ". . . whole plant aromatic"
Tanganyika / R. E. S. Tanner 1257 / 1953 / "Sukuma tribe: Roots chewed to treat cough, leaves pounded and used as poultice for sores" / "Kisukuma: Ng'oga batemi"
Tanganyika / R. E. S. Tanner 1884 / 1953 / "Leaves used as vegetable" / "Kisukuma: Lujaga"

01   *C. suffruticosa* S. W. Africa / R. Seydel 800 / '56 / "Gruen vom Vieh kaum beruehrt; die trockenen Schoten gefressen"

1302   *C. viscosa* Ceylon / Fosberg & Sachet 52974 / 1970 / "herbage strong-smelling"
Mariana Is. / F. R. Fosberg 24944 / 1946 / "Seeds eaten by birds" /
" 'mungus paluma'—bird food"

1303   *Cleomella cornuta* U.S.: Ariz. / H. C. Cutler 3120 / 1939 / "plant with a strong vile odor"

1304   *Crataeva palmeri* Mexico / H. S. Gentry 5515 / 1940 / "Fruit eaten when ripe" / "Perihuete"

1305   *C. magna* Hainan / Chun & Tso 43849 / 1932-33 / "fr . . . edible"

1306   *Maerua flagellaris* Tanganyika / R. E. S. Tanner 1253 / 1953 / "Sukuma tribe: leaves used to treat boils—as poultice with butter" / "Kisukuma: Ng'wikachiza"

1307   *M. hoehelii* Tanganyika / R. E. S. Tanner 3305 / 1956 / "Fruits edible" / "Masai: Alamalogi"
Tanganyika / R. E. S. Tanner 940 / 1952 / "flowers white, strongly aromatic" / "Kisukuma: Msangwasangwa"

1308   *M. pubescens* Tanganyika / R. E. S. Tanner 2672 / 1956 / "flowers white, aromatic"

1309   *M. triphylla* Tanganyika / R. E. S. Tanner 999 / 1952 / "Sukuma tribe: used to wash pots in butter making to improve butter; stems chewed to treat dry cough" / "Kisukuma: Kaninagu; Kalilingwe"

1310   *Steriphoma peruvianum* Peru / Hutchison & Wright 6772 / 1964 / "Fruit multiseeded pod to ½ m. long. Pulp white, deliciously sweet"

1311   *Stixis suaveolens* Hainan / S. K. Lau 60 / 1932 / "Fr. red; edible. Fr. eaten by natives"

1312   *Thylachium africanum* Tanganyika / R. E. S. Tanner 2230 / 1955 / "flowers . . . strongly aromatic"

RESEDACEAE

1313   *Caylusea hexagyna* AFR. / M. Drar 41/597 / 1930 / "Largely eaten by goat & sheep & desert animals" / "Danaban"

MORINGACEAE

1314   *Moringa oleifera* Brit. Honduras / P. H. Gentle 304 / 1931–1932 / "medicinal" / " 'wild rhuda' "
Tanganyika / R. E. S. Tanner 1088 / 1952 / "flowers . . . aromatic. Used as vegetable (seeds)" / "Kisukuma: Boyo"
Caroline Is. / F. R. Fosberg 47481 / 1965 / "leaves eaten"

5   *M. aptera* AFR. / M. Drar 42/610 / 1932 / "The seeds yield an oil" / "Mâï" "Yasser"

6   *Moringa* indet. Ceylon / V. E. Rudd 3097 / 1970 / "Flowers and leaves eaten"

NEPENTHACEAE

7   *Nepenthes mirabilis* Caroline Is. / F. R. Fosberg 25560a / 1946 / " 'ad' inner part of stem called 'malig' used to weave with bamboo to make fish traps"

PODOSTEMACEAE

8   *Apinagia richardiana* Venezuela / F. Cardona 2172 / 1947 / ". . . antiguamente usada su ceniza por los indios como sal" / "Arekuna: *uirín*" Venezuela / A. L. Bernardi 6678 / 1957 / "Muy apetecida por los bachacos (Atta sexdens) que se acercan al agua para cortarla"

9   *Marathrum elegans* Mexico / Hinton et al. 15952 / '41 / "In the river; eaten by cattle"

CRASSULACEAE

20 *Bryophyllum pinnatum* U.S.: Fla. / H. N. Moldenke 3667 / 1927 / " 'Life-plant' "
W.I.: Grand Cayman / W. Kings G.C. 117 / '38 / "Leaves heated & juice used on sores & boils" / "Cunosity Plant"
Brit. Guiana / J. S. de La Cruz 976 / 1921 / "flowers used for eye-water" / " 'Balsam' "

SAXIFRAGACEAE

21 *Heuchera americana* U.S. & CAN. / Herb. of A. Wood s.n. / no year / "Die Wurzel ist wirksam gegen scrofeln und secundaere syphilis"

22 *Saxifraga arguta* U.S.: Colo. / W. M. Rankin 152 / 1910 / "Jenny Cake"

23 *S. sarmentosa* China: Kwangtung / W. T. Tsang 20450 / 1932 / "Fl. white; for making Medic." / "So Min Dau Foo Yung"

24 *Saxifraga* indet. Korea / E. K. Tyson 4918 / 1968 / "Used by natives as food"

25 *Vahlia oldenlandioides* Tanganyika / R. E. S. Tanner 963 / 1952 / "Sukuma tribe: Roots used to treat cases of 'sudden madness' " / "Kisukuma: Malala"

GROSSULARIACEAE

1326   *Escallonia myrtilloides* Peru / J. F. MacBride 3485 / 1923 / "aromatic—citrus in fragrance"

1327   *E. myrtilloides* var. *myrtilloides* Ecuador / W. H. Camp E-2568 / 1945 / "the wood used in the manufacture of the native spoons because 'it is smooth on the lips' " / " 'Chachacoma' "

1328   *E. myrtilloides* var. *patens* Ecuador / F. Prieto P-155 / 1944 / "Wood used in making spoons" / " 'Chachacoma' "

1329   *Ribes cereum* var. *cereum* U.S.: Calif. / W. A. Dayton 483 / 1913 / "Browsed by the sheep" / "currant"
U.S.: Wash. / Hitchcock & Muhlick 21468 / 1958 / "plants . . . rather strongly aromatic . . ."

1330   *R. cognatum* U.S.: Idaho / Hitchcock & Muhlick 21723 / 1958 / "berries . . . perhaps not ripe, but pleasant tasting"

1331   *R. cruentum* U.S.: Calif. / W. A. Dayton 209 / 1913 / "Browsed limitedly by cattle" / "Gooseberry"

1332   *R. floridum* U.S. & CAN. / E. G. K. 78 / no year / " 'Wild Black Currant' "

1333   *R. hirtellum* var. *calcicola* U.S.: Mass. / MacKeever & Stockley N 834 / 1964 / "Wild Gooseberry"

1334   *R. hudsonianum* Canada: Saskatchewan / A. J. Breitung 91 / '39 / "Leaves strongly scented"

1335   *R. malvaceum* U.S.: Calif. / T. Morley 89 / 1940 / "leaves glutinous, with a pungent resinous odor"

1336   *R. nevadense* U.S.: Nev. / W. A. Archer 6043 / 1938 / "Wood has sweet fetid odor"

1337   *R. odoratum* U.S.: Conn. / M. Catherine s.n. / '40 / "Very fragrant" / "Golden . . . or Missouri Currant, Clove Current"

1338   *R. oxyacanthoides* U.S.: Mont. / Bartlett & Grayson 418 / 1952 / "Dull purplish fruit—excellent"

1339   *R. rotundifolium* U.S.: Wis. / W. N. Abbott 81 / 1893 / "Wild Gooseberry"

1340   *R. vulgare* U.S.: Conn. / A. Rosalie 18 / '40 / "American Red Current"
Alaska / Y. Mexia 2182 / 1928 / "Fruit a staple food"

1341   *R. wolfii* U.S. & CAN. / F. Fabius 151 / '46 / "Red Garden Currant"

1342   *R. magellanicum* S. AM. / Mrs. Reynolds s.n. / 1930 / "Native Currant, fruit edible, but somewhat insipid"

3 *Ribes* indet. Bolivia / W. M. Brooke 5987 / 1950 / ". . . edible but insipid fruit . . ."

4 *R. grossularioides* China: Hopei / C. F. Li 10732 / '29 / "The fruits are edible"

5 *R. manshuricum* Manchuria / P. H. & J. H. Dorsett 4005 / 1925 / "Wild currant, grown in garden"

6 *R. orientale* Himalaya / W. Koelz 679 / 1930 / "Fruit said to be yellow sweet edible eaten fresh" / "Youngke"

PITTOSPORACEAE

7 *Citriobatus pauciflorus* Australia: Queensland / C. T. White 1128 / 1946 / "Stock said to be very fond of the fruit" / "Cattleberry"

8 *Pittosporum glabratum* China: Kwangtung / S. Y. Lau 20158 / 1932 / "Used for medicine"

9 *Pittosporum* indet. Australia: Queensland / S. F. Kajewski 1352 / 1929 / "Fruit . . . when cut has a pleasing turpentine scent"

0 *P. ferrugineum* Brit. New Guinea / L. J. Brass 5136 / 1933 / "perfumed pale yellow flowers"

1 *P. pentandrum* Philippine Is. / A. D. E. Elmer 8782 / 1907 / "the wood is used to make bolo sheaths" / " 'Pasguick' "
Philippine Is. / B. Guieb 3 / 1926 / "The leaves when crushed and the bark when cut, smell like the bettle leaf (Ikmo). For medicine, toothache and stomachache" / "Buyobuyo"

2 *P. pullifolium* Brit. New Guinea / L. J. Brass 4668 / 1933 / "The women of Kuama village obtain a purple dye for their string bags from the seed pulp" / "Uti (Kuama Dialect)"

3 *P. ramiflorum* New Guinea / C. D. Sayers NGF 21645 / '65 / "Bark . . . inner cream, sweetly aromatic. Wood straw"

4 *P. resiniferum* Philippine Is. / A. D. E. Elmer 8348 / 1907 / "fruit quite juicy, turpentine odor, flamable" / " 'Keligts' "

CUNONIACEAE

5 *Weinmannia fagaroides* var. *fagaroides* Ecuador / W. H. Camp E-1996 / 1945 / "when the 'red variety', the bark used as a dye"

6 *W. multijuga* Colombia / F. R. Fosberg 21486a / 1944 / "bark used for tanning" / " 'encino' "

1357   *W. pubescens* var. *pubescens* Colombia / F. R. Fosberg 20102 / 1943 / "used as a remedy for sickness in livestock" / " 'encinillo' "

1358   *W. sorbifolia* var. *sorbifolia* Colombia / F. R. Fosberg 19088 / 1942 / "used for tanning" / " 'cascaro' "

MYROTHAMNACEAE

1359   *Myrothamnus flabelliformis* Nyasaland / L. J. Brass 16132 / 1946 / "fragrant with an odor very much like that of sandalwood oil; natives use an infusion of leaves for bathing sick children"

1360   *Myrothamnus* indet. Nyasaland / Newman & Whitmore 417 / 1956 / "Leaves . . . with sweetish aromatic smell. the plant is burnt, and the ash rubbed into the side of the head to cure headache" / "called in Chinyanga 'chison' meaning pitiful in the biblical sense presumably"

ROSACEAE

1361   *Acaena cylindristachya* Venezuela / J. A. Steyermark 55315 / 1944 / "use entire plant with root boiled with other roots to make jelly, to give women at menstruation or when they are having flow of blood" / " 'rosa rubia' "

1362   *A. ovalifolia* Ecuador / J. A. Steyermark 53001 / 1943 / "The leaves are tied around forehead for stopping headaches" / " 'huichilca' "

1363   *Alchemilla orbiculata* Ecuador / Grubb, Lloyd, Pennington & Whitmore 1003 / 1960 / "Can be eaten" / "Hor y gwela"

1364   *A. pinnata* S. AM. / Parker & Alcerreca 2-24 / 74 / "Pastoreo—vicuña—alppaca"

1365   *Amelanchier alnifolia* var. *pumila* U.S.: Tex. / D. S. Correll 13902 / 1946 / "Fruits purplish, edible"

1366   *A. laevis* U.S.: Vt. / W. H. Blanchard 2 / 1903 / "Ripe fruit. It is hard to get good specimens as the birds in all inhabited sections are ahead of you"
U.S.: Va. / H. S. Newins 8063 / 34 / "Service Berry"
U.S.: N.C. / Coher & Stewart s.n. / 1936 / "Fruit . . . fairly good to eat"

1367   *A. nantucketensis* U.S.: Mass. / F. C. MacKeever N602 / 1962 / "Nantucket Sugarplum, Juneberry"

1368   *A. utahensis* U.S.: Nev. / P. Train 407 / 1937 / "Fruit purple, mealy and insipid to taste"
U.S.: Nev. / P. Train 2787A / 1939 / "Berries edible"

1369   *Cercocarpus eximius* U.S.: N.M. / W. R. Chapline 232 / 1915 / "Grazed very freely by all classes of stock"

70 *Chamaebatiaria millefolium* U.S.: Idaho / Hitchcock & Muhlick 22742 / 1963 / "very unpleasantly odorous"
U.S.: Utah / B. R. Daly 9-BD / 1940 / "(Grazed by) S&C"

71 *Cotoneaster microphylla* India / W. Koelz 1934 / 1931 / "much cropped by cattle"

72 *Cowania stansburiana* U.S.: Utah / A. Cronquist 10101 / 1965 / "Sweetly aromatic shrub"
U.S.: Utah / A. P. Plummer 38 / 1940 / "Forage value high. (Grazed by) sheep"

73 *C. plicata* Mexico / Stanford, Retherford & Northcraft 172 / 1941 / "heavily grazed by goats"

74 *Crataegus adbita* U.S.: Fla. / W. W. Eggleston H4970 = $\frac{H4691}{2}$ = B2082 / 1901 / "Fruit . . . flesh tender, juicy, subacid edible"

75 *C. adunca* U.S.: Fla. / W. W. Eggleston H4941 = $\frac{M4063}{2}$ / 1901 / "Fruit . . . just getting ripe. Flesh orange, soft, juicy subacid edible"

76 *C. annosa* U.S.: Ala. / W. W. Eggleston $\frac{B4103}{2}$ / 1901 / "Fruit yellow, washed with red. Flesh soft & edible"

77 *C. attrita* U.S.: Ala. / W. W. Eggleston H5009 = $\frac{H4116}{2}$ / 01 / "Fruit . . . flesh rather firm but juicy, subacid, agreeable to taste"

78 *C. audens* U.S.: Fla. / W. W. Eggleston H4963 = $\frac{H4097}{2}$ / 1901 / "Fruit . . . flesh . . . soft, juicy, slillty acid. pleasant flavored edible"

79 *C. berberifolia* U.S.: La. / Biltmore Herb. H5136 = $\frac{4146}{2}$ / 1901 / "Fr. ripe . . . flesh rather firm, juicy acid pleasable taste"

80 *C. champlainensis* Canada: Québec / Bro. M.-Victorin 3 / 1914 / "Abundant in hedges, in fields and pastures. = Edible, the best species"

81 *C. crocina* U.S.: La. / W. W. Eggleston H5133 = $\frac{H4152}{2}$ / 1901 / "Fruit . . . juicy, acid pleasant flavor"

82 *C. dolosa* U.S.: Ala. / Biltmore Herb. H5030 = $\frac{H4102}{2}$ / 1901 / "Fruit . . . Flesh soft & juicy palatable"

83 *C. egens* U.S.: Fla. / W. W. Eggleston H4909 = $\frac{H4037}{2}$ / 1901 / "Fruit . . . Flesh, succulent, sweet edible"

84 *C. galbana* U.S.: Fla. / Biltmore Herb. H4966 = $\frac{H4095}{2}$ / 1901 / "Fruit . . . flesh soft, juicy, subacid, edible"

85 *C. habereri* U.S.: N.Y. / J. V. Haberer 2410 / 1905 / "Edible"

86 *C. illudens* U.S.: Fla. / W. W. Eggleston B $\frac{4055}{2}$ / 1901 / "Fruit . . . soft, yellow, edible"

87 *C. incana* U.S.: Fla. / Biltmore Herb. H4906 = $\frac{H4034}{2}$ / 1901 / "Fruit . . . flesh succulent, wild subacid, edible"

1388 *C. insidiosa* U.S.: Ala. / W. W. Eggleston H5008 = $\frac{H4115}{2}$ / 01 / "Fruit . . . sweetish but somewhat insipid, not agreeable to taste"

1389 *C. quaesita* U.S.: Fla. / Biltmore Herb. H4912 = $\frac{4038}{2}$ / 1901 / "Fruit . . . flesh soft juicy mildly acid, agreeable flavor"

1390 *C. rimosa* U.S.: Fla. / Biltmore Herb. B4862 / 1901 / "Fruit . . . soft & edible"

1391 *C. spathulata* U.S.: Miss. / S. McDaniel 3493 / 1963 / "flowers bad smelling"

1392 *C. succulenta* U.S.: Ill. / V. H. Chase 9484 / 1948 / "Anthers white, slightly pink on edges. odor offensive"

1393 *C. villaris* U.S.: Fla. / W. W. Eggleston B $\frac{4042}{2}$ / 1901 / "Fruit . . . yellow & soft, edible"

1394 *C. visenda* U.S.: Fla. / W. W. Eggleston H4903 = $\frac{H4031}{2}$ / 01 / "Fruit . . . Flesh tender, juicy and edible"

1395 *C. pubescens* Mexico / Y. Mexia 1769 / 1927 / "Fruit much used by natives in sweets"

1396 *C. scabrifolia* China: Kweichow / Y. Tsíang 7380 / 1930 / "Frt . . . edible"

1397 *C. clarkei* Kashmir / R. R. Stewart 17111 / 1939 / "fruit said to be black & edible" / " 'sinjli' "

1398 *Docynia indica* India and Burma / F. Kingdon Ward 17048 / 1948 / "Fruit . . . smelling of apple when cut"

1399 *Duchesnea indica* U.S.: Tex. / V. L. Cory 52844 / 1947 / " 'yellow strawberry' "
Colombia / E. S. Burgess s.n. / '90 / " 'Indian strawberry' "
S. AM. / R. Espinosa 1470 / 47 / "Frutos insípidos"

1400 *Eriobotrya cavaleriei* China: Kwangsi / W. T. Tsang 24709 / 1934 / "fr. black, edible"

1401 *E. deflexa* China: Kwangsi / Stewart & Cheo 105 / 1933 / "Edible Fruit"

1402 *Fallugia paradoxa* U.S.: Utah / D. W. Woodruff 1182 / 1971 / "Grazing by cattle in winter"

1403 *Fragaria insularis* Dominican Republic / Bro. A. H. Liogier 14005 / 1969 / "fruits red, edible"

1404 *F. virginiana* S. AM. / R. Espinosa 1468 / 47 / "Frutos dulces y aromáticos"

1405 *F. orientalis* China: Kweichow / Stewart, Chiao and Cheo 489 / 1931 / "Fr. reddish, edible"

1406 *Geum peruvianum* Peru / J. J. Wurdack 936 / 1962 / "aromatic root infusion used for 'stomach colic' " / " 'Canela del campo' "

1407 *Holodiscus dumosa* U.S.: Utah / B. R. Daly 3- BD / 1940 / "Grazed by . . . sheep"

8 *Horkelia bolanderi* U.S.: Calif. / E. Crum 1516 / 1933 / "herbage very aromatic"

9 *H. cuneata* U.S.: Calif. / C. B. Wolf 9556 / 1940 / "herbage fragrant"

0 *H. marinensis* U.S.: Calif. / E. Crum 1557 / 1933 / "herbage very aromatic"

1 *Malus formosana* China: Kwangsi / R. C. Ching 5932 / 1928 / "Fruit . . . edible when ripe"

2 *Osteomeles cuneata* Peru / Hutchison & Wright 4313 / 1964 / "Berries eaten as food; stomatic. Also, leaves used as tea" / " 'Umlyash' "

3 *O. heterophylla* Venezuela / J. A. Steyermark 55293 / 1944 / "Fruit edible" / " 'higuito' "

4 *O. lanuginosa* Venezuela / J. A. Steyermark 55743 / 1944 / "fruit edible" / " 'membrillo' "

5 *O. latifolia* Ecuador / W. H. Camp E-2044 / 45 / " 'Caspi-manzana' "

6 *Photinia microcarpa* Brit. Honduras / W. A. Schipp 1291 / 1934 / "Flowers white with a very delicate perfume"

7 *P. consimilis* China: Kwangsi / W. T. Tsang 24067 / 1934 / "fr. black, edible" / "Chuen Fa Muk"

8 *Physocarpus capitatus* U.S.: Calif. / W. A. Wayton 350 / 1913 / "Probably browsed, at least limitedly, by the cattle"

9 *Potentilla fruticosa* U.S.: Utah / B. R. Daly 4-BD / 1940 / "Grazed by . . . S&G"

20 *P. glandulosa* subsp. *glabrata* U.S.: Utah / Wm. D. Hurst H-8 / 1941 / "Grazed by . . . C&H"

21 *P. gracilis* subsp. *nuttallii* U.S.: Nev. / D. H. Kinnaman 41 / 1938 / "(Palatability) C&H—55 Grazed by . . . S&G—55" / "Cinquefil"

22 *P. lemmoni* U.S.: Nev. / I. W. Clokey 717 / 37 / "Grazed by . . . deer and horses (Palatability 30%)"

23 *P. pulcherrima* U.S.: Utah / Herb. of the For. Service U.S. Dept. of Agric. H11 / 1941 / "Grazed by . . . S&G (Palatability) 30%"

24 *P. supina* W. Himalayas / W. Koelz 1815 / 1931 / "Plants odoriferous"

25 *Prunus armeniaca* U.S.: Penn. / A. P. Bingaman 2835 / 1926 / " 'Apricot' "

26 *P. cuneata* U.S.: N.Y. / W. C. Ferguson 92 / 21 / "Appalachian Cherry"

27 *P. demissa* U.S.: Ore. / E. W. Hammond 103 / 1892 / " 'Choke-cherry' " U.S.: Nev. / P. Train 2763 / 1939 / "Frt . . . bitter. Foliage poisonous in fall"

28 *P. emarginata* U.S.: Idaho / Becraft & Wheeler 8247 / 36 / "Western Choke Cherry" Canada: Brit. Columbia / F. M. Knapp 8365 / 36 / "Fire Cherry, Pin Cherry"

U.S.: Nev. / W. A. Archer 5463 / 1937 / "frt. yellow when ripe; flesh bitter; bruised bark aromatic, bitter"

U.S.: Calif. / P. A. Munz 16955 / 1951 / "Much browsed by deer"

1429 *P. eximia* U.S.: Tex. / E. Whitehouse 18534 / 1947 / " 'choke cherry' "

1430 *P. melanocarpa* U.S.: Nev. / G. E. Moore M 588 / 37 / "Grazed by cattle and sheep, summer & fall" / "Chokecherry"

1431 *P. mollis* U.S.: Wash. / F. W. Johnson 1611 / 1913 / "Western Wild Red Cherry"

1432 *P. myrtifolia* U.S.: Fla. / D. H. Caldwell 8795 / 1952 / "Myrtle leaf cherry"

Brit. Guiana / D. B. Fanshawe F2437 / 45 / ". . . contains a fragrant oil with a scent like oil of almonds"

1433 *P. nana* U.S.: Penn. / H. N. Moldenke 2364 / 1925 / " 'Choke Cherry' "

1434 *P. pennsylvanica* U.S. & CAN. / F. Fabus 53 / 46 / "Wild Red Cherry"

U.S.: Penn. / H. Moldenke 3841 / 1928 / " 'Fire Cherry' "

U.S.: Mich. / H. S. Newins 8141 / 1935 / "Pin Cherry"

1435 *P. persica* U.S.: Mass. / F. C. MacKeever N984 / 1966 / "Peach"

1436 *P. prunifolia* U.S.: Wash. / F. F. Wangaard 8623 / 1941 / "Bitter cherry"

1437 *P. trichopetala* U.S.: La. / D. S. & H. B. Correll 10006 / 1938 / "fruits bitter"

1438 *P. valida* U.S.: Ariz. / Hendricks 8487 A / 38 / "Southern Black Cherry"

1439 *P. virens* U.S.: Tex. / V. L. Cory 53196 / 1946 / " 'chokecherry' "

1440 *P. virginiana* U.S.: N.J. / H. N. Moldenke 11312 / 1940 / "VOUCHER FOR PHARMACEUTICAL SPECIMEN" / " 'Bitter Cherry' "

1441 *Prunus* indet. / U.S.: Tex. / V. L. Cory 49714 / 1945 / "fruit . . . pleasantly sweet" / " 'plum' "

1442 *Prunus* indet. U.S. & CAN. / Dr. Gregg s.n. / 47 / "Durazno (peach) Characan (apricot). Bears 2 crops yearly—first (about size of plums) called *characan*—second, (smaller) *albaricoque*; literally *apricot*"

1443 *P. annularis* Panama / Stern & Chambers 54A / 1957 / " 'mamey oloroso' "

1444 *P. gentryi* MEX. & C. AM. / R. McVaugh 12017 / 1951 / ". . . bruised bark with strong (bitter) cherry odor"

1445 *Prunus* indet. Cuba / J. G. Jack 8693 / 1933 / "New Zealand Plum"

1446 *P. serotina* Bolivia / F. R. Fosberg 28579 / 1947 / ". . . fruit immature, said to be very sweet and edible when ripe"

1447 *P. sphaerocarpa* Venezuela / J. A. Steyermark 60687 / 1944 / ". . . cut bark with sweet almond or cherry-like odor; this bark used for preventing malaria; the crude is drunk in water infusion" / " 'mamga-yek' "

Venezuela / J. A. Steyermark 55412 / 1944 / " 'cereza' "

Venezuela / C. Blanco 437 / 1966 / ". . . semillas y cortezas con fuerte olor a almendra" / "Menta Berti"

1448 *P. communis* England / H. N. Moldenke 8923 / 1936 / "In cultivation" / " 'Almond' "

1449 *Prunus* indet. China: Kwangsi / Steward & Cheo 186 / 1933 / "Fruit edible" / "Ye Ying Tao"

1450 *Prunus* indet. China: Kwangsi / Stewart & Cheo 25 / 1933 / "Fruit reported edible" / "Ying T'ao"

1451 *P. acuminata* India / A. A. Bullock 906 / 1946 / "Immature fr. red with distinct odour of cyanogen but eaten by monkeys. Embryo gelatinous"

1452 *P. cerasoides* Burma / F. Kingdon Ward 355 / 1939 / " 'Carmine Cherry' "
W. Himalayas / W. Koelz 1945 / 1931 / ". . . fruit reddish, bitter"

1453 *P. divaricata* India / W. Koelz 1771 / 1931 / "Eaten" / " 'alutscha' vern. plum"

1454 *P. incana* Burma / F. Kingdon Ward 212 / 1939 / "Fruit eaten" / "Meitzu (Ch.)"

1455 *P. jacquemontii* Kashmir / R. R. Stewart 21882a / 1946 / "red, edible" / " 'tundal' "

1456 *P. undulata* TROP. ASIA / Biskam 2257 / 1933 / "This tree is poisonous to cattle"

1457 *P. gazelle-peninsulae* New Guinea / R. D. Hoogland 9073 / 1964 / "Bark with smell of bitter almond"

1458 *Prunus* indet. New Guinea / D. Frodin NGF 26525 / 66 / "Odor of sap like chewing gum"

1459 *Prunus* indet. / New Guinea / Henty & Frodin NGF 27310 / 66 / "bitter almond smell" / "Tugo"

1460 *Prunus* indet. New Guinea / D. Frodin NGF 26686 / 66 / "Odor fruity" / "Naura"

1461 *Purshia tridentata* U.S.: Utah / O. Sparrow 285 OS / 39 / "Forage value High (Grazed by . . .) C&H, S&G (Season grazed) Spring, sum. & fall"

1462 *Pygeum turnerianum* Australia / C. T. White 1221 / 1929 / "Bark . . . Has a distinctive strong almond flavour when cut"

1463 *P. fragrans* Philippines / M. D. Sulit students 167 / 1929 / "Bark when cut has bed-bug odor"

1464 *P. parviflorum* Zuid-Sumatra / W. I. Lutjeharms 4360 / 1936 / "schors bitter guer van HCN" / "tenangaoe ehoeroehoeroe"

1465 *Pygeum* indet. New Guinea / H. Streimann NGF 21085 / 65 / "Almond smell"

1466　*Pygeum* indet. Borneo / A. Kostermans 12501 / 1956 / "Bark . . . smell of cyanide"

1467　**Pyrus arbutifolia** U.S.: Conn. / J. Enequist 406 / 1932 / " 'Dogberry' "
U.S.: Mass. / F. C. MacKeever N215 / 1959 / "Red chokeberry"

1468　**P. floribunda** U.S.: Mass. / MacKeever & Stockley N922 / 1965 / "Purple Chokeberry"

1469　**P. melanocarpa** U.S.: Va. / W. H. Camp 1280 / 1936 / "The 'blueberry' of the mountaineers"
U.S.: Iowa / T. B. Croat 25009 / 1973 / " 'choke cherry' "

1470　**P. nigra** U.S. & CAN. / C. H. Peck 1894 / no year / "Black Chokeberry"

1471　**P. ussuriensis** China: Hupeh / S. C. Sun 3 / 1932 / "The fruits are preserved in sugar and eaten"
China: Hopi / J. C. Liu L.2270 / 1929 / "cultivated" / " 'Suan Li' (Sour Pear)"

1472　**Raphiolepis indica** China: Kwangsi / W. T. Tsang 23961 / 1934 / "fr. black, edible" / "Shek Pun Shue"
China: Kwangtung / To & Ts'ang (Lingnan Univ. Herb.) 12527 / 1924 / "Fruit and bark used to dye clothes" / "Shek pan muk . . . (Stone rank tree); Ts'ing li muk . . . (Green pear tree)"
China: Kwangtung / W. T. Tsang 20848 / 1932 / "Fr. black; edible" / "Sek Pun Tsai Chun Fa Muk"

1473　**Rosa californica** U.S.: Calif. / Everett, Balls & Lenz 17610 / 1952 / "leaves aromatic"

1474　**R. cinnamomea** U.S. & CAN. / F. Fabius 212 / 46 / "Cinnamon Rose"

1475　**R. gymnocarpa** U.S. & CAN. / W. A. Dayton 349 / 1913 / "Valuable browse" / "Wild Rose"

1476　**R. micrantha** U.S.: N.J. / A. Niederer s.n. / 1883 / "Small flowered Sweet Brier"

1477　**R. woodsii** var. **ultramontana** U.S.: Utah / Herb. of the For. Service U.S. Dept. of Agric. W14 / 1941 / "Use mod. (Grazed by . . .) Cattle"

1478　**R. bracteata** Jamaica / W. Harris 5711 / 1895 / "Petals . . . odor of bitter almonds"

1479　**R. barbonica** China: Kwangsi / Stewart & Cheo 162 / 1933 / "Flower . . . fragrant. Cultivated" / "Yu Yeh Hong"

1480　**R. laevigata** Hong Kong / Y. W. Taam 1827 / 1940 / "fruit yellow, edible"

1481　**R. xanthina** var. **spontanea** China: Nanking / Y. L. Keng 1327 / '28 / "Cultivated in fields"

1482　**Rosa** indet. China: Peiping / Dorsett & Morse 7345 / 1930 / "Cultivated"

1483   *R. multiflora* var. *platyphylla* Tibet / J. F. Rock 23036 / 1932 / "cultivated by natives"

1484   **Rubus alumnus** U.S.: Miss. / J. A. Steyermark 64810 / 1947 / "berries . . . sweet when ripe"

1485   *R. amicalis* U.S.: Me. / J. C. Arthur 58 / 1909 / "fruit large, and sweet"

1486   *R. argutus* Nova Scotia / Fernald & Long 24027 / 1921 / "fruit luscious"
        U.S.: Vt. / W. H. Blanchard 90 / 1906 / ". . . fruit often valuable"

1487   *R. bifrons* U.S.: Mass. / F. C. MacKeever N645 / 1962 / "Elm-leaved Blackberry"

1488   *R. canadensis* U.S. & CAN. / Herb. of W. H. Abbott 25 / 1893 / "Low Black berry or Dewberry"

1489   *R. elegantulus* U.S.: N.H. / M. L. Fernald 15834 / 1917 / "fruit good"
        U.S.: Me. / J. C. Arthur 51 / 1909 / "berries sweet"

1490   *R. frondosus* U.S.: N.J. / H. N. Moldenke 1633 / 1931 / " 'Blackberry' "
        U.S.: Vt. / W. H. Blanchard 100 / 1906 / "Very ripe and good fruit—not too ripe—hard to secure"

1491   *R. hispidus* U.S.: N.Y. / E. H. D. s.n. / 78 / "Running Swamp-Blackberry"
        U.S.: Mass. / F. C. MacKeever N64 / 1958 / "Hairy Dwarf Raspberry"

1492   *R. invisus* U.S.: N.Y. / F. W. Johnson 1500 / 1918 / " 'Dewberry' "

1493   *R. monongaliensis* U.S.: W. Va. / Mr. & Mrs. H. A. Davis 8847 / 1949 / "Abundant and fruitful; the chief source of early blackberries in Morgantown area"

1494   *R. nigricans* U.S.: N.Y. / W. C. Ferguson A-1 / 19 / "Bristly Blackberry"

1495   *R. nigricans* hybrids U.S.: R.I. / Fernald & Weatherby 843 / 1936 / "Berries sweet and of good flavor"

1496   *R. nigrobaccus* U.S.: Me. / J. C. Arthur 54 / 1909 / "berries large, sweet"
        U.S.: N.H. / Fernald & Pease 15666 / 1917 / "fruit luscious"

1497   *R. odoratum* U.S.: N.Y. / E. H. Day s.n. / '75 / " 'Purple Flowering—Raspberry' "

1498   *R. parviflorus* U.S.: Wis. / J. H. Vonuette 83 / 44 / "White Flowering Raspberry"
        U.S.: Calif. / H. E. & S. T. Parks 5644 / 1937 / "Wild pigeons in large numbers feeding on fruit"
        U.S.: Nev. / L. R. Miller 153 / 1937 / "edible frts."

1499   *R. pergatus* hybrids Canada: Newfoundland / Fernald, Long & Dunbar 26797 / 1924 / " '*English* Blackberry' "

1500   *R. plicatifolius* U.S.: Ohio / Herb. of the Ohio Wesleyan Univ. 28–80–18 / 1874 / " 'Dewberry' "

1501   *R. procumbens* U.S.: Tex. / V. L. Cory 52611 / 1947 / " 'Dewberry' "

1502   *R. pubescens* U.S. & CAN. / C. H. Peck s.n. / 1894 / "Dwarf raspberry"
U.S.: Vt. / E. T. & H. N. Moldenke 9918 / 1937 / "Fruit red; 'Dewberry' "
U.S.: & CAN. / F. Fabius 76 / 46 / "Dwarf Red Blackberry"

1503   *R. sativus* Canada: Nova Scotia / W. H. Blanchard 707 / no year / "The
fruit is good—not typica. Here was some really edible fruit"

1504   *R. stellatus* Canada: Yukon Terr. / Porsild & Breitung 11.941 / 1944 /
". . . very fragrant"

1505   *R. vermontanus* U.S.: N.H. / M. L. Fernald 11748 / 1915 / "fruit seedy &
acid"
U.S.: N.H. / Fernald & Pease 15650 / 1917 / "fruit excellent"

1506   *R. villosus* U.S.: N.Y. / E. H. D. s.n. / no year / " 'Sand Blackberry' "
U.S.: N.Y. / E. H. Day s.n. / 77 & 83 / " 'Common Blackberry' "

1507   *R. arizonicus* Mexico / G. B. Hinton 3251 / 33 / "medicine for colds" /
"Zarza de castilla"

1508   *R. coriifolius* MEX. & C. AM. / R. McVaugh 20236 / 1960 / "fruit black,
sweet"

1509   *R. pringlei* Mexico / A. J. Sharp 441042 / 1944 / "red-pink raspberry, tart
not very sweet . . ."

1510   *R. urticaefolius* Costa Rica / Molina R., Williams, Burger & Wallenta
17498 / 1966 / ". . . berries black, sour"
Honduras / A. Molina R. 10529 / 1962 / " 'Mora' "

1511   *Rubus* indet. Mexico / V. Lobato s.n. / 1888 / "Mora blanca"

1512   *R. domingensis* Dominican Republic / Bro. A. H. Liogier 12195 / 1968 /
". . . fruits red, turning black. Common in pine forest, fruit edible"

1513   *R. eggerstii* Dominican Republic / A. H. Liogier 18340 / 1972 / "fruits
fleshy, . . . edible"

1514   *R. niveus* Dominican Republic / A. H. Liogier 11630 / 1968 / ". . . fruits
. . . abundant, . . . edible"

1515   *R. selleanus* Haiti / L. R. Holdridge 902 / 1941 / "Edible . . . fruit"

1516   *R. bogotensis* Ecuador / J. A. Steyermark 54763 1st. sht. / 1943 / "fruit edible"

1517   *R. robustus* Venezuela / J. A. Steyermark 56460 / 1944 / "fruit . . . edible"

1518   *R. urticaefolius* Peru / T. R. Dudley 11834 / 1968 / fr. at first red then
turning jet black, very sweet and delicious"
Paraguay / T. M. Pedersen 4350 / 1956 / "ripe fruit black, edible"

1519   *Rubus* indet. Argentina / J. E. Montes 137 / 1944 / ". . . frutos comestibles"

1520   *Rubus* indet. Peru / B. & C. K. Maguire 60008 / 1964 / ". . . fruit edible"

1521   *Rubus* indet. S. AM. / R. F. Steinbach 557 / 1966 / "fruto dulse ácido, color rojo obscuro (comestible)" / "Zarsa mora"

1522   *Rubus* indet. S. AM. / R. F. Steinbach 638 / 1966 / "frutos guindos comestibles de gusto agri-dulse" / " 'Sarsa mora' "

1523   *R. exsuccus* Nyasaland / L. J. Brass 17706 / 1946 / "fruit black when ripe, good flavored"

1524   *R. pinnatus* Tanganyika / R. E. S. Tanner 2490 / 1955 / "Edible"

1525   *Rubus* indet. Belgian Congo / J. P. Chapin 328 / 1927 / "Fr. when ripe, black, large, and fairly good to eat"

1526   *R. corchorifolius* China: Kwangsi / Steward & Cheo 303 / 1933 / "Edible" / "san ye p'ao"

1527   *R. fenbriiferus* China: Kwangtung / W. T. Tsang 21280 / 1932 / "flower, white, edible" / "Lo Fu Bow"

1528   *R. hanceanus* China: Nanking / R. C. Ching 5369 / 1928 / "Edible"

1529   *R. lambertianus* China: Kwangtung / W. T. Tsang 21529 / 1932 / "Fr. edible" / "She Pau Lak"

1530   *R. leucanthus* China: Kwangtung / W. T. Tsang 21121 / 1932 / "fruit, red, edible" / "Wong Ngau Pau"
Hainan / C. I. Lei 690 / 1933 / "fruit red edible"

1531   *R. reflexus* China: Kwangtung / W. T. Tsang 21061 / 1932 / "flower, white, edible" / "Sze Poh Lat"

1532   *R. sumatranus* China: Kwangsi / Steward & Cheo 676 / 1933 / "Fruit Edible"
Sumatra / W. N. & C. M. Bangham 1121 / 1932 / "Fruits soft orange large; flavor sweet, good"

1533   *R. swinhoii* China: Kweichow / Y. Tsiang 5309 / 1930 / "Frt.deep purple, edible"

1534   *Rubus* indet. China: Kwangsi / Steward & Cheo 593 / 1933 / "Fruit Red—edible"

1535   *R. macilentus* India / W. Koelz 43 / 1930 / "yellow (gold), pleasant, sweet"

1536   *R. muelleri* Australia: Queensland / C. T. White 8210 / 1931 / "Ft. red, edible"

1537   *R. alceaefolius* Sumatra / W. N. & C. M. Bangham 609 / 1931 / "Fruits red. Sweet, edible"

1538   *R. ferdinandi* Brit. New Guinea / L. J. Brass 4940 / 1933 / "fr. red, palatable"

1539   *R. penetrans* Hawaiian Is. / F. R. Fosberg 33295 / 1951 / "fruit black and glossy when ripe, pleasantly tart"

1540   **R. pyrifolius** Sumatra / W. N. & C. M. Bangham 1102 / 1932 / "Fruits
. . . flavor similar to blackberry; delicious. May have value for crossing"

1541   **Rubus** indet. New Guinea / L. J. Brass 30065 / 1959 / "fruits red, palatable"

1542   **Rubus** indet. New Guinea / L. J. Brass 29876 / 1959 / "fruits red, palatable"

1543   **Rubus** indet. New Guinea / L. J. Brass 30642 / 1959 / "fruits red,
palatable"

1544   **Sorbus sitchensis** U.S.: Idaho / A. Cronquist 6062 / 1949 / "Cut stems
smelling strongly of hydorcyanic acid"

1545   **S. caloneura** China: Kwangsi / R. C. Ching 5800 / 1928 / "Edible when
ripe"

1546   **Tetraglochin pinnata** Ecuador / J. A. Steyermark 52416 / 1943 / "Fruit
white, fleshy, edible. Doctors use the leaves and branches to make an infusion for treating 'serampión' " / " 'anigua' "

1547   **Waldsteinia fragarioides** subsp. **fragarioides** U.S.: Penn. / H. N. Moldenke
2367 / 1925 / " 'Barren Strawberry' "

1548   **Rosacea** indet. Argentina / J. E. Montes 27604 / 58 / "*Uso*: Frutos comestibles" / " 'Güayabo' 'Güayabita' 'Araza' "

1549   **Rosacea** indet. Peru / R. Scolnik 1670 / 1948 / "Se toma en infusión para
los sustos" / "Hierba del susto"

1550   **Rosacea** indet. S. AM. / R. F. Steinbach 598 / 1966 / ". . . fruto comestible"

CHRYSOBALANACEAE

1551   **Acioa edulis** Brazil / Prance, Maas et al. 14015 / 1971 / "Kernel of fruit
eaten locally & crushed to add to tapioca cakes (beijú), oil used locally for
cooking and soap" / " 'Castanha de cutia' "

1552   **A. guianensis** Brazil / Pires, Rodrigues & Irvine 50490 / 1961 / "freshly cut
bark blood red . . . nice odor"
Brasil / Octavio 52237 / 1945 / "amendoa comestivel" / " 'Coco de cotia' "

1553   **Atuna racemosa** Philippine Is. / A. D. E. Elmer 11237 / 1909 / "the 'nut-meg'
like rather large seed is used by the Bogobos as a black waxy paint to cover
over such articles as sheaths of various kinds. It protects the sheaths and their
handles from the rotting influence of rains, &c" / " 'Tabumtanun' "
Philippine Is. / F. R. Bolanon 3 / 1931 / "fruit is used for condiment of
fish" / "tabontabon (Visaya)"
New Britain / J. H. L. Waterhouse 240 / 34 / "After removal of thick rind,
the flesh of the ripe fruit (of a consistency of cheese) is easily grated and
makes a most effective paste or cement which sets very hard" / "Tita"
Caroline Is. / E. Y. Hosaka 3344 / 1946 / ". . . oil extracted from seed"
Fiji Is. / O. Degener 14100 / 1941 / "Putty-like kernel used by early settlers
to cement together broken crockery"

Caroline Is. / F. R. Fosberg 26305 / 1946 / "oil from seed quick drying oil for canoes" / " 'ais' "

Fiji Is. / A. C. Smith 1713 / 1934 / "fruit yields oil scent coconut oil" / " 'Makita' "

1554 *Chrysobalanus icaco* Costa Rica / W. W. & H. E. Rowlee 120 / 1918 / "fruit good to eat" / " 'Icaco' "

W.I.: Guadeloupe / P. Duss 2212 / 1895 / "fruit édible" / "Zicac, zicaque"

W.I.: Grenada / W. E. Broadway s.n. / 1906 / " 'Fat pork' "

Brit. Guiana / For. Dept. of Brit. Guiana F 2707 / 47 / "flesh . . . edible" / "Fat Pork"

Brazil / R. Fróes 1992 / 1932 / "Fruits edible" / " 'Guajuru' "

1555 *Couepia polyandra* El Salvador / P. C. Standley 21410 / 1922 / "fr. eaten" / " 'Zapotillo, Zapote bolo' "

Brit. Honduras / W. A. Schipp 717 / 1930 / "fruits yellow, edible"

1556 *C. bracteosa* Brazil / Chajas 1956 / 55 / "Fruto comestivel, oleaginoso, doce, perfumado, carnudo" / "Papuá do mato"

1557 *C. chrysocalyx* Brazil / C. F. Baker 57 / 1908 / ". . . much cultivated in cisandine Peru for its small elliptical fruits" / "Goes under the name Parinary"

Peru / Killip & Smith 27933 / 1929 / "fruit used for ink" / " 'Parinari' "

Peru / F. Woytkowski 5059 / 1958 / "Fruto comestible" / "Parinari"

1558 *C. edulis* Brazil / J. F. Ramos P23251 / 1975 / "Fruit cotyledons edible" / " 'Castanha de cotia' "

1559 *C. glandulosa* Brasil / R. de Lemos Fróes 21102 / 1945 / "Fruto comestivel"

Colombia / Schultes & Cabrera 13145 / 1951 / "Fruit eaten" / "Puinave = Koo'-yo"

1560 *C. grandiflora* Brazil / Prance & Silva 59043 / 1964 / "Visited by humming birds"

1561 *C. habrantha* Brazil / A. Ducke 2127 / 1948 / ". . . mesocarpia albo farina adstringente"

1562 *C. longipendula* Brazil / G. T. Prance 2081 / 1965 / "Fruit edible"

Brazil / Prance, Maas et al. 16521 / 1971 / "The bark used as a hardener for pottery & the fruit eaten by Deni Indians" / " 'Tsubictsu' "

1563 *C. magnoliifolia* Brazil / Chajos 3576 / 56 / "O fruto e comestivel" / "Pajurá"

1564 *C. martianum* S. AM. / Gardner 1284 / 1838 / "The natives eat the fruit & call it oiti de Porco"

1565 *C. obovata* Brazil / G. A. Black 47-1632 / 1947 / "fruto verde, quando maduro meio alaranjado de sabor e perfume muito agradavel"

1566 *C. paraensis* Brazil / G. A. Black 47-1765 / 1947 / ". . . flor branca, muito perfumada" / " 'Uchirana' "

1567   *C. rufa* Brasil / A. Ducke 2113 / 48 / "Fruto comestivel" / " 'Oiti coró' "

1568   *C. subcordata* Brazil / Spruce 1423 / 1851 / "Frt. edible" / "Umari de Urua"

1569   *C. uiti* Brazil / Prance, Silva & Pires 59144 / 1964 / "Visited by humming birds"

1570   *Couepia* indet. Brit. Guiana / For. Dept. of Brit. Guiana F 621 / 41 / "Fr. . . . turn yellowish when ripe, fleshy; flesh sweet"

1571   *Hirtella racemosa* var. *hexandra* Brit. Honduras / S. J. Record B.H.33 / 26 / "Wild Coco Plum or Grenada"
El Salvador / P. C. Standley 20116 / 1922 / " 'Icaco montés' "
Brazil / A. Silva 254 / 1944 / "Frt. edible" / " 'ajurúrana' "
Brazil / A. Lourteig 1777 / 1966 / "Fl. rojolilácea, suavemente perfumada"

1572   *H. racemosa* var. *racemosa* W.I.: Tobago / W. E. Broadway 4457 / 1913 / "Flowers blue, fragrant"
Venezuela / J. A. Steyermark 86380 / 1960 / " 'naranjillo' "
Brazil / R. Fróes 11533 / 1939 / " 'Murta de rato' "
Venezuela / J. A. Steyermark 60979 / 1945 / "Fruit edible" / " 'carbonero' "

1573   *H. triandra* Panama / Kirkbride & Duke 1365 / 1968 / "fruits hairy, black, eaten whole, except for the seed, by the Choco Indians"
W.I.: Martinique / P. Duss 1905 / 1878 / "fruits mangeable" / "Icaque poileux"
W.I.: St. Kitts / J. S. Beard 292 / 1944 / "fruit purple, succulent, eaten by birds" / " 'Pigeon berry' "
Haiti / E. C. & G. M. Leonard 12417 / 1929 / "fruits purple, sweetish, slightly bitter"
Venezuela / J. de Bruijn 1664 / 1967 / "Flowers sweet smelling"
Venezuela / J. A. Steyermark 61058 / 1945 / "fruit edible" / " 'carbonera' "
Colombia / O. Haught 4264 / 1944 / "Fruit . . . sweet, very juicy, black with red flesh. Should be worthy of cultivation"

1574   *H. americana* Colombia / E. L. Little 7919 / 1944 / "Frt. eaten, black, gives purple dye" / " 'guarapata' "

1575   *H. bullata* Guayana / B. & C. K. Maguire 35045 / 1953 / "Fruit black, juicy, sweet"
Venezuela / A. L. Bernardi 6524 / 57 / "frutos drupaceos, comestibles" / "Calzeta de la botella"

1576   *H. guainiae* Colombia / Gasche & Desplats 154 / 1974 / "Jeunes rameaux utilisés dans les piéges a petits rongeurs" / "Nom Witoto: eĩro"

1577   *H. longipedicellata* Guayana / B. & C. K. Maguire 35043 / 1953 / ". . . fruit black, pulpy, sweet"

1578   *H. schultesii* Colombia / C. Sastre 2357 / 1973 / "Utilisé pour faire des torches" / "Nom Witoto: dofiro"

1579   *H. silicea* Venezuela / J. A. Steyermark 86384 / 1960 / " 'naranjillo' "

580 *H. tentaculata* Brazil / A. Silva 285 / 1944 / "Frt. edible" / " 'ajurúrana' "

581 *Licania michauxii* U.S.: Ga. / J. S. Harper 204 / 1930 / " 'Gopher Apple' "

582 *L. hypoleuca* Brit. Honduras / P. H. Gentle 3342 / 1940 / " 'Pigeon plum' "

583 *L. operculipetala* Costa Rica / P. H. Allen 6032 / 1951 / "Very fragrant" / " 'Sapotillo' "

584 *L. platypus* El Salvador / P. C. Standley 20532 / 1922 / "fr. very large, little eaten" / " 'Sunzapote' "

585 *L. sparsipila* Brit. Honduras / W. A. Schipp 102 / 1929 / "flowers white and sweetly perfumed"
Brit. Honduras / P. H. Gentle 3535 / 1941 / " 'bastard pigeon plum' "

586 *L. apetala* var. *aperta* Venezuela / H. Pittier 12073 / 1925 / " 'mamoncillo' "
Venezuela / L. Aristeguieta 5889 / 1965 / " 'La corteza quemada se utilizaba para mezclarla con arcilla en la fabricación de lozas' " / "N.v.: Güepe"

587 *L. apetala* var. *apetala* Brazil / W. Rodriques 289 / 56 / "a casca de madeira misturada com barro é utilizada na cerâmica de vasos" / "Caraipí"
Colombia / Schultes & Cabrera 16662 / 1952 / "Bark ground to mix with clay for pots" / "Tanimuka = ya-feé-a; Makuna = rãn-'hoo; Yukuna = a-ma-reé"
Guyane Francaise / P. Bena 1317 / 57 / "fleurs . . . odeur tres agreable se rapprochant du jasmin" / "Gaulette"

588 *L. britteniana* Colombia / Schultes & Black 8588 / 1946 / "Fl. white, very fragrant of honey"

589 *L. davillaefolia* Brazil / Prance & Pennington 2050 / 1965 / "Ash of bark used for hardening pottery" / " 'Cariperama' "
Brazil / G. A. Black 47-1975 / 1947 / "Pajurarana"

590 *L. dealbata* Brasil / Gardner 2836 / 1839 / "fruit said to be large & eatable" / " 'manja-croite'-literally Emu fruit"

591 *L. heteromorpha* var. *glabra* Colombia / H. García y Barriga 14212 / 1952 / "Frutos . . . astringente"

592 *L. heteromorpha* var. *heteromopha* Venezuela / J. A. Steyermark 88108 / 1960 / "flesh hard, slightly sweetish"
Brazil / R. Fróes 1915 / 1932 / "Bark used as a dye for ink and paint" / " 'Macucu' "
W.I.: Tobago / W. E. Broadway 4466 / 1913 / "The Widow Tree"
"W.I.: Trinidad & Tobago / R. C. Marshall 12688 / 32 / " 'wild cocoa' "

593 *L. kunthiana* Brazil / W. A. Archer 8076 / 1942 / "Bark burned and ashes mixed with clay to make jugs and other utensils" / " 'caraipé-rana' "
Venezuela / J. A. Steyermark 61093 / 1945 / "Fruit edible" / " 'carbonero' "
Brazil / Prance & Lima 4808 / 1967 / "Mature fruit . . . eaten by birds"

1594   *L. longistyla* Venezuela / J. J. Wurdack 306 / 1955 / "Fruit brown, edible when ripe" / " 'Merecurillo blanco' "

1595   *L. oblongifolia* Brazil / G. T. Prance 16799 / 1966 / "Frutos . . . comestíveis para es animais"

1596   *L. octandra* subsp. *octandra* Brazil / J. C. de Moraes 894 / 1954 / " 'Cocão' "

1597   *L. octandra* subsp. *pallida* Brazil / Prance, Maas et al. 16524 / 1971 / "the bark ash used as a hardener for pottery by Deni Indians" / " 'Cubitsa' "

1598   *L. paraensis* Brazil / Prance, Forero et al. 6532 / 1968 / ". . . flowers smelling of cooked meat"

1599   *L. steyermarkii* Venezuela / J. A. Steyermark 60465 / 1944 / "fruit edible, very tasty" / " 'macara-yek' "

1600   *L. tomentosa* Brazil / C. F. Baker 64 / 1908 / "This tree not only yields a good edible fruit called Oity, but makes a splendid shade tree, low and very broad"
Brazil / L. Coêlho 1286 / 55 / "Frutos amarelos comestíveis"

1601   *Licania* indet. Colombia / E. L. & R. R. Little 9673 / "Said to be yellow and very edible at maturity" / " 'dantamullo' (Indian name; has no Spanish name)"

1602   *Licania* indet. Colombia / Schultes & Cabrera 15456 / 1952 / "Bark used to mix with clay for making pots" / "Karapaná = noo-iñ'. Kabayari = Ka'-we. Kuripalso = Ka'-we. Puinave = wam'-chook"

1603   *Parinari pachyphylla* Venezuela / Wurdack & Monachino 41143 / 1956 / "Fruit said to be edible" / " 'Guamai' "

1604   *P. rodolphii* Brazil / Pires & Silva 4640 / 1953 / "os galhos têm ninhos de formiga, esponjosos que não pegam água e são usados para fazer fôgo"

1605   *P. romeroi* Ecuador / Little & Dixon 21014 / 1965 / " 'cuero de sapo' "

1606   *P. sprucei* Brazil / R. L. Fróes 22186 / 1947 / ". . . fruit nuts edible"

1607   *P. curatellifolia* Tanganyika / R. E. S. Tanner 1246 / 1953 / "Sukuma tribe: Roots pounded and put in porridge to treat for dropsy, water from pounded roots drunk for stomach pains" / "Kisukuma: Naji"

CONNARACEAE

1608   *Byrsocarpus orientalis* Tanganyika / R. E. S. Tanner 3872 / 1957 / "Bondei: Leaves pounded up and cooked with beans as food" / "Bondei: Kisogo"
Tanganyika / R. E. S. Tanner 3214 / 1956 / "flowers white, aromatic" / "Nyamwezi: Makaranyika"

1609   *Connarus* indet. New Guinea / D. Frodin NGF 26559 / 66 / "Odour bitter" / "Local name Asleep"

610 *Rourea glabra* var. *glabra* Mexico / Y. Mexia 1265 / 1926 / "Fr. said to be yellow when ripe. Used as poison for dogs, etc." / " 'Chilillo' "
Cuba / Leon & Roca 8088 / 1918 / "fruit eaten by aunts"

611 *R. gardneriana* Brazil / A. P. Duarte 8001 / 1963 / "flores creme perfumadas"

MIMOSACEAE

612 *Acacia baileyana* U.S.: Calif. / H. M. Hall 44 / no year / "The bark is said to be used for tanning in its native country, Australia, where it is called Cootamundra Silver Wattle"

613 *A. cornigera* El Salvador / P. C. Standley 19150 / 1922 / "pulp of fr. eaten" / " 'Iscanal, Cutupito, Iscanal de cachito' "

614 *A. goldmanii* Mexico / A. Carter 4624 / 1963 / " 'frijolillo' "

615 *A. laevis* Mexico / G. B. Hinton 4852 / 33 / " 'Root used for tanning' " / "Timbe"

616 *A. salvadorensis* El Salvador / S. Calderon 2245 / 1925 / " 'Quebracho liso' "

617 *A. unijuga* Mexico / T. MacDougall s.n. / 1970 / "N.v.: 'Huamuche' "
Mexico / Pennington & Sarukhán K. 9600 / 1968 / "Flores Fragantes"

618 *Acacia* indet. Mexico / Hinton et al. 12415 / 38 / "rots for tanning" / "Vernac. name: Timbre"

619 *Acacia* sp. nov. indet. Mexico / H. S. Gentry 5711 / 1940 / "Ant infested. Used as fuel" / "Recajadero"

620 *Acacia* indet. Brit. Honduras / C. L. Lundell 6356 / 1936 / " 'Hesmo', 'Wild tamarind' "

621 *A. macrantha* Virgin Is. / J. S. Beard 340 / 1944 / "used for charcoal" / " 'Cosher' "
Dominican Republic / E. J. Valeur 211 / 1930 / "Tr. 'Aroma or Cambrona' "
Peru / C. M. Belshaw 3374 / 1937 / "fruits green, used to make mucilage"

622 *A. maschalocephala* Cuba / E. L. Ekman 7436 / 1916 / " 'Bejuco tocino' Cub."

623 *A. tenuifolia* Cuba / A. Luna 374 / 1920 / " 'Araña gato' or 'Tocino' "

624 *A. tortuosa* W.I.: Montserrat / J. A. Schafer 109 / 1907 / " 'Cashaw' "

625 *A. adhaerans* Brazil / Y. Mexia 4494 / 1930 / "Flower cream white, fragrant. Bee plant"

626 *A. amazonica* Brazil / Spruce 1697 / 51 / "Fls. sulphur-yellow, odorif."

627 *A. farnesiana* Colombia / E. L. Little 7312 / 1944 / "Medicine from bark. (Reported at Neiva to be used as cure for typhoid fever)" / " 'pelá' "

1628   *A. flexuosa* Venezuela / F. D. Smith 229 / 1952 / "flowers yellow, unpleasant odor" / " 'Cuji hediondo' "

1629   *A. albida* S.W. Africa / R. Seydel 3168 / 1962 / "Sehr geschaetztes Futter fuer Wild und Vieh" / " 'Anaschoten' "

1630   *A. clavigera* subsp. *usambarensis* Tanganyika / R. E. S. Tanner 2656 / 1956 / "Ligua: Leaves chewed and plastered on snake bite" / "Kizigua: Mkulagembe"

1631   *A. inconflagrabilis* S. Africa / J. Gerstner 5548 / 45 / "Always a shrub, fire resistant" / " 'isikombo' "

1632   *A. mellifera* Tanganyika / R. E. S. Tanner 3085 / 1956 / "Zigua: for cough, bark chewed" / "Zigua: Msasa"

1633   *A. pennata* Tanganyika / R. E. S. Tanner 1220 / 1953 / "Sukuma tribe: Used in treating sick goats — ground with other plants to make flour, this spread on ground where goats sleep" / "Kisukuma: Igeye"

1634   *A. seyal* Tanganyika / R. E. S. Tanner 1009 / 1952 / "Sukuma tribe: Leaves used to treat fever, chewed to bring good luck, bark chewed to quench thirst, pounded bark used to give red colour to fresh skins" / "Kisukuma; Kizinza: Sese"

1635   *A. sieberiana* Tanganyika / R. E. S. Tanner 3764 / 1957 / "Bondei: For pain in the lower stomach, roots boiled and the water drunk" / "Bondei: Mguna"

1636   *A. stuhlmannii* Tanganyika / R. E. S. Tanner 3771 / 1957 / "Zigua: To cure infertility, roots pounded up and eaten with food" / "Zigua: Mgunga"

1637   *A. volkii* S.W. Africa / R. Seydel 419 / 54 / "Huelsen gefressen von Wild und Vieh"

1638   *A. xanthophloea* S. Rhodesia / G. M. McGregor 68/51 / 51 / " 'Fever Tree' "

1639   *Acacia* indet. Tanganyika / R. E. S. Tanner 1349 / 1953 / "Sukuma tribe: Roots eaten with hot water to treat stomach pains; cooked in porridge for Bilharzia" / "Kisukuma: Ghalle"

1640   *A. cambagei* Australia: Queensland / S. L. Everist 3562 / 48 / "whole plant with powerful unpleasant odour"

1641   *A. hemiteles* Australia: Queensland / C. T. White 5463 / 1927 / "Used for tanning, the whole plant being collected"

1642   *A. kempeana* Australia: Queensland / S. L. Everist 3909 / 1949 / "Local name 'Turpentine Mulga' "

1643   *A. prominens* Australia: Queensland / C. T. White 9163 / 1933 / "fls. lemon yellow, pleasantly 'wattle' scented"

1644   *A. stenophylla* Australia: Queensland / S. L. Everist 4020 / 1949 / "Gum edible" / " 'Dunthy Tree' "

645 *A. concinna* Philippine Is. / A. D. E. Elmer 8956 / 1907 / "the pods are cooked with meats as food and it gives the whole dish a sour flavor" / " 'Baiwaing' "

646 *A. spirorbis* New Caledonia / J. T. Buchholz 1409 / 1947 / "False Guiac"

647 *Adenanthera pavonina* W. I.: Grenada / R. A. Howard 10863 / 1950 / "leaves stewed and used for treating colds" / "local name 'L-eglise' "
W. I.: Martinique / P. Duss 830 / no year / ". . . ses graines rouges servent á fairé des colliers et des objets de parme"

648 *Adenanthera* indet. Ceylon / Hepper & Fernando 975 / 1972 / "Medicine tree" / "Madatiya (S.)"

649 *Albizzia caribaea* W.I.: Tobago / W. E. Broadway 3165 / 1909 / " 'Wild Tamarind' "

650 *A. carbonaria* Brazil / H. F. Leitão Filho 359 / 1968 / "Usada no sombreamento de café"

651 *A. colombiana* Colombia / A. Dugand 1143 / 1937 / "vulg. 'Guacamayo' "

652 *A. malocarpa* Colombia / W. A. Archer 530 / 1930 / " 'Piñon' "

653 *A. anthelmintica* Tanganyika / R. E. S. Tanner 3045 / 1956 / "Zigua: For cough, bark chewed" / "Zigua: Mfuleta"
S. W. Africa / R. Seydel 5a / 54 / "Blaetter u. Huelsen vom Vieh gefressen. Rinde Bandwurmmittel, omuama (herero)"

654 *A. petersiana* Tanganyika / R. E. S. Tanner 1089 / 1952 / "Sukuma: for dry cough, leaves chewed fresh. For scabies, bark cut, pounded up, soaked in water and drunk after 12 hours" / "Kisukuma: Shishiguru"

655 *A. versicolor* Tanganyika / R. E. S. Tanner 3754 / 1957 / "Bondei: For pain in lower stomach, roots boiled and the water drunk" / "Bondei: Mkingu"

656 *A. lebbeck* Mariana Is. / F. R. Fosberg 24955 / 1946 / ". . . bark smelled to make one sneeze" / " 'kalaskas' (because of noise it makes)"

657 *Calliandra grandiflora* Guatemala / J. A. Steyermark 50601 / 1942 / " 'cola de tijereta' "

658 *C. lucens* Honduras / C. & W. von Hagen 1201 / 1937 / " 'Vainilla' "

659 *C. molinae* Honduras / Standley, Williams & Allen 571 / 1946 / "Eaten by cattle"

660 *C. portoricensis* El Salvador / P. C. Standley 22958 / 1922 / " 'Guacamaya montés' "
Guatemala / P. C. Standley 76311 / 1940 / " 'Tamarindo de monte' "

661 *C. tonduzii* Panama / Cooper & Slater 318 / 1927 / " 'Aroma' "

662 *C. haematostoma* Dominican Republic / L. R. Holdridge 525 / 1940 / "Tabacuela' "

1663   *C. purpurea* W.I.: Grenada / W. E. Broadway s.n. / 1897 / "a dwarf goat-eaten shrub"

1664   *C. parviflora* Brazil / Harley, Souza & Ferreira 10993 / 1968 / "Used medicinally"

1665   *C. surinamensis* Brazil / W. A. Archer 7628 / 1942 / " 'pau de salsa' "

1666   *Dichrostachys glomerata* U.S.: Fla. / Gifford & Totten s.n. / 1940 / " 'Aroma Mala' "

1667   *D. cinerea* Tanganyika / R. E. S. Tanner 1427 / 1953 / "Sukuma tribe: Roots boiled, taken with water for Bilharzia" / "Kisukuma: Ntunduru"
Tanganyika / R. E. S. Tanner 3177 / 1956 / "Zigua: For cough, bark chewed. For snake bite, leaves chewed and put on bite as poultice" / "Zigua: Mkulagembe"
Tanganyika / R. E. S. Tanner 3760 / 1957 / "Bondei: For pain in stomach, roots boiled and the water drunk" / "Bondei: Mkulagenbe"
Tanganyika / R. E. S. Tanner 4020 / 1959 / "Leaves chewed and put on swellings for bubonic plague" / "Kizanaki: Kihuri"
Tanganyika / R. E. S. Tanner / 4424 / 1959 / "Roots and bark used to treat stomach pain" / "Kizanaki: Kihure"

1668   *Entada gigas* Brit. Honduras / W. A. Schipp 1143 / 1933 / "Flowers cream, with a sickly odor"

1669   *E. polystachya* Mexico / Y. Mexia 8883 / 1937 / "Root ground and used as hair tonic" / " 'Bejuco de Paringue' "
Mexico / J. N. Rovirosa 221 / 1888 / "Bejuco de Mandongo"
El Salvador / S. Calderon 1092 / 1922 / " 'Cola de zarrillo, Quiamol' "

1670   *Entada* indet. Venezuela / Steyermark & Velasco 100174 / 1967 / ". . . loaded with ants" / " 'caña fistula' "

1671   *E. phaseoloides* Hainan / C. I. Lei 200 / 1932 / "fruit edible" / "Lo Ah Tzam"
Marianne Is. / P. Nelson 154 / 1918 / "Roots used as a hair tonic" / "Gadye−Gogo"
Philippine Is. / A. L. Zwickey 184 / 1938 / "Seeds burned and mixed with oil as application to burns" / " 'Kirakira' "
New Hebrides / S. F. Kajewski 336 / 1928 / "Seed roasted in native oven and kernel is put in baskets in running water for two weeks, taken cooked a second time and eaten" / " 'Ou-vone-naw' "

1672   *Entada* indet. Burma / F. Kingdon Ward 17518 / 1948 / "flowers . . . with a disgusting smell"

1673   *Inga cocleensis* Honduras / A. Molina R. 18390 / 1966 / " 'Guava' "

1674   *I. coruscans* Panama / J. A. Drake 4986 / 1962 / " 'guava' "

1675   *I. endlicheria* Honduras / C. & W. von Hagen 1117 / 1937 / "Fruit eaten" / " 'Guamo pachona' "

676 *I. eriocarpa* Mexico / Y. Mexia 1842 / 1927 / "Fr. said to be long edible pod" / " 'Joaquinaquil' "

677 *I. oerstediana* Honduras / P. C. Standley 53601 / 1928 / "Fruit edible" / " 'Guama' "

678 *I. sapindoides* El Salvador / P. C. Standley 23088 / 1922 / "fr. eaten" / " 'Cujin, Cujinicuil' "

679 *I. fagifolia* W. I.: Dominica / G. P. Cooper 185 / 1933 / "Called 'Poix doux' meaning 'sweet pea' "

680 *I. vera* W. I.: Dominica / G. P. Cooper 183 / 1933 / "Called 'Poix doux marron' meaning 'wild sweet pea' "

681 *I. cinnamomea* Brazil / C. F. Baker 55 / 1908 / ". . . filled between the seeds with a copious pure white sweet edible pulp"

682 *I. codonantha* Colombia / J. M. Duque Jaramillo 4619 / 1947 / "Bueno para el sombrío" / "N.v. 'Guamo'; santafereño
Peru / J. J. Wurdack 1784 / 1962 / "fruit said to be edible" / " 'Guaba' "

683 *I. coriacea* Fr. Guiana / M. Lemoine 7805 / 1961 / "Fruits comestibles" / "(Taki-Taki): Adai Oueko"

684 *I. corymbifera* Peru / Y. Mexia 6076 / 1931 / "the pulp surrounding the seed is edible" / " 'Shimbillo' "

685 *I. cylindrica* Brasil / E. P. Heringer 11760 / 1969 / ". . . flōres perfumadas, frutos comestiveis. Cultivada"

686 *I. densiflora* Colombia / E. L. Little 7687 / 1944 / "Frts. sold in market of San Agustin" / " 'guamo machete' "

687 *I. disticha* Fr. Guiana / M. Lemoine 7890 / 1961 / "Fruits comestibles" / "(Taki-Taki): Baboun-Oueko"

688 *I. fendleriana* Venezuela / Fendler 2259 / ca. 1850 / "Fruit sweet"

689 *I. heterophylla* Brazil / A. Ducke 7602 / 1942 / "Wood has rank odor"
Brazil / A. Ducke 8048 / 1942 / "Frt. eaten by birds" / " 'ingá-i' "

690 *I. ingoides* Trinidad / R. O. Williams 12176 / 29 / "seeds sweet, edible" / "Pois doux"

691 *I. insignis* S. AM. / Hartweg 966 / no year / "The seeds are surrounded by a sweet pulp which is eaten. The pods are brought to market" / " 'Guaqua' "
Ecuador / Y. Mexia 7046 / 1935 / "pulp of fruit eaten" / " 'Guava' "

692 *I. jenmani* Brit. Guiana / For. Dept. of Brit. Guiana F674 / 41 / "seeds purple-green covered in a sweetish-juicy white wool"

693 *I. lateriflora* Fr. Guiana / M. Lemoine 7894 / 1961 / "Fruits comestibles" / "(Taki-Taki): Guebi-Oueko"
Venezuela / J. A. Steyermark 88797 / 1961 / "flowers fragrant"

1694   *I. marginata* Ecuador / Y. Mexia 7005 / 1935 / "The pulp surrounding seed is eaten" / " 'Guavo' "
Brazil / F. J. Hermann 7803 / 1946 / "Pulp around seed is edible" / " 'ingá' "

1695   *I. meissneriana* Fr. Guiana / W. E. Broadway 950 / 1921 / " 'Mango' "

1696   *I. nobilis* Peru / Killip & Smith 24734 / 1929 / "fruit edible"
Colombia / Schultes & Cabrera 16231 / 1952 / "Flowers white, fragrant" / "Makuna = ta-roó-boo-ta-ree-mě ('guama of toads')"
Venezuela / J. A. Steyermark 87240 / 1960 / "fruit edible"
Brit. Guiana / Boyan, Tillett & Tillett 44952 / 1960 / "fruit . . . sweet, white, with flavor and texture of Persimmon" / " 'mowrowai' in Akawaio"

1697   *I. pilosiuscula* Brit. Guiana / For. Dept. of Brit. Guiana F653 / 41 / "seeds covered with a sweetish juicy cottony pulp"
Fr. Guiana / W. E. Broadway 553 / 1921 / "Fruit pulpy, yellow, sweet"

1698   *I. plumifera* Colombia / Schultes & Cabrera 13538 / 1951 / "Flowers white, very fragrant" / "Puinave = choo-ee-paẃ"

1699   *I. quaternata* Colombia / Schultes & Cabrera 16261 / 1952 / "Flowers white, pungently fragrant" / "Makuna = ta-rabee-kě'-mě-nee"

1700   *I. ruiziana* Ecuador / E. Heinrichs 296 / 1933 / ". . . voller Ameisen, Blüten weiss, starker Geruch"
Peru / Y. Mexia 6334 / 1931 / "white pulp of fruit edible" / " 'Shimbillo' "
Peru / Y. Mexia 6073 / 1931 / "the pulp surrounding the seed eaten by boys and monkeys" / " 'Shimbillo' "

1701   *I. splendens* Venezuela / J. A. Steyermark 86345 / 1960 / "fruit edible" / " 'guamo pato de morrocoy' "
Venezuela / J. A. Steyermark 87298 / 1960 / "fruit terete, sweet, edible"
Brazil / Prance, Steward et al. 11064 / 1971 / " 'Oara' (Uaicá-Mucajaí)"

1702   *I. steinbachii* Bolivia / J. Steinbach 5619 / 1921 / ". . . con semilla verde envuelta en una blanca pulpa, que es dulce y sana" / "Pacaý peludo"

1703   *I. striata* Brazil / Y. Mexia 5208 / 1930 / " 'Inga banana' "

1704   *I. tomentosa* Ecuador / R. Espinosa 1818 / 1947 / "Fruto comestible" / "Arcapamba (Zaruma)"

1705   *I. umbellifera* Colombia / Schultes & Cabrera 13542 / 1951 / "Fruit edible, pulp" / "Puinave = choo'-ee"

1706   *I. umbratica* Peru / J. M. Schunke 257 / 1935 / "Fruit edible" / " 'Shymbillo' "

1707   *Inga* indet. / Brit. Guiana / Boyan, Tillett & Tillett 44952 / 1960 / "fruit . . . mucilaginous, sweet, white, with flavor and texture of Persimmon" / " 'mowrowai' in Akawaio"

08 *Leucaena glabrata* Mexico / E. Matuda 18403 / 1948 / "Cultivated for medicine in Rancho Soconusco" / " 'Guaz' "

09 *L. glauca* Mexico / G. B. Hinton 3119 / 33 / "Fruit sold in markets" / "Calguaje"

10 *L. lanceolata* Mexico / H. S. Gentry 4819 / 1939 / "Burros eat the fallen pods" / "Guaje"

11 *L. trichoides* Peru / F. Woytkowski 6870 / 1967 / "natives say that this plant crushed kills fish in the river; toxic"

12 *L. leucocephala* Mariana Is. / F. R. Fosberg 25413 / 1946 / "Used as fodder but makes horses hair fall out; beans used for making necklaces and other ornaments; 'leaves sometime eaten' " / " 'tangantangan' "

13 *Mimosa galeottii* Mexico / H. S. Gentry 5859 / 1940 / "Bark highly valued for tanning"

14 *M. pigra* Guatemala / P. C. Standley 24047 / 1922 / " 'Puta vieja, Sinvergüenza' "
Tanganyika / R. E. S. Tanner 1133 / 1952 / "Sukuma: for boils — leaves pounded up green and put on boils to draw them" / "Kisukuma: Msomeangombe"

15 *M. pudica* Dominican Republic / E. J. Valeur 212 / 1929 / " 'Muerevivir' "
Surinam / A. Kappler s.n. / no year / "Blätter und Zweige sind officinell"
Philippine Is. / A. L. Zwickey 235 / 1938 / " 'Sarambi'; 'Tamparasa' (meaning vulva?) (both Lan.; the latter name locally ambiguous)"

16 *M. scalpens* Brit. Honduras / W. A. Schipp 1306 / 1934 / "Flowers white, highly perfumed"

17 *Mimosa* indet. Mexico / Hinton et al. 7993 / 35 / "Fruit; buds edible" / "Guaje chiquito"

18 *M. arenosa* Venezuela / Aristeguieta & Montoya 2025 / 1954 / "Las hojas tienen sustancias pegajosas que se ponen de manifiesto durante la secada de la muestra"

19 *M. invisa* Brazil / Y. Mexia 4524 / 1930 / " 'Maliçia de mulher' "

20 *Mimosa* indet. / Venezuela / L. Aristeguieta 2025 / 1954 / "Las hojas tienen sustancias pegajosas"

21 *Neptunia* indet. Ecuador / C. C. Fuller 89 / 53 / "Tuber root beaten and used as soap to cleanse hair. Reported very good to remove dandruff" / "Lomo huivilan — Quichua"

22 *Neptunia* indet. Ecuador / C. C. Fuller 5 / 48 / "Tuber root beaten and used as soap to cleanse hair. Cultivated in Indian chagra" / "Lomo huivilan; Quichua"

23 *Ortholobium bubalinum* Sumatra / Kostermans 12054 / 1955 / "All part with offensive Zygia jiringa smell"

1724   *Parkia auriculata* S. AM. / Spruce 1208 / 51 / "Seeds in a small quantity of sweet-viscid pulp"

1725   *Pentaclethra macroloba* Panama / G. P. Cooper 430 / 1928 / " 'Tamarind' "
Costa Rica / R. W. Lent 1404 / 1967 / "Flowers white, strong unpleasant odor"
Surinam / W. A. Archer 2896 / 1934 / "Large tree, common *Fish Poison*, bark used; less effective than nekoe" / " 'Karoballi' (Arawak) 'Kroebara' (Negro English)"
Brit. Guiana / Jenman 3934 / 1887 / " 'Trysel' "

1726   *Piptadenia peregrina* W.I.: Grenada / J. S. Beard 164 / 1943 / "pods used for tanning"
Brit. Guiana / A. C. Smith 3217 / 1938 / "Used abundantly locally for tanning leather; bark pounded to pulp and soaked in water with skin"

1727   *Piptadenia* indet. Bolivia / O. E. White 1498 / 1921 / "Bark prized for tanning"

1728   *Piptadenia* indet. Brasil / R. de Lemos Fróes 20143 / 1942 / "casca usada para taunary"

1729   *P. africana* Liberia / J. T. Baldwin 6125 / 1947 / "Boys superstitious about cutting it" / " 'False Sasswood' "

1730   *Pithecellobium adinocephalum* Panama / J. A. Duke 8406 / 1966 / " 'frijolillo' (Spanish)"

1731   *P. austrinum* Costa Rica / P. H. Allen 5468 / 1950 / " 'Tamarindo' "

1732   *P. leucocalyx* Brit. Honduras / P. H. Gentle 3481 / 1941 / " 'wild tamarind' "

1733   *P. longifolium* Panama / J. A. Duke 8212 / no year / "En creencias las hojas sierben para baño (y dan suerte)" / "Pichinde"
Panama / J. A. Duke 14396 / 1967 / "Medicinal Plants of the Bayano Cuna (Names provided by chiefs, witch doctors or secretaries)" / " 'Apa' "

1734   *P. pallens* Mexico / Dr. Gregg s.n. / 47 / "The green beans, anthers & bulb cooked and eaten"

1735   *P. recordii* Brit. Honduras / W. A. Schipp 698 / 1931 / "Flowers . . . highly perfumed"

1736   *Pithecellobium* indet. Mexico / J. N. Rovirosa 500 / 1889 / "Quiebra hachas vel Quiebra jacha incolarum"

1737   *P. arboreum* Jamaica / W. R. Maxon 9518 / 1926 / " 'Tamarind' "

1738   *P. keyense* Bahama Is. / O. Degener 18994 / 1946 / "seed eaten" / " 'Ramhorn' in allusion to shape of legume"

1739   *P. platyloba* W.I.: Curaçao / Fr. M. Arnoldo 3013 / 1963 / "Common name: uña di gaber, Dabaruida"

40 *P. claviflorum* Brazil / Pires 396 / 1947 / "A casca raspada é usada como sabão para lavagem" / " 'Iapuúca' "

41 *P. duckei* Brazil / Prance & Prance 14762 / 1971 / "With ants associated"

42 *P. guachapele* Colombia / Bro. Elias 509 / 1828 / " 'Tabaca' "

43 *P. macrophyllum* Peru / J. Schunke V. 2635 / 1968 / " 'Virgen sisa' "

44 *P. panurense* Venezuela / E. P. Killip 37547 / 1943 / "infl. greenish white, very dense, ants often making a nest there"

45 *P. pistaceaefolium* Venezuela / Aristeguieta & Zabala 7074 / 1969 / "Se usa para estantes y horcones" / " 'Vera macho', 'Quiebrajacho' "

46 *P. polycephalum* Brazil / P. Rosa 74633 / 1934 / " 'Farinha Ceca' "

47 *P. saman* Peru / J. Schunke V. 4402 / 1970 / " 'Lluicho Vainilla' " Venezuela / Curran & Haman 1210 / 1917 / " 'Casa bicho' "

48 *Pithecellobium* indet. Brasil / A. Macedo 117 / 1945 / "Roxas aromaticas"

49 *Pithecellobium* indet. Venezuela / Steyermark & Velasco 100199 / 1967 / " 'uña de loro' "

50 *P. angulatum* Hainan / F. A. McClure 9405 / 1922 / "Lois uses the roots as medicine by steeping in water" / "T'a Hon Ts'un"

51 *P. clypearia* Hainan / Tsang, Tang & Fung 50 / 1929 / "used as medicine to cure sore eyes" / "Mah Sah"

52 *Pithecellobium* indet. Ceylon / Kostermans 23293 / 1969 / "Wood stinking (Parkia smell)"

53 *P. jiringa* Sumatra / H. H. Bartlett 8737 / 1927 / ". . . pods edible" Sumatra / B. A. Krukoff 4266 / 1932 / ". . . fruit edible" / " 'Ojekol' "

54 *Prosopis chilensis* Chile / Bowman s.n. / no year / "It produces a pod with seeds; and these are valuable for fodder, especially in the dry years" / "Algarrobo tree"

55 *P. insularum* New Guinea / D. Frodin NGF 26607 / 66 / "Odour of garlic"

56 *Stryphnodendron barbatimam* Brazil / G. Eiten 3387 / 1961 / "Sweet honey odor"

57 *Vachellia farnesiana* Colombia / J. M. Duque J. 2526 / 1945 / "Fls. amarillas muy olorosas; para perfumería" / " 'Aromo' "

58 *Mimosacea* indet. Brazil / B. A. Krukoff's 4th Exped. to Braz. Amaz. 2057 / 1933 / "Natives use the juice from leaves for injuries" / "Anjiqiunho"

CAESALPINIACEAE

59 *Adenolobus garipensis* S. W. Africa / R. Seydel 4471 / 67 / ". . . sehr guter Futterbusch"

1760   *Afzelia quanzensis* Tanganyika / R. E. S. Tanner 3772 / 1957 / "Zigua: For bilharzia, roots boiled and the water drunk" / "Zigua: Mkomba; Swaha: Mpigakofi"
Tanganyika / R. E. S. Tanner 3713 / 1957 / "Zigua: For pain in stomach, roots boiled and the water drunk" / "Zigua: Mkomba"

1761   *Amburana cearensis* Peru / R. L. Magin 22 / 1962 / "Usos: aserrfo, muebles, ebanistería. Frutos aromáticos" / " 'Ishpingo' "

1762   *Apuleia leiocarpa* var. *leiocarpa* Brasil / E. P. Heringer 10099 / 1965 / "Árvore avatife para estudo químico"

1763   *Bauhinia dipetala* Brit. Honduras / H. H. Bartlett 11966 / 1931 / ". . . green fls., odor of cabbage"

1764   *B. glabra* Brit. Guiana / Schomburgk 565 / 1838 / "Fl. highly odoriferous"
Peru / J. Schunke V. 2986 / 1969 / " 'Escalera del diablo' "
Venezuela / W. A. Archer 3098 / 1935 / "Flw. white & perfumed. Fish poison" / " 'Bejuco de cadena' "

1765   *B. guianensis* Brit. Guiana / W. A. Archer 2575 / 1934 / "Carib fish poison" / " 'Guayamufrati' (Carib) means Turtles hammock"

1766   *B. microstachya* Paraguay / A. L. Woolston 677 / 57 / "Fls. white with fragrant perfume" / "n.v.: Ca'i-escalera"

1767   *B. garipensis* S .W. Africa / R. Seydel 346 / 1954 / "gute Futterpflanze"

1768   *B. thonningii* Tanganyika / R. E. S. Tanner 3753 / 1957 / "Bondei: For ringworm, inner bark soaked and used for washing down; for cough, bark chewed" / "Bondei: Msegese"
Tanganyika / R. E. S. Tanner 1254 / 1953 / "Sukuma tribe: Roots used in diarrhoea—pounded, cooked, water drunk. Flour from roots put in drinking water of cattle with diarrhoea" / "Kisukuma: Ntindwanzagamba"

1769   *Bocoa alterna* Brit. Guiana / A. C. Smith 2828 / 1937 / "Waiwais strip bark and make into a solution for rinsing feet; said to be an effective cure for 'grounditch' "

1770   *Burkea africana* Tanganyika / R. E. S. Tanner 1363 / 1953 / "Sukuma tribe: Roots pounded, eaten in warm water to treat stomach pain" / "Kisukuma: Ngando"

1771   *Caesalpinia* indet. Mexico / T. MacDougall s.n. / 1970 / " 'granadillo' "

1772   *C. pulcherrima* Cuba / Herb. Collegii Pharmaciae Neo-Eboracensis s.n. / 1841 / "Cuban name: guacamaya"
Caroline Is. / M. Evans 352 / "Used for maremar. Informant Santos" / " 'WARIPIG' "
Caroline Is. / F. R. Fosberg 26241 / 1946 / "(Bascom 40) beans eaten" / " 'se:mutha' "

73 *C. bonduc* Colombia / R. Romero-Castañeda 5361 / 1955 / "Los cotiledones machacados, son buenos para el dolor de muela" / " 'Garranga' "

74 *C. bracteosa* Brazil / G. & L. T. Eiten 10782 / 1970 / " 'pau de rato' "

75 *C. spinosa* Colombia / J. M. Duque-Jaramillo 2709 / 1946 / "Usado por los zapateros para colorear los cueros" / " 'Dividivi serrano' "

76 *Caesalpinia* indet. Bolivia / R. F. Steinbach 23 / 1966 / "La vaina del fruto seca usan para curtido de pieles"

77 *Caesalpinia* indet. / Ceylon / R. H. Maxwell 929 / 1972 / "flowers yellow, covered with red ants"

78 *C. major* Mariana Is. / F. R. Fosberg 24978 / 1946 / "seeds crushed and used internally for medicine, very bitter; large branch with leaves used to catch flying foxes" / " 'pacao' "

79 *C. jayabo* Easter Is. / J. P. Chapin 1027 / 1935 / "Has sweet perfume" / " 'Naoho' "

80 *Campsiandra angustifolia* Peru / A. Arostegui V. 109-AA / 63 / "La corteza los nativos lo emplean para curar reomatismo y como purgante" / "Huacapurana"
Peru / A. Arostegui V. 19 / 62 / "madera para leña y carbon, corteza mezclado con cachaza lo emplean como purgante"

81 *C. comosa* var. *laurifolia* Venezuela / Wurdack & Adderley 42679 / 1959 / "cold water bark decoction said to be very efficacious for dysentery" / " 'Chigo' "
Colombia / Schultes & Cabrera 12656 / 1951 / "Bark infusion a stomachic" / "Puinave = ye'-pa. Kuripako = yĕ'-pa"

82 *C. laurifolia* Colombia / C. J. Salcedo s.n. / 1964 / "Los cotilédones se utilizan en la alimentación humana y en la de los cerdos, previa extracción de la substancia amarga que contienen" / "N. vulgar: 'Chiga' "

83 *Cassia hebecarpa* U.S.: N.J. / Herb. Columbia Univ. s.n. / 1820 / "Wild Senna"

84 *C. marginata* U.S.: Fla. / W. T. Gillis 6975 / 1968 / " 'Ceylon senna' "

85 *C. obtusifolia* U.S.: Tex. / V. L. Cory 49823 / 1945 / " 'coffeeweed' "
El Salvador / P. C. Standley 20920 / 1922 / " 'Frijolillo' "
Haiti / R. J. Seibert 1747 / 1942 / " 'Pois Pilante' "
Fr. Guiana / W. E. Broadway 38 / 1921 / "Cut parts not pleasant smelling"
Argentina / J. E. Montes 27546 / 58 / "Las hojas y flores son usadas en medicina popular como purgativas" / " 'Sen' 'Tapeyiba' "

786 *C. biflora* El Salvador / P. C. Standley 21940 / 1922 / " 'Carne asada' "
Bahama Is. / W. Cerbin C 116 / 1973 / "Used as food for livestock" / " 'Christmas bush' "
Jamaica / Houston s.n. / 1730 / "Senna Spuria minima"

1787 **C. emarginata** Mexico / T. MacDougall s.n. / 1969 / "N.v. 'vainillo' "
Honduras / A. & A. R. Molina 26030 / 1971 / " 'frijolillo' "

1788 **C. flexuosa** var. **texana** Mexico / C. & E. Seler 1977 / 1896 / " 'tamarin-dillo' "

1789 **C. grandis** El Salvador / P. C. Standley 20695 / 1922 / "pulp of fr. eaten. Lvs. & lard used to cure mange in dogs" / " 'Caragua' "
Brasil / H. S. Irwin 5034 / 1964 / "pulp surrounding seeds with consistency of very thick molasses, foetid. Cultivated"

1790 **C. leiophylla** Mexico / J. N. Rovirosa 664 / 1889 / "Hormiguera incolarum"

1791 **C. leptocarpa** var. **hirsuta** Panama / Kirkbride & Duke 1364 / 1968 / ". . . used as a vermifuge; whole plant is boiled and may be given even to children" / " 'Frijolillo de monte' "

1792 **C. nutans** Mexico / Y. Mexia 923 / 1926 / " 'Mora hedionda' "

1793 **C. reticulata** Guatemala / R. Pinkus 1159 / 1945 / "Used by natives medicinally but do not know for what" / "Barajo"
Peru / G. Klug 1107 / 1930 / "Root is boiled and used as remedy for fever" / " 'Retáma' "

1794 **C. skinneri** Mexico / G. B. Hinton 6622 / 34 / "Uses: Tanning" / "Paracata"

1795 **C. uniflora** Brit. Honduras / M. Sousa 4144 / 1973 / "Nombre maya: Xtulú-Bayá"
Mexico / H. S. Irwin 1358 / 1957 / "Foliage foetid"
Mexico / G. B. Hinton 11750 / 38 / "Frijolillo"

1796 **C. spectabilis** Honduras / C. & W. von Hagen 1060 / 1937 / " 'Frijolillo' "
Colombia / E. L. Little 7105 / 1944 / " 'vainillo' "
Tanganyika / R. E. S. Tanner 1249 / 1953 / "Sukuma tribe: Leaves cooked and water drunk to treat jaundice" / "Kisukuma: Iyolwa"

1797 **C. confusa** Haiti / E. C. & G. M. Leonard 14970 / 1929 / " 'Pois piante' "

1798 **C. ekmaniana** Cuba / E. L. Ekman 18783 / 1924 / " 'Cafecillo' "

1799 **C. leptocarpa** Cuba / Bro. León 11698 / 1922 / " 'Yerba hedionda peluda' "

1800 **C. sophera** Haiti / R. J. Seibert 1745 / 1942 / " 'Pois piante noir' "
Marianne Is. / P. Nelson 46 / 1918 / "Economic uses: Medicine" / " 'Amot Tomaga' "

1801 **C. acanthoclada** Argentina / Varela 466 / 44 / "flor amarilla con un fuerte olor a sangre"

1802 **C. alata** Brit. Guiana / J. S. de la Cruz 1091 / 1921 / "Medicinal"
Tanganyika / R. E. S. Tanner 3606 / 1957 / "Bondei: For constipation, roots boiled and the water drunk; for general restoration of health, leaves boiled and the stem inhaled" / "Bondei: Msinba"

03   *C. biflora* var. *rostrata* Brazil / Y. Mexia 5619 / 1931 / "disagreeable odor"
/ " 'Fedegosinho' "

04   *C. cathartica* var. *cathartica* Brazil / G. Gehrt 3633 / 1919 / " 'Senne do
campo' "

05   *C. chrysocarpa* Brazil / R. Fróes 1814 / 1932 / " 'Malapastinho' "

06   *C. curvifolia* var. *lucida* Brazil / J. M. Pires 3875 / 1952 / "O chá da raíz é
tido comõ remédio para gonorréa"

07   *C. ferruginea* Brazil / Y. Mexia 4356 / 1930 / " 'Banana de Macaco' "

08   *C. fruticosa* var. *benthamiana* Ecuador / H. V. Pinkley 177 / '66 /
"Medicine: for earache, make concoction from bark of twigs and hot water"
/ "Kofán: kongihi"

09   *C. hirsuta* Brazil / R. Fróes 1798 / 1932 / " 'Erva de bicho' "
Bolivia / R. F. Steinbach 336 / 1966 / "tiene un olor característico a ácido
formico y es frecuentado por hormigas chicas"

10   *C. latifolia* var. *latifolia* Brit. Guiana / Maguire, Maguire & Wilson-
Browne 45956 / 1961 / ". . . anthers eaten by ants"

11   *C. macrophylla* Ecuador / H. V. Pinkley 154 / '66 / "Medicine — wash is
made for earache and headache" / "Kofan: kongihite?ta"
Brazil / Prance, Maas et al. 13491 / 1971 / ". . . seeds fragrant" /
" 'Tamarina' "

12   *C. moschata* Panama / J. A. Duke 15192 / 1968 / "Most septae of the
'stinking' pod are riddled with insects and lack the brown seeds"
Ecuador / Little & Dixon 21076 / 1965 / " 'barbascillo' "

13   *C. occidentalis* Brazil / E. Y. Dawson 14757 / 1956 / "medicinal plant from
town of Veadeiros . . . root pounded or chopped and well boiled to stop
vomiting" / "fedegoso"
Argentina / J. Montes 27545 / 58 / "Las hojas son usadas en infusiones en
medicina popular y de sus semillas maduras y tostadas se elabora una
especie de cafe medicinal: Las hojas son purgativas"
Tanganyika / R. E. S. Tanner 1224 / 1953 / "Sukuma tribe: Roots ground
to powder, put in tea to treat fever" / "Kisukuma: Ntamba nzoka"
Tanganyika / R. E. S. Tanner 3198 / 1956 / "Zigua: For fever in children,
leaves pounded up green and mixed with oil for rubbing over child. For
stomach ache, roots boiled and the water drunk" / "Zigua: Mwangajini"
Tanganyika / R. E. S. Tanner 1014 / 1952 / "Sukuma: for fever; roots
pounded up and put in water, drunk at once without cooking" /
"Kisukuma: Tambanzoka"
Caroline Is. / D. Anderson 1088 / 1950 / "Medicine for stiff muscles, leaves
used as poultice" / " 'Kapon' = Japanese name"
Marshall Is. / F. R. Fosberg 34184 / 1952 / "Seeds said to have been
brought from Ponape and used by Japanese for coffee but make one sick"

1814   **C. papillata** Brazil / W. A. Archer 5027 / 1936 / "Used to make tea, in same manner as 'senna' "

1815   **C. quinquangulata** Brazil / Campbell, Nelson et al. F21225 / 1974 / "Paste of leaves rubbed over body & followed by a bath. Used by Paumaris Indians to lower fever" / " 'Bakadohodoho' (Paumaris)"
Venezuela / J. Jiménez Saa 1304 / 71 / "Frijolillo"

1816   **C. ruiziana** Ecuador / H. V. Pinkley 341 / '66 / "Medicine: scrape branch into hot water and then wash around ear for earache" / "Kofán: kongihi séhe?pa"
Peru / J. Schunke V. 1438 / 1966 / "Semillas tóxicas para curar sarna" / " 'Matarro' "

1817   **C. spruceana** S. AM. / R. Spruce 2558 / 1852 / "Fls. very odorifs."

1818   **C. sulcata** Brazil / Y. Mexia 4437 / 1930 / "Foliage ill smelling" / " 'Fedegosinho' "

1819   **C. abbreviata** Tanganyika / R. E. S. Tanner 3080 / 1956 / "Bondei: For sharp pains in the stomach, roots boiled and the water drunk" / "Bondei: Mhumba Mgosi"

1820   **C. afrofistula** Tanganyika / R. E. S. Tanner 3765 / 1957 / "Zigua: For gonorrhoea, roots soaked and the water drunk" / "Zigua: Mkwizinga"
Tanganyika / R. E. S. Tanner 3748 / 1957 / "Bondei: For stomach pains, roots boiled and the water drunk. Also applied externally for sores" / Bondei: Mhungba mwitu"

1821   **C. didymobotrya** Tanganyika / R. E. S. Tanner 1221 / 1953 / "Sukuma tribe: Leaves used to treat constipation—boiled in water, this drunk" / "Kisukuma: Njolwa mbogo"

1822   **C. falcinella** var. **parviflora** Tanganyika / R. E. S. Tanner 1540 / 1953 / "Sukuma tribe: Roots pounded, used as drink to treat diarrhoea" / "Kisukuma: Nyalwinze wa mwitongo"

1823   **C. goratensis** Tanganyika / R. E. S. Tanner 3081 / 1956 / "Bondei: For pain in the stomach, roots boiled and the water drunk" / "Bondei: Mhumba Mviyee"
Tanganyika / R. E. S. Tanner 1259 / 1953 / "Sukuma tribe: Roots soaked in water, mixed with white mtama flour, this drunk to treat syphilis" / "Kisukuma: Ntungulu"
Tanganyika / R. E. S. Tanner 3779 / 1957 / "Bondei: For pain in abdomen, roots boiled and the water drunk" / "Bondei: Mhumba vye"
Tanganyika / R. E. S. Tanner 3161 / 1956 / "Sambaa: For cough, leaves boiled with lemon and pepper and drunk" / "Sambaa: Mhumba"
Tanganyika / R. E. S. Tanner 977 / 1952 / "Sukuma tribe: Leaves used as hot compress for fever, roots used to treat stomach pains (boiled in water—this drunk)" / "Kisukuma: Mtunguru"

24 *C. mimosoides* Tanganyika / R. E. S. Tanner 3899 / 1957 / "Zigua: for pain near navel, roots boiled and the water drunk" / "Zigua: Lumeya" Tanganyika / R. E. S. Tanner 1496 / 1953 / "Sukuma tribe: Roots pounded, taken orally to treat deep cough; leaves chewed green and used as poultice for scorpion bite" / "Kisukuma: Nyenye gwa mumatongo" Tanganyika / R. E. S. Tanner 3454 / 1957 / "Zigua: To cause person to like you, leaves boiled and the water drunk" / "Zigua: Lusekela"

25 *C. nigricans* Tanganyika / R. E. S. Tanner 1383 / 1953 / "Sukuma tribe: Whole plant pounded, taken in water to treat fever; roots pounded, taken in water for diarrhoea" / "Kisukuma: Idingu lya mulugulu; Shishi gwa hasi"

26 *C. siamea* Tanganyika / R. E. S. Tanner 3954 / 1958 / "Zigua: For stomach pain, roots boiled and the water drunk" / "Zigua: Masaju"

27 *C. auriculata* Ceylon / V. E. Rudd 3107 / 1970 / "Used for medicine — boiled and made into tea with a little sugar, good for reducing fever and as a diuretic" Ceylon / Maxwell & Jayasuriya 818 / 1972 / ". . . fixed as tea for kidney ailments, dissolve kidney stones"

28 *Cercidium floridum* var. *floridum* U.S.: Ariz. / H. Cutler 4726 / 1941 / "Fruit ground by early Papagoes for food"

29 *C. microphyllum* U.S. & CAN. / Schott s.n. / 1855 / "Āstac of the Yumas"

30 *C. praecox* Mexico / Pennington & Sarukhán 9505 / 1968 / "Flores ligeramente perfumadas"

31 *Cercis canadensis* var. *mexicana* U.S.: N.M. / Mr. Blake s.n. / ca. 1895 / "Camels eat this"

32 *C. siliquastrum* EUR. / Herb. of Alphonso Wood s.n. / no year / "Die Blumenknospen werden wie Kappern benutzt" / "Judas-Baum"

33 *Crudia aequalis* Colombia / A. Dugand 6600 / 1964 / "vulg. 'Guamo macho' o 'Guamo de murciélago' "

34 *C. amazonica* S. AM. / Spruce 858 / 50 / "Fls . . . with odor of the fruit of Carica Papaya"

35 *Cynometra bauhiniaefolia* Brazil / E. Oliveira 2418 / 1963 / " 'Castanha pé de burro' "

36 *C. spruceana* Brazil / W. A. Archer 7943 / 1942 / "oil secured from trunk" / " 'copahyba' "

37 *Dialium guianense* Panama / Duke & Bristan 8245 / 1966 / "tree smaller than tamarindo which makes a bad flavored fruit" / " 'Tamarincillo' " Guatemala / FAO-FYDEP 15 / 1966 / " 'Tamarindo', 'palo de lacandon' " Venezuela / Wurdack & Adderley 43648 / 1959 / "Fruit dark brown, with edible seed" / " 'Tamarinda' "

1838 **Dialium** indet. Borneo / Kostermans 6587 / 1952 / "They are visited by in-
numerable insects, but have hardly any smell" / " 'Kerandji' "

1839 **Dicorynia paraensis** var. *paraensis* Brasil / R. de Lemos Fróes 21539 / 1945
/ "frutos comestiveis, si bem que amargos um pouco"

1840 **Dimorphandra macrostachya** Venezuela / B. Maguire 33563 / 1952 / "Seed
decoction good for stomach trouble"

1841 **Elizabetha leiogyne** Brazil / J. A. Steyermark 104038 / 1970 / "Tree bark
used locally by Guaica Indians as ingredient of 'yopo' "

1842 **E. princeps** Brazil / R. Souza s.n. / no year / "Cascas torradas são usadas
no preparo de Rapé de Indio pelos indios Uaica. Voucher for materials
under chemical studies by Prof. Norman A. LeBel"
Brazil / J. A. Steyermark 104038 / 1970 / "Tree bark used locally by
Guaica Indians as ingredient of 'yopo' "

1843 **Eperua purpurea** Colombia / Schultes, Baker & Cabrera 17955 / 1952 /
"Bark used for yuruparí horns. Sticky resin (colourless) in wood used to
catch birds" / "Venezuelan = yé-ba; Kuripako = wá-pa; Puinave = poó-
koó"

1844 **Gleditsia triacanthos** U.S. & CAN. / Herb. of A. Wood s.n. / no year /
"Das süsse Mark des Hülsen wird als Mittel gegen Katarrhe benutzt, auch
siedet man einen Meth daraus"

1845 **G. amorphoides** Argentina / J. E. Montes 27668 / 58 / "Frutos en medic.
popular—para lavar e higienisar el cabello" / " 'Espina de corona' "

1846 **G. delavayi** China: Yunnan / W. Hancock 549 / '96 / "used by the natives
as soap"

1847 **Haematoxylon campechianum** W.I.: Martinique / Columbia College of
Pharmacy Herb. s.n. / no year / "Dieser Baum liefert das Campecheholz.
Lignum campechianum, das ausser als Farbholz auch als mildes Adstrin-
gens in Erschlaffung des Darmcanals, bei Colliquationen Schwindelsüch-
tiger, chronischen Blutflüssen, gebraucht wird. Die Rinde ist schweis-
streibend" / "Westindisches Blutholz"

1848 **Hoffmanseggia glauca** U.S.: Tex. / V. L. Cory 53035 / 1946 / " 'camote de
raton' "

1849 **Hymenaea oblongifolia** Brazil / W. A. Archer 7877 / 1942 / "Frt. edible" /
" 'jutahy-miry' "

1850 **H. oblongifolia** var. *davisii* Brit. Guiana / Davis 943 / 29 / "Gives locust
gum when dead" / "LOCUST OR SIMIRI ARAWAK"

1851 **H. oblongifolia** var. *palustris* Peru / J. Schunke V. 4976 / 1971 / " 'Azucar
huayo' "

1852 **H. parvifolia** Peru / T. B. Croat 19968 / 1972 / "seeds black, covered by a
powdery, sweet substance"

Brazil / R. Fróes 1900 / 1932 / "Exudes a valuable resin" / " 'Jatoba' "

53 *Julbernardia globiflora* Tanganyika / R. E. S. Tanner 1562 / 1953 / "Sukuma tribe: Stems pounded, smoked in pipe to treat leprosy" / "Kisukuma: Ntundu"

54 *Libidibia coriaria* Mexico / Y. Mexia 8928 / 1937 / "Fruit crushed and used for tanning" / " 'Cascalote' "
El Salvador / P. C. Standley 19478 / 1921 / "Used to dye black & blue" / " 'Nacascol, Tinaco' "

55 *Macrolobium acaciaefolium* Brasil / Irwin, Pires & Westra 48634 / 1960 / ". . . the hollow twigs inhabited by ants ('tachi'); fruits green (immature), said to be edible" / " 'Arapari' "

56 *M. stenosiphon* Ecuador / Little & Dixon 21084 / 1965 / " 'dormilón' "

57 *Parkinsonia aculeata* El Salvador / P. C. Standley 97 / 1922 / " 'Sulfato' "

58 *P. africana* S.W. Africa / R. Seydel 4453 / 66 / ". . . von Kuddu gern gefressen"

59 *Peltogyne floribunda* Venezuela / J. A. Steyermark 88531 / 1961 / "fruit eaten by birds, mammals, and fish"

60 *P. venosa* var. *densiflora* Brit. Guiana / C. S. & C. L. Tillett 45403 / 1960 / "used by natives to make 'wood-skins' " / "Suru-a-gu-puh"

61 *Poeppigia procera* El Salvador / S. Calderon 2139 / 1924 / " 'Frijolillo' "

62 *Swartzia ochnacea* Mexico / Carter & Chisaki 1239 / 1959 / " 'naranjillo' "

63 *S. sumorum* Nicaragua / A. Molina R. 2393 / 1949 / " 'Uvillo' "

64 *Swartzia* indet. Panama / Duke & Kirkbride 14053 / no year / "Unpleasantly aromatic twigs"

65 *Swartzia* indet. Nicaragua / Proctor, Jones & Facey 97 / 1966 / " 'Guavo colorado' "

66 *S. acutifolia* var. *ynesiana* Brazil / Y. Mexia 5069 / 1930 / " 'Sangue de burro' "

67 *S. littlei* Ecuador / Little & Dixon 21171 / 1965 / " 'canelón' "
Ecuador / Little & Dixon 21194 / 1965 / " 'naranjillo' "

68 *S. sprucei* Colombia / R. Romero-Castañeda 9298 / 1962 / " 'Tamarindo de mico' "

69 *Swartzia* indet. Venezuela / J. de Bruijn 1602 / 1967 / "Tree visited by black ants" / "Frisolillo"

70 *Tamarindus indica* Tanganyika / R. E. S. Tanner 2232 / 1955 / "For diarrhoea bark strips pounded up, cooked and eaten" / "Kiswaheli: Mkwaju Kizigua: Mkwaju"
Tanganyika / R. E. S. Tanner 3953 / 1958 / "Zigua: For constipation, roots boiled and the water drunk" / "Zigua: Mkwazu"

Ceylon / C. F. Baker 131 / 1907 / "The pulp is also used in medicine, being rich in formic and butyric acids. It is laxative and refrigerant and is also used to prepare a gargle for sore throat"

FABACEAE

1871   *Abrus africanus* Tanganyika / R. E. S. Tanner 1361 / 1953 / "Sukuma tribe: Roots boiled, mixed with white mtama flour, taken with water to treat syphilis" / "Kisukuma: Bula wa subi"

1872   *A. precatorius* Tanganyika / R. E. S. Tanner 3392 / 1956 / "Bondei: For asthma, roots cooked and the water drunk" / "Bondei: Mfyambo"

1873   *A. precatorius* subsp. *africanus* Brit. Honduras / P. H. Gentle 5 (collected for H. H. Bartlett) / 1931–1932 / "medicinal" / " 'John — crow bead' "

1874   *Aeschynomene uniflora* Tanganyika / R. E. S. Tanner 1567 / 1953 / "Sukuma tribe: Pounded roots used internally to treat stomach pain" / "Kisukuma: Nyalwinzi wa kulanda"

1875   *Alysicarpus glumaceus* Tanganyika / R. E. S. Tanner 3591 / 1957 / "Zigua: For impotence, roots soaked in water and drunk regularly with sugar added" / "Zigua: Mzalingala"

1876   *A. rugosus* Tanganyika / R. E. S. Tanner 3687 / 1957 / "Bondei: For asthma, leaves pounded up and the sap drunk" / "Bondei: Meso ya pasi" Tanganyika / R. E. S. Tanner 1951 / 1955 / "Roots chewed to treat swollen testicles" / "Swaheli: Muhufia"

1877   *Apurimacia michelii* Peru / A. Weberbauer 7508 / 1926 / "Roots used as fishing narcotic" / " 'Chacanhuay' "

1878   *Astragalus americana* Yukon Terr. / M. W. Gorman 1091 / 1899 / "This plant has a very disagreeable odor like decaying meat, when drying"

1879   *A. bisulcatus* Canada: Manitoba / H. Marshall 1 / 1939 / "Unpleasant smell persisting in root after drying" Canada: Saskatchewan / Boivin & Dore 7564 / 1951 / "Flowers . . . smelling of old urine"

1880   *A. caricinus* U.S. & CAN. / F. Tweedy 615 / 1883 / " 'Stock eat this . . . greedily' "

1881   *A. flavus* var. *argillosus* U.S.: Utah / A. Cronquist 9081 / 1961 / "Plants smelling of selenium"

1882   *A. flavus* var. *candicans* U.S.: Utah / A. Cronquist 10051 / 1965 / "Plants with a mild seleniferous odor"

1883   *A. gracilis* U.S.: Mo. / C. A. Geyers 184 / 1839 / "The root has a taste like liquorice"

84 *A. hornii* U.S.: Calif. / E. C. Twisselmann 4904 / 1958 / "Poisonous to livestock, especially sheep"

85 *A. kentrophyta* var. *implexus* U.S.: Mont. / A. Cronquist 8084 / 1955 / "Leaves pungent"

86 *A. lentiginosus* U.S. & CAN. / F. Tweedy 604 / 1883 / " 'Said to be poisonous to sheep' "

87 *A. miser* var. *oblongifolius* U.S.: Wyo. / C. L. Porter 3483 / 1944 / "poisonous"

88 *A. moencoppensis* U.S.: Utah / A. Cronquist 9198 / 1961 / "Plants smelling of selenium"

89 *A. oxyphysus* U.S.: Calif. / E. C. Twisselmann 8048 / 1963 / "Occasionally troublesome as a livestock poisoner"

90 *A. preussii* var. *preussii* U.S.: Utah / J. Anderson 694 / 1968 / "Plants have strong silenium odor"
U.S.: Utah / A. Cronquist 8993 / 1961 / "Plant with the characteristic odor of organic selenium"

91 *A. purshii* var. *tinctus* U.S.: Calif. / E. C. Twisselmann 6245 / 1961 / "Questionably poisonous to livestock"

92 *A. reventus* U.S.: Wyo. / C. R. Gunn 2782 / 1963 / "Flowers yellow, selenium smell"

93 *A. spaldingii* U.S. & CAN. / Tweedy 614 / 1883 / " 'Greedily eaten by stock' "

94 *A. insularis* Mexico / R. Moran 12835 / 1966 / " 'locen los caballos' " / " 'Garboncillo' "

95 *A. mollissimus* Mexico / C. V. Hartman 605 / 1891 / "causing intoxication or madness when eaten, if boiled" / " 'Loco weed' "
Mexico / Gregg s.n. / no year / "This plant is said by the Indians to be very poisonous killing horses & cattle"

96 *A. garbancillo* Bolivia / W. Brooke 5269 / 1949 / ". . . poisonous to animals"

97 *Astragalus* indet. Peru / Iltis & Ugent s.n. / 1963 / "Eaten by sheep sparingly" / " 'salka' "

98 *A. chrysopterus* China / C. F. Li 10820 / 1929 / "of some medicinal value"

99 *Baphia kirkii* Tanganyika / R. E. S. Tanner 3721 / 1957 / "Zigua: For epilepsy, roots boiled and the water drunk" / "Zigua: Mkoma vidare"

00 *Cajanus cajan* Grand Bahama Is. / W. Lewis 7129 / 1968 / "pigeon pea cultivated & eaten by African residents"
Cuba / C. F. Baker 87 / 1907 / "furnishes a good edible pea" / "Pigeon Pea or Gandul"

Cuba / H. A. Van Hermann 382 / 1904 / "Fruit edible"

Peru / C. M. Belshaw 3209 / 1937 / "fruits edible" / " 'Puspoporoto' "

Brazil / A. Silva 157 / 1944 / "Cult . . . Frt. cooked as green beans" / " 'feijão cuandú' "

Brazil / W. A. Archer 8019 / 1942 / "Cult. for the edible frt." / " 'feijão quandú' "

Tanganyika / R. E. S. Tanner 1234 / 1953 / "Sukuma tribe: Roots boiled, mixed with white mtama flour, taken orally in water to treat syphilis; leaves chewed for cough . . . Seeds used as vegetable bean" / "Kisukuma: Mbaazi"

Hainan / S. K. Lau 1065 / 1933 / "fruit, edible"

Hainan / C. I. Lei 361 / 1933 / "Leguminous fruit, the seeds of which serve to make delicious soup"

Hainan / C. I. Lei 522 / 1933 / "fruit edible. The seeds gives best taste by boiling with water and sugar"

1901   *C. bicolor* Peru / Y. Mexia 6505 / 1932 / "cultivated green; choice eating" / " 'Puspo-poróto' (lungs beans)"

Peru / Killip & Smith 24735 / 1929 / "fruit edible"

1902   *Canavalia brasiliensis* Peru / C. M. Belshaw 3326 / 1937 / "Edible"

1903   *C. gladiata* Tanganyika / R. E. S. Tanner 3163 / 1956 / "Zigua: For stomach pains, roots boiled and the water drunk" / "Zigua: Tuha"

Hainan / F. A. McClure 708 / 1929 / "Seeds eaten as food" / "To Shiu Tau"

1904   *Canavalia* indet. Tanganyika / R. E. S. Tanner 1294 / 1953 / "Sukuma tribe: Seeds used as famine food; leaves made into poultice with sheep's urine to treat smallpox" / "Kisukuma: Ihoho"

1905   *C. gladiolata* China: Kwangsi / W. T. Tsang 24070 / 1934 / "fr. edible" / "To Pan Tau"

1906   *C. maritima* New Hebrides / S. F. Kajewski 304 / 1928 / "Ripe seed boiled and eaten by natives" / " 'No sorrae' "

1907   *C. microcarpa* Marshall Is. / F. R. Fosberg 39439 / 1958 / "used for medicine, for headaches and pains" / " 'Marlap' "

1908   *Cantharospermum scarabaeoides* Guam / F. R. Fosberg 25423 / 1946 / "leaves used medicinally" / " 'kakaguates aniti' (devil's peanuts)"

1909   *Clathrotropis brachypetala* Brit. Guiana / W. A. Archer 2452 / 1934 / "reported as *Fish poison* . . . Wood has strong odor" / " 'Arumatta' "

1910   *C. macrocarpa* Brazil / R. Fróes 12160/70 / 1941 / "Bark is used for poisoning fish together with Cashanby" / " 'Timborana' "

1911   *Crotalaria purshii* U.S. & CAN. / P. P. Sheehan s.n. / 1919 / "Used to cure sore-throat. The pods are boiled, and the liquor then drunk, or used as a gargle" / "Rattlebox"

12 *C. pumila* Mexico / G. B. Hinton 1636 / 32 / "Edible boiled with ripeseed"
/ "Chipil"

13 *C. nitens* Peru / Killip & Smith 23601 / 1929 / "applied with salt to
abcesses or swellings" / " 'Japincho' "

14 *C. velutina* Brazil / Ratter & Bertolda 1009 / 1968 / "Foliage smelling of
cucumbers"

15 *C. vespertilio* Brazil / J. A. Ratter 90 / 1967 / "The plant has a strong scent
of cucumbers when crushed" / "Xica-Xica"

16 *C. dawei* Tanganyika / R. E. S. Tanner 1560 / 1953 / "Sukuma tribe:
Roots pounded, used as drink to treat swollen spleen" / "Kisukuma:
Mazebazeba"

17 *C. polysperma* Tanganyika / R. E. S. Tanner 1275 / 1953 / "Sukuma tribe:
Pounded leaves used as poultice to treat sores" / "Kisukuma: Ipondo;
Nchekecheke"
Tanganyika / R. E. S. Tanner 2982 / 1956 / "Used as a vegetable" /
"Zigua: Cheluwe"

18 *C. retusa* Tanganyika / R. E. S. Tanner 3193 / 1956 / "Zigua: For sores,
leaves pounded up and applied as paste" / "Sambaa: Koma nguku"
China: Kwangtung / Groff & Siu 00501 / 1931 / "Leaves used to make tea"
/ "Kau Tsai Tau"

19 *C. vasculosa* Tanganyika / R. E. S. Tanner 1418 / 1953 / "Sukuma tribe:
Used to treat scabies—whole plant pounded and used as poultice" /
"Kisukuma: Nzegezege swa hasi"

20 *C. linifolia* Hainan / H. Fung 20416 / 1932 / "When woman gives birth
then use it to cook with rice to nurish the blood" / "Po Pek Ts'o"

21 *Dalbergia foliolosa* Bolivia / J. Steinbach 7134 / 1925 / "Estando en flor se
percive cerca de los arboles un olor á maní" / "Manicito"

22 *D. stuhlmanii* Tanganyika / R. E. S. Tanner 978 / 1952 / "Sukuma tribe:
roots used to treat gonorrhoea and syphilis—used in fermented drink with
mtama flour" / "Kisukuma: Mfifi"

23 *D. vacciniifolia* Tanganyika / R. E. S. Tanner 3351 / 1956 / "Zigua: For
asthma, roots boiled and the water drunk" / "Zigua: Mhaha nkuku"

24 *Dalea carnea* var. *carnea* U.S.: Fla. / Isely & Wemple 9229 / 1964 /
"Crushed foliage with parsley-like odor"

25 *D. lumholtzii* U.S.: Ariz. / Gould, Darrow & Haskell 2720 / 1944 / "Herbage with lemon odor"

26 *D. bicolor* var. *orcuttiana* Mexico / A. Carter 4396 / 1962 / "Locally . . .
used for tea" / " 'damiana' "

1927 **D. foliolosa** var. *foliolosa* Guatemala / A. F. Skutch 1605 / 1934 / "Foliage, stems & roots lemon-scented"

1928 **D. coerulea** var. *coerulea* Colombia / F. R. Fosberg 20286 / 1943 / "Said to be very useful as an insecticide"

1929 **D. cylindrica** var. *nova* Ecuador / F. Prieto P-174 / 1945 / "fluid from boiled fls taken as diuretic; same decoction thrown about on floors of houses before sweeping to get rid of fleas" / " 'Zhordán' "

1930 **Derris amazonica** Brazil / Pires, Rodrigues & Irvine 50832 / 1961 / "Stem used as fish-killing agent" / " 'Tinbo' "

1931 **D. latifolia** Brazil / Prance, Maas et al. 13930 / 1971 / "Roots beaten out and used as fish poison. Cultivated by Indians in their fields" / " 'Cuña' "

1932 **Derris** indet. / Brazil / Prance, Steward et al. 9639 / 1969 / "Used by Mayagão Indians as a fish poison"

1933 **Derris** indet. China / A. Henry 13820 / no year / "The fruit is pounded & used for poisoning fish — (thrown into water)"

1934 **Derris** indet. Burma / F. Kingdon Ward 00472 / 1939 / "Roots, (or stems?) pounded to extract a poison used to intoxicate fish"

1935 **D. mindorensis** Philippine Is. / A. D. E. Elmer 8678 / 1907 / "the stems are pounded to a pulp put in the water for killing fish — it is said to be sour" / " 'Upe' "

1936 **D. trifoliata** Fiji / A. C. Smith 3 / 1933 / "Stem used as a fish poison" / " 'Nduva'; 'Raurau' "
Fiji / O. Degener 15038 / 1941 / "Fijians take root and mash it to liberate milky juice. A coconut shell full is enough to poison fish in a small stream. Fish good to eat. Root is very poisonous but tea made from fresh leaves is used as purgative" / " 'Nduva' (Serua)"

1937 **Desmodium nicaraguensis** Honduras / P. C. Standley 13929 / 1943 / " 'Engordacaballos' "

1938 **D. canum** Bahamas / W. Cerbin C129 / 1973 / "With midwives, it is popular to give a woman in labor Wild Granite tea which induces faster labor. Some use it after birth to regain strength" / " 'Wild Granite' "

1939 **D. molliculum** Ecuador / F. Prieto E-2508 / 1945 / "the roots are ground and an infusion made which is used to assist during child-birth ; also, the same concoction is said to 'purge one of bad hysterics' " / " 'Yerba del Infante' "

1940 **Desmodium** indet. Venezuela / Steyermark & Velasco 100034 / 1967 / "Introduced from Mexico for forage" / " 'mermelado de caballo' "

1941 **D. barbatum** Tanganyika / R. E. S. Tanner 1423 / 1953 / "Sukuma tribe: Roots pounded and eaten with salt to treat anklyostomiasis" / "Kisukuma: Kalandi"

1942   *D. salicifolium* Tanganyika / R. E. S. Tanner 1600 / 1953 / "Sukuma tribe: Roots pounded, used in water orally to treat Bilharzia" / "Kisukuma: Ng'weyo gwa hisanga"

1943   *D. pseudotriquetrum* China: Kwangtung / T. W. Tak 55 / 31 / "Drug plant" / "Paak Lo Sit"

1944   *D. triquetrum* China: Kwangtung / W. T. Tsang 21397 / 1932 / "used as medicine" / "Paak Lo Sit"

1945   *Desmodium* indet. Philippine Is. / A. L. Zwickey 673 / 1938 / "infusion of lvs. for stomachache" / " 'Tinuk a amo' "

1946   *Dioclea apiculata* Peru / D. McCarroll 113 / 1940 / "Used for cough remedy" / " 'Flor de arco' "

1947   *Diphysa robinioides* El Salvador / P. C. Standley 19255 / 1922 / "Remedy for chills & malaria" / " 'Guachipilín, Uachipilín' "

1948   *Dipteryx emarginatus* Brazil / G. & L. T. Eiten 5416 / 1963 / "Seeds containing an oil used for toothache" / " 'sucupira amarela' "

1949   *Dolichos lablab* China: Kwangtung / F. Hom A-405 / 1930 / "For food & medicine" / "Ngo Mei Tau"

1950   *D. purpureus* Ceylon / R. H. Maxwell 1020 / 1972 / "Seeds eaten"
Ceylon / R. H. Maxwell 1016 / 1972 / "flowers not beans eaten, mixed with chili, pepper, and salt, fried and eaten" / "Called Wild Bean by Sinhalese"

1951   *Eriosema psoraleoides* Tanganyika / R. E. S. Tanner 3388 / 1956 / "Bondei: Male aphrodisiac, roots chewed fresh with ground nuts" / "Swaheli: Mbaazi mwitu"
Tanganyika / R. E. S. Tanner 3478 / 1957 / "Bondei: For blood in urine, roots boiled and the water drunk. For male impotence, roots chewed" / "Bondei: Kibaazi mzitu"
Tanganyika / R. E. S. Tanner 3105 / 1956 / "Bondei: To increase male potency, roots chewed fresh" / "Swaheli: Kibazi Mwitu"
Tanganyika / R. E. S. Tanner 3950 / 1958 / "Sambaa: for stomach ache, roots boiled and the water drunk" / "Sambaa: Kibazi mwitu"
Tanganyika / R. E. S. Tanner 3027 / 1956 / "Zigua: For T.B. whole plant pounded up very finely and drunk in water" / "Zigua: Mbalazi Luwala"

1952   *Erythrina herbacea* subsp. *herbacea* U.S.: Fla. / P.P. Sheehan s.n. / 1919 / "Used for most canine diseases. The plant is boiled and the infusion used as an emetic for dogs" / "Coral-bean"

1953   *E. berteroana* El Salvador / P. C. Standley 21826 / 1922 / "fls. red, eaten boiled, fried with eggs etc.; lvs. cooked in bean soup" / " 'Quilite, Pito' "
Panama / C. & W. von Hagen 2125 / 1940 / "Bark used for poisoning dogs and wild animals / " 'Palo Santo' "

1954   *E. costaricensis* Panama / J. A. Duke 14503(2) / 1967 / "Use for female problems . . . Natives distinguish the orange and white varieties with qualifying adjectives for the genus '*parsu*' this being parsu sipu"

1955   *E. folkersii* Guatemala / P. C. Standley 24031 / 1922 / "fls. cooked & eaten" / " 'Pito' "

1956   *E. horrida* Mexico / R. E. Schultes 644a / 1939 / "Seeds regarded as poison by natives" / " 'Sompantle' "

1957   *E. lanceolata* Honduras / C. & W. von Hagen 1115 / 1937 / "Tea drunk for kidneys" / "Pito (Gualaquemo)"

1958   *E. mexicana* Mexico / Schultes & Reko 687 / 1939 / "Seeds regarded as poisonous by natives" / "Chinantec name = 'ma-nya' "

1959   *E. poeppigiana* Ecuador / W. von Hagen 13 / 1934 / "Bark used for strained ligaments—Make paste and apply hot to injured parts" / "Borotillo-Shukei"

1960   *E. sacleuxii* Tanganyika / R. E. S. Tanner 3124 / 1956 / "Bondei: For gonorrhoea; roots boiled in water and drunk" / "Bondei: Mbambangoma; Zigua: Mlumgumagoma"

1961   *E. subumbrans* Philippine Is. / A. D. Elmer 8666 / 1907 / "green limbs are used at times in the house for fuel in order to smoke out the insects" / " 'Sabdañg' "

1962   *Eysenhardtia punctata* Mexico / Gregory & Eiten 134 / 1956 / "Orange-like odor when bruised"

1963   *E. texana* Mexico / Ripley & Barneby 13258 / 1963 / "Foliage tar-scented"

1964   *Fordia filipes* Brit. N. Borneo / B. Evangelista 894 / 1929 / "Natives use the root for poisoning fish. Not commonly used" / "Maripari"

1965   *Gastrolobium bilobum* Australia / C. T. White 5306 / 1927 / "A bad poison bush but not usually touched by stock" / "Heart-leaf Poison Bush"

1966   *Geoffraea spinosa* Venezuela / Aristeguieta & Zabala 7072 / 1969 / "Las semillas cocinadas son comestibles. También la parte externa del fruto es comestible" / " 'Taque' "

1967   *Glycyrrhiza uralensis* China: Hopei / C. F. Li 10643 / 1929 / ". . . of some medicinal value"

1968   *Gourliea decorticans* Argentina / V. Martínes 73 / 44 / "hojas medicinales" / "Chañar"

1969   *Hedysarum americanum* Canada: Yukon Terr. / M. W. Gorman 1018 / 1899 / "The roots of this plant form a staple article of food amongst the Ayans, Nahónces, and allied of Athabascun stock. It is eaten both raw and boiled in conjunction with other food"

1970   *Hymenolobium heterocarpum* Brazil / A. Ducke 246a / 1936 / "sementes comestives" / " 'Carámate' "
Brazil / R. L. Fróes 22237 / 1947 / "edible by Bominas indian"

1971　*Indigofera lespedezioides* Honduras / C. & W. von Hagen 1081 / 1937 / "Used for diarrhoea. The root is mashed & made into tea" / "Escorcienera" Brazil / Prance, Forero et al. 4603 / 1967 / "Venenous to horses but not to cattle"
Brazil / Prance, Forero et al. 4265 / 1967 / "Poisonous to horses"

1972　*I. pascuorum* Brazil / A. G. Amora s.n. / 1946 / "Reputed to cause death of cattle, especially horses, which graze the plant" / " 'mata zambando' "

1973　*I. suffruticosa* Surinam / W. A. Archer 2819 / 1934 / "Poison for small fish . . . Used by Carib children. Said not to be strong" / " 'Basra wiriri nekuru' (Carib)"

1974　*I. colutea* Tanganyika / R. E. S. Tanner 3977 / 1958 / "Zigua: for success in hunting, leaves rubbed on nets or gun barrel" / "Zigua: Mzange"
Tanganyika / R. E. S. Tanner 1421 / 1953 / "Sukuma tribe: Roots pounded and mixed with ghee as poultice to treat bruises or cuts . . . aromatic like ripe mangoes" / "Kisukuma: Lubungwa lukuma"

1975　*I. drepanocarpa* Tanganyika / R. E. S. Tanner 1488 / 1953 / "Sukuma tribe: Roots boiled, used as mouth rinse to treat sores in mouth" / "Kisukuma: Itama lya bagikulu"

1976　*I. garckeana* Tanganyika / R. E. S. Tanner 1591 / 1953 / "Sukuma tribe: Roots cooked in food and eaten to treat sharp pains in side" / "Kisukuma: Lwamu Iwa mlutogoro"

1977　*I. hirsuta* Tanganyika / R. E. S. Tanner 1372 / 1953 / "Sukuma tribe: Roots pounded, used orally with water to treat diarrhoea; leaves pounded, heated, mixed with butter as poultice for boils" / "Kisukuma: Lwamu lugasha"

1978　*I. kirkii* Tanganyika / R. E. S. Tanner 2851 / 1956 / "Swaheli: For general debility. Leaves boiled and drunk hot which causes light fever" / "Swaheli: Muhato"

1979　*I. vohemarensis* Tanganyika / R. E. S. Tanner 3338 / 1956 / "Sambaa: For asthma, roots boiled and the water drunk" / "Zigua: Lumuye"

1980　*Indigofera* indet. Hawaiian Is. / W. H. Hatheway 195 / 1950 / "Poisonous to fowl. Hair of white rabbit fed this plant turned blue, according to Dr. G. D. Sherman" / " 'Creeping indigo' "

1981　*Lonchocarpus velutinus* Panama / Herb. of The New York Botanical Garden s.n. / 1949 / " 'The local people there believe that it is the plant used for extraction of rotenone. They say that the natives use this plant for making a poisonous bait for sharks' "

1982　*L. domingensis* W.I.: Martinique / F. E. Egler 39–268 / 1939 / "The fish poison tree"

1983　*L. chrysophyllum* Surinam / W. A. Archer 2908 / 1934 / "Cultivated *Fish poison*" / " 'Nekoe' "

Brit. Guiana / A. C. Smith 2838 / 1938 / "used as fish-poison by Wai-wai Indians"

1984   *L. denudatus* Brazil / R. Fróes 1958 / 1932 / "Roots are very poisonous . . . Roots used to poison fish" / " 'Timbo' "

1985   *L. floribundus* Surinam / W. A. Archer 2790 / 1934 / "Fish poison. Roots used" / " 'Wiriri mekuru' (Carib)"

1986   *L. latifolius* Venezuela / Killip & Tamayo 37048 / 1943 / "Used as fish poison" / " 'Barbasco de Palo' "

1987   *L. martynii* Brit. Guiana / Ex Herb. Hort. Bot. Reg. Kew. 477 / 1926 / "Stem pounded in water and extract used for poisoning fish" / "Vernacular A-YA; Dialect Patamona"
Brit. Guiana / Brit. Guiana For. Dept. 927 / 1929 / "Stems and roots used as a fish poison" / "Haiari"

1988   *L. rariflorus* Brit. Guiana / A. C. Smith 2161 / 1937 / "used as fish poison" / " 'Black haiari' "

1989   *L. rufescens* Brit. Guiana / W. A. Archer 2519 / 1934 / *"Fish poison.* The stems are used" / " 'Wakorokoda' "

1990   *L. urucu* Brazil / Killip & Smith 30585 / 1929 / "roots used as a fish poison" / " 'timbo urucu'; 'timbo legitimo' "
Brazil / Prance, Maas, Woolcott et al. 15573 / 1971 / "Leaves & stems pulverized for fish poison" / " 'Timbo' (Portuguese), 'Doouí' (Makú)"

1991   *L. utilis* Ecuador / Y. Mexia 7305 / 1935 / "Roots used to poison fish" / " 'Barbasco'; 'Timu-ambi' "
Ecuador / W. von Hagen 98 / 35 / "Cultivated . . . Bark stripped from roots—pounded—use in river for poising fish"

1992   *L. eriocalyx* Tanganyika / R. E. S. Tanner 578 / 1952 / "Bark used for rope, also to concoct medicines for girls who delay in marrying" / "Kisukuma: Male"

1993   *L. stipularis* Australia: Queensland / S. F. Kajewski 1348 / 1929 / "A native told me they use this vine for poisoning fish"

1994   *Lotononis dichotoma* AFR. / Fouad I. Agric. Mus. Herb. 46/637 / 1932 / "Largely eaten by cattle" / "Arab. Name 'Eshb.' "

1995   *Lotus arabicus* Egypt / Fouad I. Agric. Mus. herb. 47/706 / 1928 / "Said to be poisonous to cattle" / "Arab. Name Qabd."

1996   *Lupinus fraxinetorum* U.S.: Calif. / R. Hopping 7031½ / 1911 / "Cattle eat it readily" / "Lupine"

1997   *L. mutabilis* Peru / Killip & Smith 22014 / 1929 / "Cultivated in sandy field. Seeds are soaked in water, and eaten, the water then being used as an insecticide and fish poison"

1998   *Marina scopa* Honduras / Acuñz 9112 / 1944 / "Considered as a medicinal plant . . . Cult.—Cuba" / "common name Conchalagua"

1999   *Medicago polymorpha* U.S.: La. / Gregg s.n. / 47 / "Eaten by cattle"

2000   *Millettia lasiopetala* Hainan / C. I. Lei 334 / 1933 / "fruit, poisonous. Roots give off poisonous substance to kill fish"

2001   *Millettia* indet. Indochina / F. A. McClure 790 / 1921 / "Seeds used to make a poison to stupefy fishes"

2002   *Mucuna* indet. Panama / J. A. Duke 14507(2) / 1967 / "Medicinal Plants of the Bayano Cuna"

2003   *M. aterrima* Venezuela / Steyermark & Nilsson 842 / 1960 / "grown as commercial plant with other crops as coffee substitute; sedds roasted and drink used cold or hot; has coffee flavor" / " 'nescafé' "

2004   *M. urens* Colombia / H. H. Smith 286 / 1898–1899 / "The use of the stinging hairs of this species as a remedy for worms is well know here"

2005   *M. rubro-aurantiaca* Tanganyika / R. E. S. Tanner 1568 / 1953 / "Sukuma tribe: Roots pounded, taken with water to treat anklyostomiasis, diarrhoea" / "Kisukuma: Jundu"

2006   *Mucuna* indet. Ceylon / Maxwell, Hepper & Fernando 970 / 1972 / "In garden. Beans eaten, packing around the seed removed, and the rest cooked and eaten" / "called broadbean"

2007   *Mundulea sericea* Tanganyika / R. E. S. Tanner 1273 / 1953 / "Sukuma tribe: Roots used in medicine added to porridge to treat impotence; flour from roots taken orally for stomach pains" / "Kisukuma: nkolomije"

2008   *M. suberosa* India / K. Yeshoda 57 / 32 / "Use—A positive insectcide"

2009   *Neorautanenia* indet. Tanganyika / R. E. S. Tanner 1238 / 1953 / "Sukuma tribe; Roots cooked and used as poultice on sores, or ground into flour and put on scabies sores" / "Kisukuma: Izumangiji"

2010   *Ormocarpum kirkii* Tanganyika / R. E. S. Tanner 3160 / 1956 / "Sambaa: Leaves used as a vegetable" / "Zigua: Kiumbu"
Tanganyika / R. E. S. Tanner 3212 / 1956 / "Swaheli: For diarrhoea, leaves pounded up and mixed with water and drunk" / "Swaheli: Mlambuzi"

2011   *O. trichocarpum* Tanganyika / R. E. S. Tanner 1252 / 1953 / "Sukuma tribe: Roots cooked with fat meat, juice drunk to cure sexual impotence" / "Kisukuma: Mkondwa mpuli"

2012   *Ostryoderris stuhlmanii* Tanganyika / R. E. S. Tanner 993 / 1952 / "Sukuma tribe: bark used to treat diarrhoea—pounded, soaked in water, this drunk" / "Kisukuma: Njundo"

2013   *Oxytropis lambertii* var. *bigelovii* U.S.: Ariz. / W. N. Clute 12 / 1920 / "Many horses and cattle were seen dead from eating this" / "loco weed of the Tuba region"

2014   *O. microphylla* W. Himalayas: Kashmir / W. Koelz 2153 / 1931 / "Plant has a rare delightful fragrance"

2015   *Pachyrhizus erosus* China: Kweichow / Y. Tsiang 7098 / 1930 / "Frt. pod green, poison"

2016   *Parryella filifolia* U.S.: Ariz. / R. & M. Spellenberg 3510A / 1973 / "plant with a lemonny odor"

2017   *Periandra dulcis* Brazil / Fróes 2081 / 1933 / "wood very sweet and greatly used as medicine . . . roots are used for cough"

2018   *Phaseolus formosus* Mexico / G. B. Hinton et al. 8245 / 35 / "Vernac. Name Frijol de ratón"
Mexico / King & Soderstrom 5048 / 1961 / "Common name 'frijol de venado' "

2019   *Phaseolus* indet. Mexico / G. B. Hinton 11684 / 37 / "Vernac. name: Frijol de raton"

2020   *Phaseolus* indet. Hainan / S. K. Lau 1132 / 1933 / "fruit, edible" / "Muk Dou"

2021   *P. sublobatus* Philippine Is. / A. D. E. Elmer 11830 / 1909 / "The fruits and leaves are used by the natives for food" / " 'Caratan' "

2022   *Piscidia grandifolia* Honduras / C. & W. von Hagen 1152 / 1937 / "Jicaque Indians use crushed bark as fish poison" / "Sapilote"

2023   *P. carthagenensis* W.I.: Martinique / P. Duss 120 B / no year / "Le fruit contient un principe tonique—narcotique. Les feuilles enivrent les poisons, quand on les froisse et les met dans l'eau" / "Vulgo: bois enivrant"

2024   *Pongamia pinnata* Philippine Is. / T. Simbajon 6 / 1931 / "Medicinal for women sickness"

2025   *Pseudarthria hookeri* Tanganyika / R. E. S. Tanner 3685 / 1957 / "Used for cleaning teeth" / "Bondei: Mswaki mwitu"
Tanganyika / R. E. S. Tanner 1939 / 1955 / "Roots used to treat stomachache; boiled and water drunk" / "Kibondei: Mbazi Mwitu"

2026   *Psoralea pentaphylla* Mexico / Wilson, Johnston & Johnston 8588 / 1972 / "tea cures flu and colds" / "contra yerba"

2027   *P. maleolens* Peru / MacBride & Featherstone 50203 / 1922 / "A stomach remedy" / " 'Coling Imbra' "

2028   *P. mexicana* Ecuador / W. H. Camp E-2477 / 1945 / "an infusion used for diarrhoea and for 'a bad state of the stomach' " / " 'Culin' or 'Trinitaria' "
Ecuador / W. H. Camp E-2165 / 1945 / "An infusion of the leaves given to children when they have a stomach-ache" / " 'Culen' or 'Trinitaria' "

Ecuador / B. & C. Maguire 44278 / 1959 / "used as purgative" / " 'Culin ' "

029   *P. mexicana* var. *trianae* Ecuador / F. Prieto E-2706 / 1945 / " 'good for a bad stomach' " / " 'Trinitaria' or 'Culin' "

030   *P. eriantha* Australia: Queensland / S. L. Everist 5909 / 1957 / "Eaten readily by sheep and cattle and highly esteemed as a fodder plant"

031   *Pterocarpus hayesii* Costa Rica / Alfonso Jiménez M. 3096 / 1965 / " 'Congos' comian flores" / " 'Chajada amarilla' "

032   *P. amazonum* Peru / J. J. Wurdack 2176 / 1962 / "Clear red bark exudate (exudate used for 'uta' [leishmaniasis] treatment), . . . flowers yellow, fragrant" / " 'Unjunche-numa' (Aguaruna) or 'Sangre de grado' "

033   *P. santalinoides* Colombia / Schultes & Cabrera 16237 / 1952 / " 'fish fruit' " / "Makuna — wý-see-mee-o"

034   *P. angolensis* Tanganyika / R. E. S. Tanner 1362 / 1953 / "Sukuma tribe: Roots pounded, mixed with butter, used as poultice for headache" / "Kisukuma: Ngwininga"

035   *P. tinctorius* Tanganyika / R. E. S. Tanner 600 / 1952 / "Roots used to treat fever in children (ground into flour, mixed with water, this drunk); used by bees in honey making" / "Kisukuma: Mkulungu"

036   *P. indicus* N. Borneo / A. Cuadra A. 3260 / 1951 / "The red sap from the bark said to cure toothache"

037   *Rhynchosia monticolor* Guatemala / J. A. Steyermark 42689 / 1942 / "Pulverized fruit a remedy for snake bite" / " 'frijol de casanpulga' "

038   *R. precatoria* Mexico / H. S. Gentry 1261 / 1935 / "Med. seeds ground and mixed with grease making an ointment for various afflictions. Mayos reported to use the seeds for necklaces and bracelets" / "Chanae pusi, Mayo"
El Salvador / P. C. Standley 21289 / 1922 / " 'Huevos de Casapulga' "

039   *R. pyramidalis* Mexico / J. N. Rovirosa 697 / 1890 / "Ojitos-de picho"
Cuba / Fr. Leon 423 / 1920 / " 'Bejuco culebra' "

040   *R. erythrinoides* Jamaica / W. Harris 8775 / 1904 / " 'Wild Liquorice' "

041   *R. phaseoloides* Brazil / F. C. Hoehne 268 / 1917 / " 'Olho de pombo', 'Favinha' "

042   *R. rojasii* Brazil / Duarte & Pereira 1790 / 1949 / "considerada toxica para ogado" / "Feija de veado"

043   *R. minima* var. *prostrata* Tanganyika / R. E. S. Tanner 1327 / 1953 / "Sukuma tribe: Roots pounded into flour stirred into water, taken orally to treat diarrhoea" / "Kisukuma: Kagobi"

044   *Sesbania exaltata* U.S.: Tex. / V. L. Cory 50808 / 1945 / " 'peatree' "

2045   *Spartium junceum* Ecuador / W. H. Camp E-2025 / 1945 / "The dried flowers are smoked as a cigarette for asthma" / " 'Retama' "
Ecuador / W. H. Camp E-2501 / 1945 / ". . . an infusion of the roots is taken by women who do not want children' (i.e. it is used as an abortificant)"

2046   *Sphenostylis stenocarpa* N. Rhodesia / A. Angus 2583 / 61 / "Vine with edible tuberous roots. (Kind of yam)"

2047   *Stylosanthes viscosa* Jamaica / W. Harris 12899 / 1919 / "An infusion of the plant is used at 'tea' in cases of headache, fever etc." / " 'Poor man's friend' "

2048   *S. mucronata* Tanganyika / R. E. S. Tanner 3881 / 1957 / "Zigua: male aphrodisiac, roots chewed" / "Zigua: Msekera"

2049   *Sutherlandia frutescens* S. Africa / R. D. A. Bayliss 4981 / 1971 / "Said to delay toe progress of cancer. Infusion used for chest colds" / "Local name 'Cancer Bush' "

2050   *Tephrosia heydeana* Guatemala / J. A. Steyermark 51312 / 1942 / "Root used for getting rid of fleas and lice, and also for killing fish" / " 'barbasco' "

2051   *T. multifolia* Honduras / C. & W. von Hagen 1236 / 1937 / "Used for fish poisoning" / "Varvasco"

2052   *T. emarginata* Brazil / Killip & Smith 30234 / 1929 / "roots used as a fish poison" / " 'Timbo de cayenne' "

2053   *T. sinapon* Peru / J. V. Schunke 1585 / 1967 / "La raiz toxico utilizan para pescar" / " 'Firano barbasco' "
Brit. Guiana / C. D. Cook 251 / 1957 / "Roots crushed for poison"

2054   *T. bracteolata* Tanganyika / R. E. S. Tanner 1406 / 1953 / "Sukuma tribe: Roots pounded and drunk with water twice daily for pregnant women infected with syphilis" / "Kisukuma: Ngwininga gwa hasi"

2055   *T. heckmanniana* Tanganyika / R. E. S. Tanner 1393 / 1953 / "Sukuma tribe: Leaves and roots pounded and used as poultice on sores" / "Kisukuma: Ghukuka mbuli gwa mlugulu"

2056   *T. noctiflora* Tanganyika / R. E. S. Tanner 3756 / 1957 / "Bondei: Aphrodisiac; roots boiled and the water drunk" / "Bondei: Utupa Mwitu"
Tanganyika / R. E. S. Tanner 3125 / 1956 / "Bondei: For gonorrhoea, roots boiled in water for one day and the water drunk" / "Bondei: Utupa Mwitu"

2057   *T. polystachyoides* Tanganyika / R. E. S. Tanner 1261 / 1953 / "Sukuma tribe: Whole plant pounded, soaked, patient bathed in this to treat fever, scabies" / "Kisukuma: Lwalamu"
Tanganyika / R. E. S. Tanner 1267 / 1953 / "Sukuma tribe: Whole plant pounded to flour used as poultice to treat boils on cattle and humans" / "Kisukuma: Luduta"

)58 **T. radicans** Rhodesia / C. Lennard 32460 / 51 / "Eaten by sheep"

)59 **T. rigida** Tanganyika / R. E. S. Tanner 1566 / 1953 / "Sukuma tribe: Leaves pounded, mixed with water, drunk twice daily for swollen testicles" / "Kisukuma: Lwamu lwa hisanga"

)60 **T. villosa** var. **incana** Tanganyika / R. E. S. Tanner 3159 / 1956 / "Sambaa: As a male aphrodisiac, roots dipped in water and chewed" / "Sambaa: Kitupa"

)61 **Vataireopsis speciosa** Surinam / W. A. Archer 2924 / 1934 / "Fish poison. Rank, overpowering odor in cut cortex or roots. Sap said to be extremely poisonous even to man." / " 'Man jongo' (Djuka)"
Brazil / R. Fróes, under the direction of B. A. Krukoff 12149 / 1941 / "Used by Indians to cure leprosy" / " ' Jaré' "

)62 **Vicia hyrcanica** Afghanistan / E. Bacon 92 / 1939 / "Used medicinally" / "bankakedeona (This name may be incorrect)"

063 **Vigna marina** Caroline Is. / E. Y. Hosaka 3435 / 1946 / "plant fed to pigs" / "suwakimo"
Caroline Is. / F. R. Fosberg 25776 / 1946 / "used medicinally" / " 'keldelel' "
Marshall Is. / F. R. Fosberg 26739 / 1946 / "used medicinally" / " 'merkunen chojo' "
Mariana Is. / F. R. Fosberg 25342 / 1946 / "medicine for many ailments" / " 'akanakan mahlos' "
Caroline Is. / M. Evans 497 / 1965 / "Used for medicine" / " 'HOLU' "
Samoa / D. W. Garber 700 / 1922 / "Used medicinally for circumcision and eyes" / " 'fia fia tai' "

064 **Zornia latifolia** Brazil / Prance, Rodrigues et al. 8917 / 1968 / "Leaves dried and smoked as a halluceogenic substitute for *Cannabis*" / " 'Maconha Branca' "

065 **Fabacea** indet. Honduras / C. & W. von Hagen 1029 / 1937 / "Indians eat fruit" / " 'Arisillo' "

PANDACEAE

066 **Microdesmis puberula** Liberia / J. T. Baldwin, Jr. 9080 / 1947 / "Tea from leaves for abortion and for inducing menstruation"
Liberia / J. T. Baldwin, Jr. 9058 / 1947 / "Concoction for inducing abortion made from leaves . . . beverage from leaves caused resumption of menstruation in one case after a lapse of two years"

GERANIACEAE

2067 **Biebersteinia emodi** Kashmir / W. Koelz 2071 / 1931 / "Pleasant spicy odor to plant. Rupshu people use it for horse load-wounds"

2068 **Erodium moschatum** Ecuador / F. Prieto 2448 / 1945 / "An infusion is used for gas on the stomach" / " 'Cuchi-agujilla' (Pig's needles)"

2069 **E. tibetanum** Kashmir / W. Koelz 2322 / 1931 / "rhubarb Animals dig it in fall & eat it"

2070 **Geranium** indet. Ecuador / W. H. Camp E-2511 / 1945 / "an infusion used when 'one is filled with air' " / " 'Agujillas del puerco' "

2071 **Geranium** indet. / Ecuador / W. H. Camp E-2469 / 1945 / "an infusion used for 'wind in the intestines' " / " 'Agujillas del cercas' "

2072 **Pelargonium anceps** U.S.: Calif. / J. T. Howell 19842 / 1944 / "Herbage with smell of turpentine"

2073 **P. graveolens** Ecuador / R. Espinosa 2455 / 48 / ". . . hojas muy aromáticas . . . Se le atribuyen propiedades medicinales" / "Esencia de rosa"

2074 **P. inquinans** S. Africa / R. D. A. Bayliss 4676 / 1971 / "Used as cold and headache cure by tribespeople"

2075 **Geraniacea** indet. Argentina / Rodriguez 616 / 1944 / "usados contra infecciones a las timaduras"

OXALIDACEAE

2076 **Averrhoa bilimbi** Caroline Is. / F. R. Fosberg 32067 / 1950 / "leaves pounded and applied as poultice to wounds" / " 'Kamin' "

2077 **Hypseocharis pimpinellifolia** Bolivia / W. M. A. Brooke 5114 / 1949 / "An extraction is used for making soap and washing hair"

2078 **Oxalis rufescens** Ecuador / F. Prieto E-2564 / 1945 / "used as a tonic in convalescence from various diseases in the belief that the plant contains iron" / " 'Sacha recaida' "

2079 **O. scandens** Colombia / J. M. Duque Jaramillo 3170 / 1946 / "Medicinal" / " 'Acedera' 'Chulco' "

2080 **O. spiralis** Colombia / F. R. Fosberg 20410 / 1943 / "roots yield tubers which are eaten like potatoes" / " 'oca' "

2081 **O. corniculata** China: Kwangsi / W. T. Tsang 24623 / 1934 / "fr. edible" Hainan / C. I. Lei 364 / 1933 / "fr. green edible"

2082 **Sarcotheca** indet. Borneo / Kostermans 4703 / 1948 / "fr. darkred, edible" / "krumbai merah"

HUMIRIACEAE

083 *Humiria balsamifera* var. *attenuata* Colombia / Schultes & Cabrera 16893 /
1952 / "Small black edible fruit" / "Yukuna = wa-toó-moo-ko"

084 *H. balsamifera* var. *balsamifera* Brit. Guiana / D. B. Fanshawe F715 / 42 /
"ripe fr . . . edible" / "Tauaranru"
Brazil / J. M. Pires 989 / 1947 / "frutos comestiveis" / " 'Umiri' "

085 *H. balsamifera* var. *coriacea* Venezuela / J. A. Steyermark 93218 / 1964 /
"fruit . . . edible" / " 'amuri-yek' (arekuna)"
Venezuela / Steyermark & Nilsson 282 / 1960 / "fruit . . . edible"

086 *H. floribunda* Brit. Guiana / D. Davis 141 / 1967 / "fruit . . . edible"

087 *Saccoglottis amazonica* Venezuela / J. J. Wurdack 293 / 1955 / "fruit
green, edible, used by Guaraos for diarrhea treatment" / " 'Nabaru'
(Guarauno)"

088 *Schistostemon retusum* Brazil / R. Fróes 21411 / 1945 / "frutos
comestiveis"

ERYTHROXYLACEAE

089 *Erythroxylon carthagenense* Colombia / R. Romero Castaneda 10449 / 1966
/ " 'Jayo', 'Coca' "

ZYGOPHYLLACEAE

090 *Larrea* indet. Argentina / J. Semper 31 / 1944 / "desinfectante"

091 *Tribulus cistoides* Dutch W. Indies / J. G. v. d. Bergh s.n. / no year /
"Root used as a cure"
Marquesas Is. / Mumford & Adamson 426 / 29 / "3 spined fruits cause ab-
cesses on feet of wild dogs—probably a factor in reducing their number" /
"Tataaotuna"

RUTACEAE

092 *Acronychia pedunculata* China: Kwangsi / W. T. Tsang 23815 / 1934 / "fr.
edible" / "Ye Yau Kam Shue"
Hong Kong / Y. W. Taam 1644 / 1940 / "fruit yellow, edible"

093 *Agathosma ovata* S. Africa / R. D. A. Bayliss BS 4683 / 1971 / "Leaves
used for making 'Tea' "

094 *Atalantia disticha* Philippine Is. / O. Ladia 25 / 1929 / "fruit edible" /
"agoyhoy"

2095 **Citrus** indet. China: Kwangtung / G. W. Groff 00493 / 1931 / "Fruits eaten" / "Kom?"

2096 **Citrus** indet. China: Kwangtung / S. K. Lau 987 / 1933 / "fr . . . edible"

2097 **C. maxima** Hainan / F. A. McClure 20074 / 1932 / "Fruits eaten" Hainan / S. K. Lau 409 / 1932 / "Fr. yellow & edible" / "Tai Kwat"

2098 **Clausena anisata** Tanganyika / R. E. S. Tanner 3896 / 1957 / "Zigua: for pain in testicle, roots boiled and the water drunk" / "Zigua: Ndavu Kali"

2099 **C. excavata** Hainan / H. V. Swa 7 / 1932 / "Fr. yellow. edible. Cure for indigestion" / "Kai Mo Wong"
Hainan / H. Fung 20073 / 1932 / "fr . . . edible" / "Che Tiu"

2100 **Dictyoloma peruvianum** Peru / J. V. Schunke 1950 / 1967 / "Las hojas son muy tóxicas" / " 'Huamansamana' "

2101 **Glycosmis citrifolia** China: Kwangtung / S. K. Lau 643 / 1932 / "fr . . . edible" / "Kom Kat Tsz"

2102 **G. cochinchinensis** Hainan / S. K. Lau 306 / 1932 / "fruit, pink, edible" / "Tim To"

2103 **Micromelum falcatum** China: Kwangtung / S. I. Nin 18424 / 1928 / "Medicine" / "A Tam"
Indochina / Poilane 694 / 1919 / "les feuilles son employées pour préserver les plaies de l'infection et des insectes montagne" / "Annamite: Sang ot"

2104 **M. integerrimum** China: Kwangtung / Y. Tsiang 2234 / 1929 / "Fruit . . . edible"

2105 **M. minutum** Fiji / A. C. Smith 7035 / 1953 / "leaves crushed in coconut oil and used as liniment" / " 'Ngginggila' "
Fiji / O. Degener 15000 / 1941 / "Leaf used as medicine by Fijians. Dry leaves are then used as tea by Fijians" / " 'Karakarakuro' (Serua)"

2106 **Murraya paniculata** Ecuador / W. von Hagen 2 / 1934 / "Used by the natives for snake bite. Leaves are mashed into a paste and put into alchol. Efficacy doubtful" / "Merto (Sp.-Ind.)"

2107 **Skimmia japonica** India / W. Koelz 79 / 1930 / "Said to be eaten by Muskdeer" / "Nehr (Kulu)"

2108 **Spiranthera odoratissima** Brazil / Philcox, Ramos & Sousa 3076 / 1967 / "Used medicinally for blood disorders" / " 'manaca' "

2109 **Teclea glomerata** Tanganyika / R. E. S. Tanner 1334 / 1953 / "Sukuma tribe: Roots pounded, cooked in porridge to treat anklyostomiasis" / "Kisukuma: Nungubashite"

2110 **T. nobilis** Tanganyika / R. E. S. Tanner 952 / 1952 / "Sukuma tribe: Leaves boiled in water to produce steam and turn fever to sweat" / "Kisukuma: Mjuv"

111 *T. simplicifolia* Tanganyika / R. E. S. Tanner 3463 / 1957 / "Zigua: For sharp stomach pain, roots boiled and the water drunk" / "Zigua: Ndizi"

112 *Toddalia asiatica* Tanganyika / R. E. S. Tanner 3143 / 1956 / "Bondei: For snake bite, leaves pounded up and drunk in water" / "Nyamwezi: Mlamata; Bondei: Mdadai or Mgosi"

113 *T. aculeata* China: Kwangsi / R. C. Ching 7605 / 1928 / "Fruit . . . edible"

114 *Toddaliopsis sansibarensis* Tanganyika / R. E. S. Tanner 3967 / 1958 / "Zigua: for pain in the stomach, roots boiled and the water drunk" / "Zigua: Mdizi"

115 *Triphasia trifoliata* U.S. Virgin Is.: St. Croix / E. L. Little, Jr. 21512 / 1966 / "Fruit . . . edible" / "sweet-lime"

116 *T. trifolia* Vietnam / no collector s.n. / 1920 / "small bush to tree medicinal"

117 *Vepris lanceolata* Tanganyika / R. E. S. Tanner 3900 / 1957 / "Swaheli: for fever, twigs boiled and patient sits over pot covered with blanket" / "Mti mkuu (Swaheil)"

118 *Zanthoxylum caribaeum* Honduras / C. & W. von Hagen 1165 / 1937 / "Bark & stem chewed as narcotic for toothache" / "Duerme lengua"

119 *Z. culantrillo* Peru / R. Scolnik 1203 / 1948 / "fl. blanco verdosas para el reuma" / "shapilleja"?
Peru / Killip & Smith 25400 / 1929 / "Leaves much used by Indians as remedy for snake bites and cuts"
Peru / Killip & Smith 23597 / 1929 / "leaves put in alcohol or aguardiente and drunk for snake bites" / " 'Contravenosa' "

120 *Z. pterota* Peru / J. V. Schunke 1071 / 1966 / "Leaves medicinal; drinking in infusion for rheumatism, and picadura of culebras" / "Contra veneno"

121 *Z. ruiziana* Peru / J. V. Schunke 984 / 1965 / "The leaves are anti-rheumatic" / "Huajala' "

122 *Z. chalybea* Tanganyika / R. E. S. Tanner 3894 / 1957 / "Zigua: for pain in stomach, roots pounded up and boiled and eaten" / "Zigua: Mkunungu"
Tanganyika / R. E. S. Tanner 1350 / 1953 / "Sukuma tribe: Pounded roots used in water orally to treat tuberculosis and fever, as poultice for sores and pain in wisdom teeth" / "Kisukuma: Nungunungu"

123 *Z. bungei* China / C. Y. Chiao 21354 / 1929 / "Shoots edible, as a condiment"

124 *Z. nitidum* China: Kwangtung / T. W. Tak 19276 / 31 / "For drug plant" / "Ye Oo Ts'in"
Philippine Is. / P. Afalla 30213 / 1926 / "For poisoning eel & mud fish (dalag). The vine is to be pounded thoroughly and to be inserted into the holes of eel or dalag" / "SUAL"

2125   *Z. simulans* China: Kweichow / Steward, Chiao & Cheo 17 / 1931 / "seeds shiny black, used for seasoning"

2126   *Rutacea* indet. Bolivia / G. Bejerano s.n. / 1964 / "used by natives of Beni; infusion of ground bark for amoebiasis" / " 'ebanta' "

SIMAROUBACEAE

2127   *Ailanthus altissima* U.S.: Penn. / E. M. Gress 10076 / 1937 / "Seedlings 'suspected of being a narcotic plant' "

2128   *Eurycoma longifolia* Brit. N. Borneo / Evangelista & Orsat 977 / 1929 / "Leaves are used by natives as medicine for skin disease"

2129   *Harrisonia abyssinica* Tanganyika / R. E. S. Tanner 1223 / 1953 / "Sukuma tribe: Leaves mixed with ghee for poultice used to treat boils; pains in stomach treated with flour made from roots and drunk in tea" / "Kisukuma: Soma"
Tanganyika / R. E. S. Tanner 3455 / 1957 / "Zigua; For pain in stomach, roots boiled and the water drunk" / "Zigua: Mkusi"
Tanganyika / R. E. S. Tanner 3322 / 1956 / "Bondei: For stomach ache, roots soaked in water and drunk" / "Bondei: Mdadai"

2130   *H. perforata* Hainan / C. E. Lei 116 / 1932 / "Boil ground leaves with water, wash the body to heal skin disease" / "Goong Chi"

2131   *Picramnia allenii* Panama / C. & W. von Hagen 2126 / 1949 / " 'amargocito' "

2132   *P. andicola* El Salvador / J. A. Steyermark 51261 / 1942 / " 'cafeyillo' "

2133   *P. antidesma* Brit. Honduras / W. A. Schipp 1079 / 1932 / "Indians use bard for malaria"
Brit. Honduras / P. H. Gentle 3306 / " 'Pasa embra', 'wild raisin' "

2134   *P. latifolia* Ecuador / Grubb, Lloyd et al. 1637 / 1960 / "Fruits . . . astringent"

2135   *Picrasma excelsa* Brit. Guiana / no collector 5744 / 48 / "Tree whose bark is ingredient of 'Wourali' poison" / "Urariwong (M)"

2136   *Picrolemma sprucei* Brazil / Prance, Maas et al. 16489 / 1971 / "the bark used as an ingredient of Deni Indian arrow poison" / " 'Mapizuuba' "

2137   *Simarouba glauca* Nicaragua / Bunting & Licht 688 / 1961 / "Sweet tasting, edible" / " 'Aceituna' "

SURIANACEAE

2138   *Suriana maritima* Bahama Is. / O. Degener 18777 / 1946 / "boil bark of root or stem, add salt, and hold in mouth for remedy for boil in mouth" / " 'Bay Cedar' "

BURSERACEAE

39 *Bursera diversifolia* Mexico / Hinton et al. 10072 / 37 / "Bark and leaves smell like ripe oranges" / "Copal"

40 *B. trifoliolata* Mexico / Hinton et al. 6340 / 34 / "Poisonous" / "Quicanchire"

41 *Bursera* indet. / Mexico / G. B. Hinton 10851 / 37 / "gum collected for incense"

42 *B. simaruba* Bahama Is. / S. R. Hill 2346 / 1974 / "Used as tea and incense" / " 'Gum-elemi' "
Bahama Is. / O. Degener 19011 / 1946 / "tea from leaves drunk as remedy for backache" / " 'Gemalamey' "
Bahama Is. / W. Cerbin C 150 / 1973 / "Bark popular to treat ailments. Tree used to feed livestock" / " 'Camalami' "

43 *Canarium asperum* var. *asperum* Philippine Is. / M. Oro 31072 / 1929 / "fruit eaten, seeds roasted & made into coffee" / "Mayagyat (Tag)"

44 *C. dichotomum* N. Borneo / A. Cuadra A.1322 / 48 / "edible by natives" / "Kandis daham (Malay)"

45 *Canarium* indet. New Guinea / D. Frodin NGF 26695 / 66 / "Fruit . . . edible"

46 *Commiphora boiviniana* Tanganyika / R. E. S. Tanner 3342 / 1956 / "Zigua: For lasting sores, roots pounded up and applied fresh" / "Zigua: Dandasindi"

47 *C. laxiflora* Tanganyika / R. E. S. Tanner 1255 / 1953 / "Sukuma tribe: Roots used to treat snake-bite" / "Kisukuma: Missusu"

48 *C. pilosa* S. Africa / H. G. Faulkner 20 / 1944 / "The natives des—it as a 'Great Medicine'. The part used is the outer skin of the roots. This is removed, cut up and boiled. For children cut up small and mixed with food. It is a remedy for stomach and head" / "Kukulu"
Tanganyika / R. E. S. Tanner 3775 / 1957 / "Zigua: For bringing on delayed birth, roots boiled and the water drunk" / "Zigua: Mti-ntwi"

49 *C. ptelaefolia* Tanganyika / R. E. S. Tanner 3904 / 1957 / "Bondei—for gonorrhoea, roots boiled and the water drunk" / "Bondei—Mtwitwi"

50 *Dacryodes peruviana* Brazil / R. Fróes 21430 / 1945 / "Planta alimenticia entre os Indios Baniuas; frutos comestiveis ou preparados em bebidas" / " 'Iaá-pixuna grande' "

51 *Protium panamense* Panama / T. B. Croat 4235 / 1967 / "odor of mango"

52 *P. tenuifolium* subsp. *sessiliflorum* Panama / N. Bristan 1082(2) / 1967 / "fruit green, EDIBLE when ripe"
Panama / Cooper & Slater 259 / 1927 / " 'Comida del mono' "

2153 **P. poeppigianum** Peru / Poeppig 2830 / 1831 / "remedium contra dolores arthriticos indigenis celeberrimum" / " 'Balsamo de Caraña' "

2154 **Protium** indet. Venezuela / J. A. Steyermark 90578 / 1962 / "fruit deep red, edible"

2155 **Santiria** indet. E. Borneo / A. Kostermans 4838 / 1951 / "Fruit edible" / " 'Kumbajau tëta' "

2156 **Tetragastris balsamifera** Dominican Republic / J. J. Jimenez 5192 / 1967 / "From the fruits an aromatic oil is extracted used in different diseases (Aceite de Palo = Tree Oil)" / " 'AMACEY' "

2157 **Burseracea** indet. Borneo / Kostermans 7113 / 1952 / "Fr. dull yellow, edible" / " 'Keramu' (Benua—Dajak language) or 'Merasam' (Kutei)"

2158 **Burseracea** indet. Borneo / Kostermans 6275 / 1952 / "Fr. acid, edible . . ." / " 'Asam' or 'Rarawa pipit' (pipit = bird)"

MELIACEAE

2159 **Aglaia** indet. China: Kwangsi / W. T. Tsang 24088 / 1934 / "fr. brown, edible" / "Shap Man Taai Shan"

2160 **A. roxburghiana** Ceylon / R. G. Cooray 69111624R / 1969 / "fruits . . . edible but astringent"

2161 **A. eusideroxylon** Java / Martati 123 / 1960 / "The smell like a ripe papaya"

2162 **A. vitiensis** Fiji / A. C. Smith 7042 / 1953 / "bark used for medicine" / " 'Thawaru' "

2163 **Aglaia** indet. Borneo / Kostermans 4701 / 1948 / "Fruit brown, edible" / "Bunjauw"

2164 **Aglaia** indet. Philippine Is. / R. S. Williams 3017 / 1905 / ". . . pulp eaten by natives"

2165 **Aphanamixis** indet. / New Guinea / D. Frodin NGF 26587 / 66 / "Odour of cabbage" / "Aumamia"

2166 **Carapa guianensis** Venezuela / J. J. Buza 7589 / 1959 / "Su aceite es de valor medicinal" / "Carapa"
Venezuela / J. A. Steyermark 61338 / 1945 / "take the inside part of the seed and mash it, cook in water with aceite de castilla and drink for stomach ache; also crushed seed scraped raw placed over cuts, also with acite de castilla heals the wound"

7   *Dysoxylum hornei* var. *hornei* Fiji / O. Degener 15308 / 1941 / "For diarrhoea and bloody stools Fijians drink liquid by squashing the fresh leaves in water and straining" / " 'Viviniura' in Serua dialect"

8   *D. richii* Fiji / Degener & Ordonez 14181 / 1941 / "leaves boiled in water, the liquid being used as a shampoo or hair-dye" / " 'Manawi' "

9   *Guarea bilibil* Colombia / E. L. Little, Jr. 7469 / 1944 / "Purgative for hogs made from bark" / " 'bilibil' "

0   *G. trichilioides* Venezuela / J. A. Steyermark 61037 / 1945 / "fleshy aril of fruit eaten by birds and is edible for man" / " 'cabimbo' "
Peru / C. M. Belshaw 3324 / 1937 / "Said to be a cure for lunatics. Rub the leaves on the person or crush in water for a bath" / " 'Requia' "
Brit. Guiana / C. W. Anderson 25 / 1908 / "Bark used as an emetic" / " 'Buck (sp.?) vomit' "

1   *Guarea* indet. Argentina / J. E. Montes 14881 / 1956 / "Su corteza es medicinal" / " 'Cedrillo', 'Catigua-pora' "

2   *Guarea* indet. Argentina / J. E. Montes 14761 / 1955 / "corteza empleada en medicina popular" / " 'Cedrillo' "

3   *Lansium kostermansii* Indonesia / Kostermans 19215 / 1961 / "the undepeloped seed edible, sweetish" / "narab"
Indonesia / Kostermans 19117 / 1961 / "soft (undeveloped seed), edible, slightly sweet" / " 'kaju narab' "

4   *Trichilia havanensis* Guatemala / J. A. Steyermark 48519 / 1942 / "Bark reputed to be used as antidote for malaria" / " 'quina sylvestre' "

5   *Trichilia* indet. Panama / T. B. Croat 11711 / 1970 / "fruit green, the interior red, being eaten"

6   *T. catigua* Argentina / J. E. Montes 27362 / 1958 / "corteza tánica — curtiente y medicinal . . . (Usada en medicina popular)" / " 'Catigu á-guazú' "

7   *Trichilia* indet. Tanganyika / R. E. S. Tanner 1243 / 1953 / "Sukuma tribe: Pounded roots mixed with mtama flour and drunk to treat sores; used as poultice in syphilis" / "Kisukuma: Njundu"

78   *Turraea fischeri* Tanganyika / R. E. S. Tanner 1614 / 1953 / "Sukuma tribe: Roots pounded and drunk with water at meals for ankylostomiasis or sharp pain in stomach" / "Kisukuma: Ningiwe nkima"

79   *Walsura robusta* Hainan / H. Fung 20154 / 1932 / "fr. red, edible"

80   *Xylocarpus mollucensis* Tanganyika / R. E. S. Tanner 3759 / 1957 / "Swaheli: For circulating pain in stomach, roots boiled and the water drunk" / "Swaheli: Mkelenge"

81   *Meliacea* indet. Argentina / J. E. Montes 14825 / 1955 / "cortesa (sp.?) = medicina popular" / " 'Quina' "

MALPIGHIACEAE

2182 **Byrsonima karwinskiana** Mexico / H. H. Rusby 32 / 1910 / "commercial edible fruit" / " 'Nanchi agria' "

2183 **B. spicata** Jamaica / W. Harris 9093 / 1905 / "Fruits edible"
Colombia / R. Romero-Castañeda 8429 / 60 / "frutos . . . comestibles" / " 'Pajarito' "

2184 **B. basiloba** Venezuela / Steyermark & Dunsterville 105695 / 1972 / "fruit . . . edible, but Indians do not appear to know about it"
Brazil / E. P. Heringer 13.069 / 1974 / "frutos alimento de passarinho"

2185 **B. chrysophylla** Peru / C. M. Belshaw 3395 / 1937 / "Used to stop flow of blood in a cut" / " 'Indano' "

2186 **B. hypoleuca** Venezuela / Steyermark, Bunting & Wessels-Boer 100364 / 1967 / "fruit edible"

2187 **B. intermedia** Brazil / E. P. Heringer 13.006 / 1973 / "fruto comestivel por passarinho"

2188 **B. poeppigiana** Peru / J. V. Schunke 2656 / 1968 / "utilizado por los indios para lavado vaginal y anti-tuberculosis" / " 'Indano' "

2189 **Malpighia angustifolia** W.I.: Montserrat / J. A. Shafer 102 / 1907 / " 'Wild Coffee' "

2190 **Malpighia** indet. Venezuela / H. M. Curran 2M71 / 49 / "edible red fruit"

2191 **Tetrapterys discolor** Venezuela / J. A. Steyermark 61108 / 1945 / "Monkeys eat flowers / " 'bejuco de mono' "

2192 **Triaspis macropteron** Tanganyika / R. E. S. Tanner 1285 / 1953 / "Sukuma tribe: Roots pounded, dried, mixed with white mtama flour and water, taken orally to treat syphilis" / "Kisukuma: Njundu gwa mulugulu"

2193 **Tristellateia africana** Tanganyika / R. E. S. Tanner 3725 / 1957 / "Fruits edible" / "Swaheli: Mtole"

VOCHYSIACEAE

2194 **Vochysia** indet. Argentina / J. E. Montes 14945 / 1956 / "Frutos comestibles" / " 'Agüa-hi' "

POLYGALACEAE

2195 **Carpolobia alba** Liberia / J. T. Baldwin, Jr. 10699 / 1947 / "Fruit yellow edible"

96 *Monnina xalapensis* Guatemala / P. C. Standley 65753 / 1939 / "Eaten by deer and sheep" / " 'Zacate de venado' "

97 *M. salicifolia* Ecuador / F. Prieto 2676 / 1945 / "Roots used to wash the head" / " 'Iguala' "
Ecuador / W. H. Camp E-2217 / 1945 / "The ground root used as soap" / " 'Iguilan' "
Ecuador / W. H. Camp E-2484 / 1945 / "the bark of the roots used as a substitute for soap, especially for washing the head to get rid of dandruff" / " 'Iguila' "

98 *Moutabea guianensis* Venezuela / C. Blanco 1093 / 1971 / "Semilla con árilo blanco, comestible, sabor dulce" / " 'Ojo de zamuro' "
Venezuela / J. A. Steyermark 99921 / 1967 / ". . . acid-sweet pulp; edible pulp"

99 *Muraltia macrocarpa* S. Africa / R. Story 4504 / 1954 / "excellent stock feed"

00 *Polygala boykinii* U.S.: Fla. / P. P. Sheehan s.n. / 1919 / "Used as a head-wash to correct dizziness. Plant is soaked in cold water, and the decoction is used externally on the head" / "White-milkwort"

01 *P. aparinoides* Brit. Honduras / W. A. Schipp 822 / 1931 / "whole plant when crushed gives of the odor of menthol"

02 *P. timoutou* Brit. Honduras / W. A. Schipp 553 / 1930 / "roots when crushed have odor like Menthol"

03 *P. acuminata* Peru / J. V. Schunke 1092 / 1966 / "Leaves utilizan para curar heridas infectadas" / "Irgapirina Sacha"

04 *P. amboniensis* Tanganyika / R. E. S. Tanner 2852 / 1956 / "Swaheli: For delayed birth leaves pounded up and drunk cold" / "Swaheli: Mzalia Nyuma"

05 *P. erioptera* Tanganyika / R. E. S. Tanner 1306 / 1953 / "Sukuma tribe: Roots chewed to treat stomach pains: roots pounded to flour and sniffed for the head pains in child; pounded roots used as poultice on breasts to increase lactation" / "Kisukuma: Kasseni"
Tanganyika / R. E. S. Tanner 1500 / 1953 / "Sukuma tribe: Roots pounded up, used as drink to treat stomach pains" / "Kisukuma: Kahekalla"

06 *P. multiflora* Ivory Coast / F. R. Fosberg 40600 / 1959 / "roots with wintergreen odor"

07 *P. sphenoptera* Tanganyika / R. E. S. Tanner 3493 / 1957 / "Bondei: For sores, pounded up and put on as a poultice" / "Bondei: Kitutu shaba"
Tanganyika / R. E. S. Tanner 3640 / 1957 / "Zigua: For sores and scabies, leaves pounded up and put on sores as a paste" / "Zigua: Mlutlutu"

2208   *P. arillata* China: Kwangsi / W. T. Tsang 24014 / 1934 / "used for medicine" / "To Tiu Wong Shue"

2209   *P. aureocauda* China: Kwangtung / W. T. Tsang 20628 / 1932 / "for medical value" / "To Dew Wong"

2210   *P. tenuifolia* China: Hopei / J. C. Liu L.2537 / 1930 / "root used Chinese medicine" / " 'Yuan Chih' "

2211   *Securidaca virgata* Dominican Republic / R. A. & E. S. Howard 8955 / 1946 / "Roots used medicinally" / " 'Maraveli' "

2212   *S. longepedunculata* Tanganyika / R. E. S. Tanner 1139 / 1952 / "Sukuma: for gonorrhoea—roots cut up into small pieces and boiled with white mtama flour and then drunk. Causes much irritation" / "Kisukuma: Nengonengo"
Tanganyika / R. E. S. Tanner 2669 / 1956 / "Bondei: Roots boiled and drunk for stomach ache" / "Kiswaheli: Mkurukatani"

XANTHOPHYLLACEAE

2213   *Xanthophyllum hainanense* Hainan / S. K. Lau 35 / 1932 / "The ripe fruit is edible" / "Laiziyu (Lois)"

DICHAPETALACEAE

2214   *Dichapetalum hainanense* Hainan / Tsang, Tanh & Fung 117 / 1929 / "edible" / "Wan Tsu"

2215   *Tapura peruviana* Peru / E. Ancuash 665 / 1974 / "Comestible su fruto" / " 'yumpig' "

EUPHORBIACEAE

2216   *Acalypha phleoides* Mexico / Stanford, Retherford & Northcraft 231 / 1941 / "heavily grazed by goats"

2217   *A. triloba* Guatemala / J. A. Steyermark 51959 / 1942 / " 'yerba cancer' "

2218   *A. umbrosa* Mexico / R. McVaugh 10030 / 1949 / " 'Hierba cáncer' "

2219   *Acalypha* indet. Guatemala / Jones & Facey 3510 / 1966 / " 'Hierba del cáncer' "

2220   *A. macrostachya* Venezuela / J. A. Steyermark 61009 / 1945 / "crushed leaves mashed up give soapy sap used to wash clothes when there is no soap; also crushed leaves used in treating sores and scratches; (lepra of Bergantin)" / " 'cura sana' "

21   *A. fruticosa* Kenya / B. Mathew 6371 / 1970 / "Used to cure sores on animals - stems ground up in water" / " 'Etetelait' (Turkana)"
Tanganyika / R. E. S. Tanner 3337 / 1956 / "flowers . . . slightly aromatic. Sambaa: For scabies, leaves pounded up and put on affected areas" / "Sambaa: Msagati kizumba Kizuga: Mfulwe"
Tanganyika / R. E. S. Tanner 3902 / 1958 / "flowers . . . slightly aromatic. Bondei: for pains in the left side of the stomach, roots boiled and the water drunk" / "Zigua: Mfulwe. Bondei: Msagati kivumba"
Tanganyika / R. E. S. Tanner 3472 / 1957 / "flowers . . . aromatic. Sambaa: For extensive irritating sores, leaves pounded up and put on as a paste" / "Sambaa: Msgati Kizumba"

22   *A. oranta* Tanganyika / R. E. S. Tanner 3491 / 1957 / "flowers . . . aromatic. Bondei: For scabies in children, leaves soaked in water and used for washing down" / "Bondei: Msagati Kivumba: Ngosi"

23   *A. volkensii* Tanganyika / R. E. S. Tanner 4247 / 1959 / "Whole plant used to treat scabies" / "Kizanaki: Kiherehre"

24   *Acalypha* indet. Tanganyika / R. E. S. Tanner 1122 / 1952 / "flowers . . . aromatic. Used for making fish traps" / "Kisukuma: Ngesha"

25   *Adelia ricinella* Venezuela / J. A. Steyermark 62796 / 1945 / "wild doves eat the fruit" / " 'fruta de paloma' "

26   *Alchornea castaneifolia* Venezuela / Aristeguieta & Zabala 7083 / 1969 / "De la corteza se raspa y se hacen cataplasmas contra la picada de raya. Ligado al curso de agua" / " 'Mangle' "

27   *Alchornea* indet. Venezuela / J. A. Steyermark 99955 / 1967 / "fruit eaten by birds . . ." / " 'arbol de cucaracho' "

28   *Aleurites triloba* Jamaica / H. A. Wood 3180 / 1908 / " 'Jamaica Walnut' "

29   *A. moluccana* Philippine Is. / A. L. Zwickey 151 / 1938 / "nuts edible; their oil used on hair, in making cakes, etc." / " 'Togaya' "

30   *A. saponaria* Philippine Is. / no collector 145 / 1914 / "The oil yielded by the seeds . . . is somewhat caustic, causing eruptions when applied to the skin" / "balocand, baguilumbang, etc."

31   *Antidesma venosum* Tanganyika / R. E. S. Tanner 3918 / 1958 / "Bondei: for bilharzia, roots boiled and the water drunk" / "Bondei: Mnamiaziwa"
Tanganyika / R. E. S. Tanner 3924 / 1958 / "Bondei: for removing afterbirth, roots boiled and the water drunk" / "Bondei: Mnamiaziwa"
Tanganyika R. E. S. Tanner 3419 / 1957 / "Zigua: For sharp stomach ache, leaves boiled and the water drunk. For toothache, leaves boiled and the water rinsed through the mouth" / "Zigua: Mgwejameno"
Tanganyika / R. E. S. Tanner 3958 / 1958 / "Zigua: For infections veneral, roots boiled and the water drunk" / "Zigua: Msele"

2232 *A. fordii* China: Nanking / R. C. Ching 8258 / 1928 / "Purplish edible when ripe"

2233 *A. gracile* China: Kwangtung / W. T. Tsang 21272 / 1932 / "fruit, black edible" / "Yuen Ip Suen Mei Tsz"

2234 *Antidesma* indet. China: Kwangsi / Steward & Cheo 854 / 1933 / "Fruit edible"

2235 *A. pleuricum* Philippine Is. / J. G. Gojar 8 / 1932 / "fruit is edible like Binayuyu"

2236 *Aporosa alvarezii* Philippine Is. / M. Oro 275 / 1929 / "for small horses" / "Kanaga-ay. Dialect Tag"

2237 *Baccaurea griffithii* Malay Peninsula / M. R. Henderson 21974 / 1929 / "Fruit edible" / " 'Taban' "

2238 *B. membranacea* Brit. N. Borneo / Md. Tahir 799 / 1928 / "Fruit can be eaten"

2239 *Bischoffia javanica* Hainan / C. I. Lei 208 / 1932 / "Use for medicine" / "Tzou Fung"
Philippine Is. / A. D. E. Elmer 8968 / 1907 / " 'Tuol'—the young leaves are used in drawing sores and boils to maturity"

2240 *Breynia fruticosa* China: Kwangtung / W. T. Tsang A-664 / 30 / "drug plant"
Hainan / H. V. Swa 4 / 1932 / "used as a dye" / "Sik Koi Ying Fa"

2241 *B. acuminata* New Britain / J. H. L. Waterhouse 244 / 1934 / "A shrub with small fruit used by native children as 'shot' for small bamboo pop-guns" / "Pipil"

2242 *B. cernua* Philippine Is. / A. D. E. Elmer 9983 / 1908 / "berries . . . soft juicy and sweet when ripe"

2243 *Bridelia cathartica* Tanganyika / R. E. S. Tanner 3043 / 1956 / "Zigua: Bark boiled and the water drunk for cough" / "Zigua: Mkunde"
Tanganyika / R. E. S. Tanner 3013 / 1956 / "Ngindo: Aphrodisiac, leaves and roots pounded up and boiled and cooked with beans" / "Ngindo: Nongomela Bondei: Msimisi"
Tanganyika / R. E. S. Tanner 2848 / 1956 / "Swaheli: For pain in upper stomach leaves lightly burnt and eaten in porridge" / "Swaheli: Namya Kwale"
N. Rhodesia / W. R. Bainbridge 730 / 63 / "The Tonga chew leaves, then drink water, this gives water a sweet taste" / "Tonga—MUNOHYA-MENDA"
Tanganyika / R. E. S. Tanner 1947 / 1955 / "Roots used to treat stomach-ache; boiled and water drunk" / "Kibondei: Msisimisi"

4   *B. scleroneura* Tanganyika / R. E. S. Tanner 4160 / 1959 / "Leaves used in treating headaches; roots for sores" / "Kizanaki: Ekihatyo"

5   *B. scleroneuroides* Tanganyika / R. E. S. Tanner 1209 / 1953 / "Sukuma tribe: Roots used to treat sores on gums-pounded up, mixed in cold water, this used as mouth rinse" / "Kisukuma: Sorwanghuru"

46   *Caperonia palustris* Surinam / J. Lanjouw 615 / 1934 / "Leaves used for tea"

47   *Claoxylon carolinianum* Caroline Is. / R. Kanehira 647 / 1929 / "latex sometimes poisons the skin"

48   *C. polot* Melanesia / J. H. L. Waterhouse 88 / 1932 / "Leaf used as spinach" / "Maritu"

49   *Cleistanthus megacarpus* Philippine Is. / A. D. E. Elmer 12381 / 1910 / " 'Cacao' "

50   *Codiaeum variegatum* Fiji / Degener & Ordonez 14160A / 1941 / "Some kinds planted by natives about their graves" / " 'Sathasatha' "
Caroline Is. / F. R. Fosberg 32060 / 1950 / "leaves green with white spots. Said to be used medicinally for fever, juice of pounded leaves and stems used" / " 'Kasuk' "

51   *Croton ciliatoglandulosus* Honduras / S. F. Glassman 1591 / 48 / "Glandular hairs blind cattle" / " 'Ciegavista' "

52   *C. ovalifolius* Mexico / T. MacDougall s.n. / 1970 / "Infusion of root is remedy for 'trisia' — yellowing of skin with fever"

53   *C. nitens* Jamaica / G. R. Proctor 23667 / 1963 / "inner bark aromatic"

54   *C. poitaei* Dominican Republic / Bro. A. H. Liogier 11596 / 1968 / ". . . the foliage very aromatic"

55   *C. rigidus* Brit. Virgin Is. / E. L. Little, Jr. 23764 / 1969 / "Crushed leaves have pungent odor"

56   *C. stenophyllus* Cuba / Fre. Leon 12077 / 1924 / "much used medicinal plant" / " 'Té' "

257   *C. cajucana* Brazil / G. T. Prance 1262 / 1965 / "Leaves used for cure of liver ailments" / " 'Sacaca' "

258   *C. draconoides* Peru / A. G. Ruíz 126 / 1963 / "Usos: savia para cicatrizar heridas" / " 'Sangre de grado' "

259   *C. glabellus* Colombia / J. M. Idrobo 2197 / 1956 / "Le atribuyen varias propiedades, entre éllas hipotensor y colagogo' " / " 'Almizclillo' "

260   *C. gossypifolius* Venezuela / J. A. Steyermark 61266 / 1945 / "reddish sap in bark used by children in raw state as a tooth paste or mouth wash" / " 'mestre' "

2261　*C. malambo* Colombia / R. Romero Castañeda 9688 / 1963 / "Madera olorosa, utilizada en afecciones estomacales" / " 'Malanbo' "

2262　*Croton* indet. Colombia / E. L. Little, Jr. 7108 / 1944 / "Medicine for stomach ache, dysentery" / " 'sangre gallo' "

2263　*Croton* indet. Peru / P. C. Hutchison 1493 / 1957 / "Juice stains clothes and causes skin inflammation"

2264　*Croton* indet. Argentina / J. E. Montes 14765 / 1955 / "Uso: raíz medicina popular" / " 'Belame colorado' "

2265　*Croton* indet. Argentina / Herb. del Inca Huasi 68 / 42 / "El fruto de uso medicinal"

2266　*Croton* indet. Argentina / J. E. Montes 14784 / 1955 / "Uso: raiz medicinal" / " 'Belame blanco' "

2267　*Croton* indet. Argentina / J. E. Montes 789 / 45 / "Uso: raices medicina popular" / " 'Belame guazu' "

2268　*Croton* indet. Argentina / J. E. Montes 3 / 1944 / "Medicinal"

2269　*Croton* indet. Argentina / J. E. Montes 808 / 45 / "Uso Medicina popular" / " 'Belame enano' "

2270　*C. dichogamus* Tanganyika / R. E. S. Tanner 3302 / 1956 / "Masai: For syphilis, roots with their outside removed are boiled and the water drunk" / "Masai: Engataroo"
Tanganyika / R. E. S. Tanner 3650 / 1957 / "For colds in head, roots pounded up, dried, and used as snuff" / "Zigua: Mghwati Mkomba"
Tanganyika / R. E. S. Tanner 1066 / 1952 / "Costume made of leaves for a dance of this name" / "Kisukuma: Ifubo"

2271　*C. macrostachyus* Tanganyika / R. E. S. Tanner 3588 / 1957 / "Zigua: For severe cough, leaves boiled in pot and the patient sits over it for inhalent" / "Zigua: Msindusi"

2272　*Croton* indet. Borneo / J. & M. S. Clemens 5083 / 1929 / "Somewhat irritating to the skin"

2273　*Dalechampia scandens* Brazil / Philcox & Fereira 4066 / 1968 / "Stinging properties of leaves" / " 'Feijão-ortiga' "
Surinam / Lanjouw et Lindeman 1114 / 1948 / "Hairs burning on sensitive skin" / "brongwirie (Sur.)"

2274　*Drypetes lateriflora* Mexico / F. Ventura A. 3553 / 1971 / "comestible"

2275　*D. battiscombei* Tanganyika / R. E. S. Tanner 1326 / 1953 / "Sukuma tribe: Roots and leaves pounded, taken orally in water, used to treat gonorrhoea" / "Kisukuma: Shonwanguku"

2276　*D. gerrardii* Tanganyika / R. E. S. Tanner 3959 / 1958 / "Zigua: For stomach pain. Roots boiled and the water drunk" / "Zigua: Kisimpanda"

7   *D. gossweileri* Nigeria / J. P. M. Brenan 8474 / 1947 / "Slash smelling strongly of horseradish" / "OKHUABA (Beni)"

8   *Drypetes* indet. China: Kwangsi / R. C. Ching 8299 / 1928 / "Fruit . . . edible . . ."

9   *D. sepiaria* Ceylon / Nowicke & Jayasuriya 306 / 1973 / "fruits red, edible; monkeys nearby. Termite nest at the base"

0   *Drypetes* indet. Sumatra / W. N. & C. M. Bangham 750 / 1932 / "Somewhat edible, a little sweet"

1   *Erythrococca kirkii* Tanganyika / R. E. S. Tanner 3147 / 1956 / "Bondei: For pain in heart, roots cooked in water which is drunk when cool" / "Bondei: Mnyembeuwe"
Tanganyika / R. E. S. Tanner 3916 / 1958 / "Bondei: for syphilis, roots boiled and the water drunk" / "Bondei: Mnyembeuwe"
Tanganyika / R. E. S. Tanner 3485 / 1957 / "Bondei: For stomach pain, roots boiled and the water drunk when cool" / "Bondei: Mnyembuewe"
Tanganyika / R. E. S. Tanner 3210 / 1956 / "Leaves used as vegetable" / "Bondei: Mnyembeuwe"

82   *E. menyharttii* Tanganyika / R. E. S. Tanner 1301 / 1953 / "Sukuma tribe: Roots pounded and eaten with honey to treat cough; sap from leaves used to treat snake poison in eye" / "Kisukuma: Soleleja"

83   *Euphorbia fulva* Mexico / G. B. Hinton 731 / 32 / "The sap is milky., and is used as an adhesive on the bandages with which broken bones are tied, whether of men or animals"
Mexico / R. McVaugh 17307 / 1958 / "juice copious, milky, said to cause blindness" / "called by boys 'Escalón' "

84   *E. heterophylla* Guatemala / R. Pinkus 1158 / 1945 / "Plant juices used lactogenetically by natives" / "Ixbut"

85   *E. hirta* Mexico / C. D. Mell s.n. / 1923 / "The Indians crush it and prepare a tea to cure internal disorders" / " 'Memella' "
Tanganyika / R. E. S. Tanner 4049 / 1959 / "For styes in the eye, sap squeezed into the eye" / "Kizanaki: Kinymaurere"
Mariana Is. / F. R. Fosberg 24968 / 1946 / " 'golondrina' used with 'abas' or Psidium guajava and Phyllanthus niruri for vaginal astringent"

86   *E. radians* Mexico / G. B. Hinton 3483 / 33 / "eaten cooked"
Mexico / G. B. Hinton 666 / 32 / "Young buds are mashed and eaten" / " 'Cuaresma' "

87   *E. segoviensis* Mexico / Y. Mexia 1272 / 1926 / "Thick milky juice. Said to be remedy for bite of arlomo" / " 'Yerba de Arlomo' "

88   *E. steyermarkii* Guatemala / J. A. Steyermark 50777 / 1942 / "Reputed to be effective in curing pain arising from broken bones or ligaments; use the crude milky sap in the plant to effect cure" / " 'contra ratura' "

2289   *E. buxifolia* W.I.: Cayman Is. / W. Kings C-B-26 / 38 / "White latex—used for warts warts made to bleed & latex applied" / "Tittie Molly"

2290   *E. petiolaris* Puerto Rico / Little, Jr., & Wadsworth 16472 / 1954 / "Latex . . . poisonous"

2291   *E. orbiculata* Colombia / J. M. Duque Jaramillo 2737 / 1946 / "Medicinal" / " 'Teología' "

2292   *E. papillosa* Paraguay / P. Jorgensen 1676 / no year / ". . . poisonous to cattle"

2293   *E. thymifolia* Brazil / Prance, Maas, Woolcott, Monteiro & Ramos 15595 / 1971 / "Latex used as cure for eye infections by Makú" / " 'Tubiden' (Makú)"

2294   *Euphorbia* indet. Argentina / J. E. Montes 14740 / 1955 / "Uso: medicinal?" / " 'Rompepiedra' "

2295   *E. enormis* AFR. / W. G. Barnard 357 / 1935 / ". . . used medically" / " 'Sahlokoane' "

2296   *E. grantii* Tanganyika / R. E. S. Tanner 1303 / 1953 / "Sukuma tribe: Roots pounded, mixed with porridge to treat constipation" / "Kisukuma: Sululu"

2297   *E. systyloides* Tanganyika / R. E. S. Tanner 2007 / 1955 / "Seeds rubbed on skin infected with ringworm" / "Kizigua: Lubasasuwa"

2298   *E. tirucalli* Tanganyika / R. E. S. Tanner 1147 / 1952 / "Sukuma: for constipation—small pieces of meat liberally covered with sap and cooked. 2–3 pieces causes diarrhoea. Planted with tobacco plants, symbolically to encourage it to have strong flavor" / "Kisukuma: Inala"

2299   *E. chrysocoma* China: Kwangsi / R. C. Ching 6097 / 1928 / "Cultivated by Miu used as medicine"

2300   *E. drummondi* S. Australia / M. Koch 63 / no year / "Poison weed"

2301   *E. atoto* Philippine Is. / M. Ramos 1013 / 1930 / "poisonus plants for bad the age"

2302   *Flueggea microcarpa* Tanganyika / R. E. S. Tanner 3115 / 1956 / "For bilharzia, roots boiled and the water drunk" / "Bondei: Kwamba"

2303   *Garcia nutans* Colombia / W. A. Archer 430 / 1930 / "Said to be poisonous" / " 'Avellano' "

2304   *Gelonium zanzibariense* Tanganyika / R. E. S. Tanner 3188 / 1956 / "Zigua: For stomach ache, gonorrhoea and bilharzia, roots boiled and the water drunk" / "Zigua: Mdimu mbago"

2305   *Glochidion macrophyllum* Hainan / C. I. Lei 310 / 1932 / "Boil ground leaves to dye nets" / "Ngai Goong Tsou"

2306   *G. oblatum* Hainan / C. I. Lei 366 / 1933 / "fruit poisonous" / "Shui Lou"

)7   *G. cordatum* Fiji / A. C. Smith 8624 / 1953 / "leaves chewed for stomach troubles and other ailments" / " 'Molau vulua' "

)8   *G. marianum* Guam / P. Nelson 139 / 1918 / "Sometimes eaten" / " 'abas duendes' "

)9   *Glochidion* indet. Philippine Is. / A. L. Zwickey 657 / 1938 / "frts. edible raw" / " 'Karimog' (Lan.)"

10   *Grimmeodendron jamaicense* Jamaica / G. R. Proctor 9945 / 1955 / "Tree . . . with latex. 'Burn-eye' "

11   *Hieronyma colombiana* Colombia / J. M. Idrobo 2222 / 1956 / "Los frutos comestibles (se venden en Pasto en julio)" / " 'Motilón' "

12   *Hieronyma* indet. Ecuador / Grubb, Lloyd, Pennington & Whitmore 1221 / 1960 / "Fruits black when ripe, favoured by humans and birds; rather insipid drupes" / "Motilon"

13   *Homalanthus alpinus* Philippine Is. / A. D. E. Elmer 8759 / 1907 / " 'Buta'—The green smoke is said to make the eyes run unusually"

14   *H. stokesii* Austral Is. / J. P. Chapin 913 / 1934 / "Fruit eaten by green pigeons"

15   *Homalanthus* indet. Samoa / A. J. Eames 127 / '21 / "Lvs. used for dressings in circumscission and on other wounds" / "Fogamamala"

16   *Hura crepitans* Colombia / R. Romero-Castañeda 9716 / 1963 / " 'Ceiba de leche' "
Brazil / Prance, Maas, Woolcott, Monteiro & Ramos 16342 / 1971 / "The latex & inner bark put on gums & in cavities as a cure for toothache by Deni Indians" / " 'Asacú' (Port.) 'Inupupu' (Ina = tooth)(Deni)"
Peru / F. R. Fosberg 29034 / 1947 / "latex said to be poisonous on contact, especially to eyes. Used as fish poison. Fish said to sink when killed by this plant" / " 'catahua' "
Peru / Killip & Smith 25397 / 1929 / "juice in petioles & stems used as syphilis cure"

\$17   *Hymenocardia mollis* Tanganyika / R. E. S. Tanner 1373 / 1953 / "Sukuma tribe: Roots pounded, cooked, used in porridge to treat problems of lactation" / "Kisukuma: Lukarangwa"

\$18   *Jatropha arizonica* Mexico / Parry, Bigelow et al. 1321 / no year / "known to the Mexicans as possessing purgative properties"

\$19   *J. sessiliflora* Mexico / Parry, Bigelow et al. s.n. / 1852 / "Juice by chewing in use to clean or preserve the teeth & to clean the hair"

\$20   *J. urens* Brazil / A. Silva 113 / 1944 / "The latex is capable of burning the skin to produce serious sores" / " 'ortiga branca' "
Colombia / H. H. Smith 1471 / 1898–99 / "The poisonous hairs produce great irritation of the skin which may last for several days"

2321 *J. curcas* Tanganyika / R. E. S. Tanner 584 / 1952 / "Leaves used for snakebite, sap for sore eyes, roots for sore gums, oil from seeds for lamps" / "Kisukuma: Ihale"

2322 *Julocroton* indet. Paraguay / F. Rojas 12633 / 1945 / 'tóxica para las ovejas"

2323 *Mallotus contubernalis* China: Kwangtung / T. W. Tak 16 / 31 / "Drug plant" / "Wong Ip Fa"

2324 *Manihot grahami* U.S.: La. / Rogers, Thieret & Reese 498 / 1963 / "Also said to have insect-repellant qualities for those who sit in its shade. Plant has strong odor of HCN in all parts"

2325 *M. leptophylla* Peru / Carneiro & Dole s.n. / 1960 / "Sometimes used as food" / " 'yushiwá asti' = spirit's manioc, Amahuaca Indian name"

2326 *M. peruviana* Peru / Carneiro & Dole s.n. / 1960 / "Sometimes used as food" / " 'yushiwá asti' = spirit's manioc. Amahuaca Indian name"

2327 *M. pilosa* Brazil / Y. Mexia 4109 / 1929 / "Thick milky juice. Said to be poisonous" / " 'Mandioca bravo' "

2328 *Manniophyton africanum* Liberia / Daniel & Okeke 25 / 1950 / "dried leaves powdered and put on sores"

2329 *Mareya micrantha* Nigeria / P. W. Richards 3193 / 1935 / "poisonous" / "OJOJO. OMODE. KOTAKE (Yoruba)"

2330 *Margaritaria nobilis* Peru / Y. Mexia 6140 / 1931 / "green fruit eaten by Indians" / " 'Loroñaúe' "

2331 *Micrandra sprucei* Venezuela / Schultes & López 9284 / 1947 / "Seed said to be eaten"

2332 *Phyllanthus grandifolius* Brit. Honduras / P. H. Gentle 204 / 1931–32 / "medicinal"

2333 *P. nobilis* Mexico / R. McVaugh 15868 / 1957 / "fruit green, acid, said to be eaten when immature for this quality" / " 'Barudo' "

2334 *P. acidus* Puerto Rico / P. R. Jarru (?) 11 / 1911 / "taste is like a green gooseberry" / "called 'cherries' and 'gooseberries' "

2335 *P. brasiliensis* Brit. Guiana / Aelson (?) 449 / 1926 / "a fish poison"
Fr. Guyana / R. A. A. Oldeman B-775 / 1966 / "Usage: Utilisation comme poison de peche" / "Nom creole: nivoué"
Brazil / Prance, Maas et al. 15556 / 1971 / "Leaves mashed for fish poison by Makú Indians" / " 'Cantibuã' (Makú)"

2336 *P. orbiculatus* Brazil / Prance, Maas et al. 15853 / 1971 / "Used for kidney infections"

2337 *P. piscatorum* Colombia / W. A. Archer 3416 / 1935 / "leaves used as fish poison . . ." / " 'Tinta Barbasco' "

38 *P. pseudo-conami* Peru / F. Woytkowski 5019 / 1958 / "Toxic—used for fishing" / " 'Culi' "
Peru / J. Schunke V. 4841 / 1971 / "Los indios usan las hojas para pescar, en quebradas de poca agua" / " 'Cúli' "

39 *Phyllanthus* indet. Brazil / G. & L. Eiten 10773 / 1970 / "Used for acid stomach & heartburn, also for kidney stones. Chew leaves and drink water, or make tea" / " 'quebra pedra' "

40 *P. discoideus* Tanganyika / R. E. S. Tanner 3774 / 1957 / "Zigua: For heart burn, roots boiled and the water drunk" / "Zigua: Mchakuzi"

41 *P. maderaspatensis* Tanganyika / R. E. S. Tanner 1431 / 1953 / "Sukuma tribe: Used to treat scabies—whole plant pounded, made into solution to wash in" / "Kisukuma: Igagate igosha"
Tanganyika / R. E. S. Tanner 3416 / 1957 / "Zigua: For constipation, roots boiled and the water drunk" / "Zigua: Luhahaso"

42 *P. muelleranus* Tanganyika / R. E. S. Tanner 3022 / 1956 / "Zigua: Roots pounded up and drunk in water, as cure for diarrhoea" / "Zigua: Mkwambamazi"

43 *P. reticulatus* Tanganyika / R. E. S. Tanner 1478 / 1953 / "Sukuma tribe: Roots pounded, taken with water to treat headache" / "Kisukuma: Minzwandimi"
Tanganyika / R. E. S. Tanner 3720 / 1957 / "Zigua: Male aphrodisiac, roots boiled and the water drunk" / "Zigua: Mkwambasi"
Tanganyika / R. E. S. Tanner 3200 / 1956 / "Zigua: To increase fertility, roots boiled in water and drunk" / "Zigua: Mkamba Mazi"

44 *P. pectinatus* Borneo / J. & M. S. Clemens 22496 / 1929 / "used by Chinese for preserves"

45 *Phyllanthus* indet. S. Pacific: Halmaheira Is. / G. A. L. de Haan 1729 / 1949 / "Use by natives as a depurative" / "Gosao madina (Sawai)"

46 *Pseudolachnostylis maprouneifolia* Tanganyika / R. E. S. Tanner 2788 / 1956 / "Zigua: For stomach-ache; boiled and drunk when cool. Fruits edible" / "Kizigua: Msempele"

47 *Pycnocoma zenkeri* Tanganyika / R. E. S. Tanner 3960 / 1958 / "Zigua: For delay in delivery of afterbirth, the roots boiled and the water drunk" / "Zigua: Mbelubelu"

48 *Richeria* indet. Colombia / H. García y Barriga 18044 / 1964 / "Frutos comestibles, de color morado cuando maduros" / " 'MULATO' "

49 *Sapium laurocerasus* Puerto Rico / Stimson & Montalvo 3746 / 1966 / "Native states that sap from tree very poisonous. Collector felt no discomfort when sap touched his skin but others are reported to be subjected to 'terrible itching' when exposed"

2350   *S. pallidum* Brazil / W. A. Archer 8361 / 1943 / "Seed eaten by fish" / " 'tartaruguinha da praia' "

2351   *Sauropus androgynus* Sumatra / W. N. & C. M. Bangham 612 / 1931 / "Natives cook and eat leaves. Very sweet"

2352   *Sauropus* indet. Sumatra / R. S. Boeea 5800 / 1933 / "leaves edible" / "nasi-nasi"

2353   *Savia sessiliflora* Venezuela / H. Pittier 12120 / 1926 / "fruit edible"

2354   *Sebastiana bilocularis* Mexico / P. V. LeRoy s.n. / '84 / "used by Apache Indians to poison their arrows / " 'Yerba de fleche' "

2355   *S. longicuspis* Brit. Honduras / P. H. Gentle 6755 / 1949 / " 'White poison wood' "

2356   *S. standleyana* Brit. Honduras / P. H. Gentle 5292 / 1945 / " 'White Poisonwood' "

2357   *S. obtusifolia* Peru / P. C. Hutchinson 1091 / 1957 / "poisonous. Juice blinds" / " 'Chiche' "

2358   *Spirostachys africana* S.W. Africa / R. J. Rodin 2674 / 1947 / ". . . fish toxicant" / "Ovambo name: 'Omuhongo' "

2359   *Stillingia acutifolia* Guatemala / L. M. Andrews 512a / 1965 / "milky juice, poisonous"

2360   *Victorinia regina* Cuba / Hno. Leon 11912 / 1924 / "Latex abundante e irritante" / " 'Sabrosa' "

2361   *Euphorbiacea* indet. Honduras / C. & W. von Hagen 1031 / 1937 / "fruit eaten by Quetzal bird" / " 'Aguacatillo' "

2362   *Euphorbiacea* indet. Ecuador / C. Fuller 88 / 52 / "Crushed pod intensely stains hands a lasting dark blue. Juice in eyes, burning & painful. Nuts cracked, meats toasted as food . . . Remindful of roasted peanuts" / "Hauchansi, Quichua; Maní del monte, Span."

2363   *Euphorbiacea* indet. Ecuador / C. Fuller 63 / 49 / "Leaf type barbasco used crushed as fish poison" / "Cajali, Jivaro"

2364   *Euphorbiacea* indet. Peru / Killip & Smith 27502 / 1929 / "Tea brewed from leaves used for headache" / " 'Toque' "

2365   *Euphorbiacea* indet. Peru / Killip & Smith 27474 / 1929 / " 'Corteza de oje medicina"

2366   *Euphorbiacea* indet. Brazil / Prance, Maas, Woolcott et al. 15593 / 1971 / "Root bark scraped, added to water as cure for dysentery" / " 'Ocochugbiden' (Maku)"

2367   *Euphorbiacea* indet. Argentina / J. E. Montes 27732 / 58 / "medicina popular" / " 'Rompepiedras' (Variedad)"

68 *Euphorbiacea* indet. Argentina / J. E. Montes 14765 / 1955 / "Raíz medicina popular" / " 'Belame colorado' "

69 *Euphorbiacea* indet. N. Rhodesia / W. R. Bainbridge 432 / 1960 / "Latex said to irritate eyes" / "SIBWALAKATA (Valley Tonga name)"

70 *Euphorbiacea* indet. Tanganyika / R. E. S. Tanner / 3941 / 1958 / "Sambaa: for painful eyes, root juice squeezed into them" / "Zigua: Msofu"

71 *Euphorbiacea* indet. China: Kwangsi / R. C. Ching 7682 / 1928 / "edible when cooked"

DAPHNIPHYLLACEAE

72 *Daphniphyllum calycinum* Hainan / C. I. Lei 61 / 1932 / "Leaves emersed in wine may heal wounds" / "Ngou Sun Fung"

CALLITRICHACEAE

73 *Callitriche deflexa* var. *subsessilis* Colombia / F. R. Fosberg 19852 / 1943 / "Infusia made with corn silk and Equisetum for liver" / " 'berro' "

CORIARIACEAE

74 *Coriaria thymifolia* Ecuador / F. Prieto E-2488 / 1945 / "fruits . . . eaten by birds and sometimes by small children. If the children eat a considerable amount they 'act like drunken persons' — i.e. it has some narcotic effect" / " 'Pinan' "

75 *C. nepalensis* Punjab / R. R. Stewart 1591 / 1917 / "a child died after eating the fruit others recovered" / " 'masuri' "

EMPETRACEAE

76 *Empetrum rubrum* Africa / R. A. Dyer 3548 / 37 / "berries sour, used for puddings" / "Island berry"

ANACARDIACEAE

77 *Allospondias chinensis* China: Kwangsi / R. C. Ching 7593 / 1928 / ". . . seed . . . edible"
Hainan / H. Fung 20065 / 1932 / "fr. edible"
Hainan / S. K. Lau 285 / 1932 / "Used to make bed, for the fleas cannot live there" / "Sai Sa Dack (Lai)"

78 *Anacardium pumilum* Brazil / Prance, Silva & Pires 59086 / 1964 / "Edible, sweet"

2379   *Buchanania nitida* Philippine Is. / M. Oro 172 / 1919 / "Fruit . . . said to be edible"

2380   *Cotinus coggygria* Punjab / W. Koelz 1824 / 1931 / "wood used for cleaning teeth" / "Kulu name 'Tunt' "

2381   *Dracontomelum vitiense* Fiji / A. C. Smith 943 / 1934 / "fruit edible" / " 'Tarawau ndina' "

2382   *Heeria insignis* Tanganyika / R. E. S. Tanner 3343 / 1956 / "Zigua: For diarrhoea, roots boiled and the water drunk" / "Zigua: Msempele"

2383   *H. reticulata* Tanganyika / R. E. S. Tanner 1087 / 1952 / "Sap used as a gum. Sukuma: for poisoning by witchcraft; roots cut up, pounded green, dried, ground down and cooked in food and eaten" / "Kisukuma: Kalakala"

2384   *Lannea fulva* Tanganyika / R. E. S. Tanner 1213 / 1953 / "Fruits edible. Sukuma tribe: Bark chewed for cough" / "Kisukuma: Sellya"

2385   *L. stuhlmannii* Tanganyika / R. E. S. Tanner 3345 / 1956 / "Sambaa: For sore gums, bark boiled and the water rinsed through mouth" / "Sambaa: Mumbu"
Tanganyika / R. E. S. Tanner 3903 / 1957 / "Bondei—for ulcers in nose (yaws?), roots boiled and the water drunk. Soft hair on root smoked via cigarette for same disease" / "Bondei—Mumbu"
Tanganyika / R. E. S. Tanner 3763 / 1957 / "Bondei: For asthma, roots boiled and the water drunk" / "Bondei: Mumbu"

2386   *Loxopterygium sagotii* Venezuela / J. A. Steyermark 86346 / 1960 / "Sap considered irritating, therefore name for tree" / " 'picatón' "
Brit. Guiana / Jenman 4094 / 1887 / "Milk so acid that it will blister outer (?) skin like fire" / " 'Hoobaballi' "

2387   *Metopium linnaei* U.S.: Fla. / H. N. Moldenke 778a / 1930 / "Juice poisonous" / " 'Poisonwood' "

2388   *Pistacia chinensis* China: Hupeh / S. C. Sun 35 / 1932 / "From the roots an infusion is derivated for medicine"

2389   *Pseudosmodingium perniciosum* Mexico / G. B. Hinton 1993 / 32 / "Fair people waking under this tree get a rash" / "Cuajiote"

2390   *Rhus natalensis* Tanganyika / R. E. S. Tanner 1328 / 1953 / "Sukuma tribe: Bark ground, used as poultice to treat sores" / "Kisukuma: Sese bulika"

2391   *R. chinensis* China: Kwangsi / W. T. Tsang 24155 / 1934 / "fr. edible" / "Im Sheung Pak Shue"

2392   *Schinus weinmannifolius* Paraguay / W. A. Archer 4929 / 1937 / "Used as gargle and to wash wounds"

2393   *Schinus* indet. Argentina / J. E. Montes 27652 / 58 / "usada en medicina popular" / " 'Molle colorado' "

2394 **Schinus** indet. Argentina / J. E. Montes 27440 / 1957 / "cortesa usada en medicina popular" / " 'Molle gris' "

2395 **Schmaltzia trilobata** U.S.: Ariz. / W. N. Clute 105 / 1920 / "Indians use the berries" / " 'Squaw berry'; 'Lemonade berry' "

2396 **Sclerocarya birrea** Tanganyika / R. E. S. Tanner 1266 / 1953 / "Sukuma tribe: Roots ground up, mixed with water, drunk to treat Bilharzia, used to bathe in for scabies" / "Kisukuma: Ngongwa"

2397 **Semecarpus atra** PACIFICA: Berar / G. H. Wittenbaker 19 / 1889 / ". . . resembling almonds edible nuts and tasty much eaten, an oil is expressed from them" / "Seeds called Chironji; Char or Chironji"

2398 **S. merrilliana** Philippine Is. / A. D. E. Elmer 7117 / 1906 / "poisonous causing large red itching blotches in the skin which is very irritable"

2399 **S. vitiensis** Fiji / A. C. Smith 9676 / 1953 / "the leaves containing a very poisonous oil" / " 'Kaukaro' "

2400 **Sorindeia madagascariensis** Tanganyika / R. E. S. Tanner 3138 / 1956 / "Sambaa: For sharp stomach pains, roots cooked with maize and eaten" / "Bondei: Mbwakabwaka; Zigua: Mkwelengala"

2401 **Spondias mombin** Mexico / Hinton et al. 6599 / 34 / "Fruit edible" / "Chacumo"
Panama / T. B. Croat 12291 / 1970 / "being eaten by spider monkeys"
Peru / A. G. Ruiz 162 / 1964 / "frutos comestibles" / " 'Itahuba' "

2402 **S. axillaris** China: Kwangsi / Ching & Chun 5803 / 1928 / "Fruit . . . edible when ripe"

2403 **S. pinnata** Hainan / S. K. Lau 1614 / 1933 / "fruit, edible" / "Pan Long Ching"
Philippine Is. / R. S. Williams 498 / 1904 / "Fruit cooked with fish by natives"

2404 **Swintonia schwenckii** Siam / H. S. Cunniff 97 / 1931 / "poison"

2405 **S. elmeri** Borneo / native collector 5230 / 1927 / "skin poisonous" / "Kraus or Rengas utan (Malay)"

2406 **Tapiria guianensis** Venezuela / J. A. Steyermark 60729 / 1944 / "Fruit edible" / " 'acorga-yek' "

2407 **Thyrsodium spruceanum** Brazil / B. A. Krukoff 5017 / 1933 / "Seeds are edible" / "Castanha de Porco"

2408 **Toxicodendron lobatum** U.S.: Calif. / J. H. Redfield 45 / 1872 / " 'Poison-Oak' "

2409 **T. radicans** subsp. **radicans** U.S.: N.J. / H. N. Moldenke 1661 / 1931 / "poisonous to touch" / " 'Poison-ivy' "

2410 **T. toxicarium** U.S.: Penn. / Herb. Collegii Pharmaciae Neo-Eboracensis s.n. / 1861 / "Poison vine"

2411   *T. vernix* U.S.: Wis. / J. H. Schuette s.n. / 36 / "Poisonous to the touch for most persons. Juice poisonous" / "Poison Sumach or Dogwood"

2412   *T. striata* Costa Rica / Burger & Ramirez B. 3980 / 1966 / "Tree caused severe facial swelling of Wm. Ramirez who apparently got some resin near his eye"
Honduras / C. & W. von Hagen 1033 / 1937 / "Not used by natives who fear its sap which is supposed to burn" / " 'Palo sarrno' "

2413   *Anacardiacea* indet. Honduras / C. & W. von Hagen 1164 / 1937 / "fruit edible" / "Joco-mico"

2414   *Anacardiacea* indet. Argentina / C. L. Schulz 614 / 45 / "La raíz se usa como tonico" / "Quena"

AQUIFOLIACEAE

2415   *Ilex coriacea* U.S.: Va. / Fernald & Long 13677 / 1941 / "the fruit said . . . to be edible" / "Sweet Gallberry"

2416   *I. asprella* China: Kwangtang / W. T. Tsang 21752 / 1932 / "root is edible" / "Fu Moot Kan"

2417   *I. rotunda* Hainan / C. I. Lei 1 / 1932 / "The bark is used to heal stomach ache" / "Paak Ngaan"

CELASTRACEAE

2418   *Elaeodendron capensis* Tanganyika / R. E. S. Tanner 3677 / 1957 / "Zigua: For sores, leaves pounded up and pasted on green"

2419   *Gymnosporia senegalensis* Tanganyika / R. E. S. Tanner 960 / 1952 / "Sukuma tribe: leaves used to treat fever by pounding into flour, putting in water, person bathing in this" / "Kisukuma: Luweja"

2420   *G. undata* Tanganyika / R. E. S. Tanner 1293 / 1953 / "Sukuma tribe: Roots and bark boiled, steam used to treat fever" / "Kisukuma: Ngazu; Kiukerewe: Nyabwele"

2421   *Maytenus* indet. Brazil / R. Fróes 20995 / 1945 / "cascas usadas para sifilis" / " 'chichuasca' "

2422   *M. senegalensis* Tanganyika / R. E. S. Tanner 3638 / 1957 / "Zigua: For sores, leaves put on sore green" / "Zigua: Mngamea"
Tanganyika / R. E. S. Tanner 2072 / 1955 / "Leaves pounded and used as poultice on sores"

2423   *M. thompsonii* Mariana Is. / F. R. Fosberg 25354 / 1946 / "used medicinally" / " 'lalukut' "
Mariana Is. / F. R. Fosberg 24981 / 1946 / "tips used as medicine" / " 'luluhod' "

2424   *Celastracea* indet. Bolivia / Scolnik & Luti 612 / 1947 / "fruto comestible"

HIPPOCRATEACEAE

2425   *Cheiloclinium cognatum* Peru / J. Schunke V. 1544 / 1967 / "Corteza medicinal" / " 'Chuchuhuasha' "

2426   *Hemiangium excelsum* Guatemala / J. A. Steyermark 51013 / 1942 / "Reputed to be used in killing fleas in the human hair" / " 'matapioja' "

2427   *Salacia oblonga* Ceylon / N. Balakrishnan NBK 1057 / 1971 / "green fruit edible"

2428   *S. prinoides* Sarawak / J. W. Purseglove P.4961 / 56 / ". . . orange-red fruits . . . with sweet edible flesh"

STAPHYLEACEAE

2429   *Euscaphis japonica* China: Szechwan / W. P. Fang 2138 / 1926 / "The buds may be eaten as food"

2430   *Turpinia arguta* China: Kwangtung / F. Hom A290 / 1930 / "Medicine" / "Shang Yuk Yauk"

ICACINACEAE

2431   *Apodytes dimidiata* Tanganyika / R. E. S. Tanner 3909 / 1957 / "Bondei: for pain in left side of stomach, roots boiled and the water drunk" / "Bondei — Msungura"

2432   *Poraqueiba paraensis* Brazil / B. A. Krukoff 7064 / 1934 / "fruit . . . edible" / " 'Mary' "
Brazil / W. A. Archer 7740 / 1942 / "frt. edible" / " 'Mari' "

2433   *P. sericea* Brazil / R. Fróes 21169 / 1945 / "frutos comestiveis"

2434   *Icacinacea* indet. Melanesia / J. H. L. Waterhouse 87 / 1932 / "Leaf edible, used as spinach" / "Rarui"

ACERACEAE

2435   *Acer nigrum* U.S.: N.Y. / O. P. Phelps 1175 / 1915 / "The 'black maple' is much prized by owners of sugar bushes, it being thought to produce sweeter sugar and finer than the ordinary maple" / " 'black maple' "

HIPPOCASTANACEAE

2436   *Billia columbiana* Colombia / Idrobo & Jaramillo 1594 / 1954 / "En fruto. Semilla farinosa but amarga. La cocción es similar a la castaña europea" / " 'Manzano' "

Venezuela / Steyermark & Fernández 99613 / 1967 / "Seed when dried placed in infusion, used as drink to help mothers in child birth" / " 'cobalonga' "

SAPINDACEAE

2437 *Allophylus edulis* Argentina / T. Meyer 3425 / 1940 / "fruto comestible" Argentina / J. E. Montes 14770 / 1955 / "hojas medicinales: Frutos comestibles" / " 'Conhcú' 'Cokú' 'Picarú-Rembiú' "

2438 *A. alnifolius* Tanganyika / R. E. S. Tanner 1216 / 1953 / "Sukuma tribe: Roots used to treat gonorrhea" / "Kisukuma: Mpilimisha"
Tanganyika / R. E. S. Tanner 3318 / 1956 / "Bondei: For swellings leaves pounded up green and put on as a poultice" / "Bondei: Mbangwe"

2439 *A. fulvo-tomentosus* Tanganyika / R. E. S. Tanner 1256 / 1953 / "Sukuma tribe: Roots used to treat diarrhoea, sharp stomach pains, syphilitic sores" / "Kisukuma: Mpalanyonga ngosha"

2440 *A. viridus* Hainan / S. K. Lau 454 / 1932 / "Use to cure stomachache by cooking the root" / "Jia lun"

2441 *A. timorensis* New Hebrides / S. F. Kajewski 309 / 1928 / "Leaves baked in native oven by means of hot stones & when steaming applied to swellings" / "Nau-nom-pe-val"

2442 *Allophylus* indet. Indonesia / G. A. L. de Haan 1784 / 1950 / "Pulped leaves used against sprain" / "Kohilèwès (Weda)"

2443 *Cardiospermum halicacabum* Brit. Guiana / H. A. Gleason 567 / 1921 / "An Indian fish poison"

2444 *Cupania latifolia* Brit. Honduras / S. J. Record 8803 / 26 / "Leaves medicinal" / "Grande Betty"

2445 *C. vernalis* Brazil / Y. Mexia 5219 / 1930 / "fruits fed upon by large parrots"

2446 *Cupaniopsis serrata* Australia / S. F. Kajewski 1258 / 29 / "Bark medium grey and when cut has a faint odour resembling that of a cut watermelon"

2447 *C. leptobotrys* Fiji / O. Degener 15371 / 1941 / "Bark squashed in water—drink liquid for stomach trouble" / " 'Malawathe' (Ra)"

2448 *Deinbollia borbonica* Tanganyika / R. E. S. Tanner 3723 / 1957 / "Zigua: For pain in spleen, roots boiled and water drunk" / "Zigua: Mbwakabwaka"

2449 *Dialiopsis africana* Tanganyika / R. E. S. Tanner 1219 / 1953 / "Sukuma tribe: Used with Acacia Pennata in treating sick goats—roots ground up, spread on ground where goats sleep" / "Kisukuma: Nkewe"

2450 *Dipterodendron venezuelense* Venezuela / Bernardi 2093 / 1955 / "semillas con pulpa externa blanco-vidriosa, agri-dulce, comestibles" / "Machirió tamarindo"

2451 *Erioglossum rubiginosum* Hainan / H. Fung 20030 / 1932 / "fr. red, edible"

2452 *Euphoria gracilis* Philippine Is. / N. Santos 31321 / 1930 / "fruit—edible, sweet" / "Alupagamo (Dialect Y)"

2453 *Euphoria* indet. Borneo / A. Kostermans 13326 / 1957 / "Fruit . . . sweet, edible" / " 'Buku' "

2454 *Euphoria* indet. Borneo / A. Kostermans 13625 / 1957 / ". . . seedcoat jelly like, sweet, edible" / " 'Djuring' (Bassap)"

2455 *Litchi philippinensis* Philippine Is. / L. Logan 4 / 1930 / "edible" / "Baet"

2456 *Magonia pubescens* Brazil / Y. Mexia 5618 / 1931 / "Cinnamon-brown fruits . . . used to make soap" / " 'Tingui' "

2457 *Matayba guianensis* Venezuela / J. A. Steyermark 61390 / 1945 / "Fruit makes a soapy lather used for washing (outside part of fruit); fruit poisonous or rather makes one 'loco' if eat it; . . . scratch up bark until white and put over cuts in conjunction with aceite de castilla" / " 'para-para' "

2458 *Matayba* indet. Argentina / J. E. Montes 755 / 45 / "Fr. rojos comestibles"

2459 *Melicocca lepidopetala* Paraguay / T. M. Pedersen 9419 / 1969 / "fruit green, the pulpy arillus edible"

2460 *Melicocca* indet. Argentina / T. M. Pedersen 1833 / 1952 / "fruit edible" / " 'Coq vibo de Sem Juan' "

2461 *Nephelium tuberculatum* Malaysia / M. R. Henderson 22107 / 1929 / "Fruit edible" / "Nati, name 'Mutan Sipuk' "

2462 *Nephelium* indet. Singapore / Sinclair & Kadim 10447 / 1960 / "Fruit red, sweet, edible . . ." / "Pulasan hutan"

2463 *Pappea capensis* Tanganyika / R. E. S. Tanner 1231 / 1953 / "Sukuma tribe: Roots used in treating ankylostomiasis—made into a flour which is put in food" / "Kisukuma: Ntanguta"

2464 *Paullinia pachycarpa* Brazil / B. A. Krukoff 1687 / 1931 / "wood (stem) is used as a fish-poison" / " 'Timbo preto' "

2465 *P. tarapotensis* Peru / Y. Mexia 6297 / 1931 / "fruit green . . . when mature pulp is eaten" / " 'Ycánchem', (Indian), 'Lúcumia' "

2466 *Paullinia* indet. Colombia / G. Klug 1937 / 1931 / " 'Taruca (poison) yoco' "

2467 *Serjania glabrata* Peru / Killip & Smith 25401 / 1929 / "Stems mashed and thrown in water to stupify fish" / " 'Verap' "

2468 *S. leptocarpa* Brazil / B. A. Krukoff 7880 / 1936 / "used as fish poison by Indians Parintintins" / "Tirbo—pucú"

2469 *S. paucidentata* Brit. Guiana / A. C. Smith 3589 / 1938 / "stems used as fish poison"

2470 *S. rubicaulis* Peru / J. Schunke V. 2221 / 1967 / "El tallo y las hojas utilizan en infusión para lavado vaginal" / " 'Cashahuasca' "

2471 *S. rubicunda* Peru / Killip & Smith 25376 / 1929 / "Stems mashed and thrown into water to stupify fish" / " 'Verap' "

2472 *Serjania* indet. Peru / J. Schunke V. 3354 / 1969 / "Toda la planta es tóxica. Los nativos lo utilizan para matar peces en las quebradas" / " 'Macota blanca' "

2473 *Talisia olivaeformis* Mexico / C. L. Lundell 1376 / 1932 / "edible fruit" / " 'Guaya' "
Guatemala / Ortíz 90 / 1965 / "Frutos comestibles" / "Guaya"
W.I.: Trinidad / W. E. Broadway 5230 / no year / "Fruits yellow, edible" / " 'Yellow chenip' "
Venezuela / Curran & Haman 781 / 1917 / "Edible fruit"

2474 *T. cavinata* Guyana / Granville 376 / 1970 / "Fruits oranges . . . comestible" / "Créole: 'bois flambeau'; Oyampi: 'touliatan' "

2475 *T. cerasina* Brazil / Harley, Souza et al. 10982 / 1968 / "Fruit yellow with single seed covered with edible juicy pale pink flesh" / " 'Pitomba' "

2476 *T. hexaphylla* Venezuela / J. A. Steyermark 86624 / 1960 / "fruit . . . yellow with edible sweet white mesocarp" / " 'cotoperiz' "
Venezuela / L. Aristeguieta 5994 / 1966 / "frutos comestibles" / "Cotopalo"

2477 *T. microphylla* Brit. Guiana / For. Dept. of Brit. Guiana F 3402 / 52 / "edible pulp tasting of mangoes"

2478 *T. pulverulenta* Brazil / Maguire, Murça Pires et al. 56058 / 1963 / "fruits edible, slight acid taste, a little astringent"

2479 *T. reticulata* Guyana / C. Blanco 303 / 1965 / "Fruto comestible" / "Cotoperíz montañero"

2480 *T. squarrosa* Brit. Guiana / N.Y. Sandwith 333 / no year / "Has a poisonous sap which is used as a fish-poison"

2481 *Talisia* indet. Bolivia / H. H. Rusby 1409 / 1921 / "Fruit edible"

2482 *Talisia* indet. Colombia / J. A. Duke 11074(3) / 1967 / "fruits yellow, edible (much eaten in Darien)" / " 'mamon de monte' "

2483 *Talisia* indet. Fr. Guiana / Fr. Guiana For. Service s.n. / 1954 / "pulpe entre l'ecorce et le noyau trés comestibles sucre gout de raisin" / "TATOU (Idiome Paramaka)"

2484 *Tristira triptera* Philippine Is. / R. S. Williams 2898 / 1905 / "Fruit mostly inhabited by ants, outer coat eaten by wild pigs"

2485 *Sapindacea* indet. Brit. Guiana / A. C. Smith 2829 / 1937 / "Bark made into tea and drunk to reduce swellings"

2486 *Sapindacea* indet. Brazil / Fróes 2063 / 1933 / "Grease is taken from the seed for soap" / "Jingui (sp.?)"

2487 *Sapindacea* indet. Argentina / J. E. Montes 14881 / 1956 / "Su corteza es medicinal" / " 'Cedrillo' "; 'Catigua-pora' "

BALSAMINACEAE

2488 *Impatiens gigantea* India / W. Koelz 1436 / 1930 / "Seeds considered edible"

RHAMNACEAE

2489 *Alphitonia philippinensis* Hainan / C. I. Lei 373 / 1933 / "fruit poisonous" / "Shan Muk Min"

2490 *Alphitonia* indet. / New Guinea / D. Frodin NGF26629 / 66 / "Odour same as root beer"

2491 *Ampelozizyphus* indet. Brazil / B. A. Krukoff 5812z / 1933 / "Bark of roots is scraped off placed in water and used as a beverage (beer like taste!)" / "Saracuramurá"

2492 *Berchemia franchetiana* China: Kwangtung / W. T. Tsang 21574 / 1932 / "Fr. edible" / "Kau Shi Pak T'ang"

2493 *B. lineata* China: Kwangtung / S. K. Lau 615 / 1932 / "fruit, black, edible" / "Lo Sue Shi"

2494 *Colubrina arborescens* Bahamas / Fr. W. Cerbin C103B / 1973 / "Years ago the plant was used by the island people in place of soap. It becomes sudsy in water and was used in washing floors etc." / " 'Soapbush' "

2495 *C. asiatica* Tanganyika / R. E. S. Tanner 3449 / 1957 / "Zigua: For swellings, leaves boiled and applied as a poultice"
Fiji / O. Degener 15359 / 1941 / "To cure man, Colubrina leaves are taken and squashed with leaves of sama, of lolo, of bombo, of dawa, of wasasala in water. Then give to victim to drink. If Fijian is to be killed, leaves from several kinds of plants (kept secret by informant) are put in bottle or bamboo joint with water. Container is buried secretly in or about victim's house. If he gets sick from this, he drinks cure. This method of murder is not practiced nor known in all districts but only in a few as about Tavua, for example" / " 'Verelailai' (Ra)"
Guam / F. R. Fosberg 25343 / 1946 / "used medicinally for many ailments" / " 'gasoso' "

2496 **Condalia hookeri** Mexico / Gregg s.n. / no year / "fruit black . . . edible — sweet and pleasant" / "capul"

2497 **Gouania lupuloides** Peru / J. Schunke V. 978 / 1965 / " 'Jaboncillo' "

2498 **G. polygama** Peru / F. Woytkowski 6510 / 1961 / "wood has odor of Barbasco (cube), pungent"

2499 **Karwinskia humboldtiana** Mexico / H. S. Gentry 4812 / 1939 / "Decocted as a wash for sores" / "Kakachila"

2500 **K. oblongifolia** Bolivia / H. H. Rusby 1724 / 1921 / "Fruit edible"

2501 **Kentrothamnus weddellianus** Bolivia / W. M. A. Brooke 5114 / 1949 / "An extraction is made for soap and shampoos"

2502 **Noltea africana** S. Africa / R. D. A. Bayliss 6334 / 1974 / "Soap bush"

2503 **Reissekia smilacina** Brazil / Y. Mexia 4248 / 1930 / "greenish flower much visited by small ants. Tea made of leaves for blood purifier" / " 'Amora Lisa' "

2504 **Rhamnidium caloneura** Panama / G. P. Cooper 434 / 1928 / "Wood . . . has odor like peanuts"

2505 **R. elaeocarpum** Paraguay / T. M. Pedersen 4230 / 1956 / "fruit said to be edible ('dulce')"
Brazil / R. Fróes 2050 / 1933 / "Fruit eatible" / "Bitrinho"

2506 **Rhamnus tomentella** U.S.: Ariz. / J. C. Blumer 2317 / 1906 / "berries power fully purgative"
U.S.: Calif. / E. L. Greene s.n. / 1889 / "black, edible"

2507 **R. prinoides** Kenya / Glover, Gwynne et al. 2582 / 1961 / "Roots used for soup" / "Kosisitiet (Kisp.); Orkonyel (Masai)"

2508 **R. procumbens** Punjab / W. Koelz 1935 / 1931 / "attractive plant with disagreeable odor"

2509 **Sageretia elegans** Mexico / G. B. Hinton 7720 / 35 / "Fruit edible" / "Techalo"

2510 **Scutia spicata** Peru / Simpson & Schunke V. 595 / 1968 / "fruit . . . edible" / " 'Negrito' "

2511 **Zizyphus amole** Mexico / G. B. Hinton 3795 / 33 / "Fruit edible" / "Corongoro"

2512 **Z. taylori** W.I.: Caicos Is. / D. S. Correll 43240 / 1974 / "Leaves relished by iguanas"

2513 **Z. joazeiro** Brazil / H. C. Cutler 12439 / no year / "Make a syrup with water and sugar by heating slightly with the chips, then strain to remove chips and boil down the liquid. Used for coughing. the chips are often used as they are in washing hair and for dyed materials" / "Juá"

514   *Z. saeri* Venezuela / Aristeguieta & Zabala 7011 / 1969 / "Los frutos se comen pero producen vómitos" / " 'Chica' "
Venezuela / Aristeguieta & Zabala 7059 / "Es utilizado como estantes, viguetas, soleras" / " 'Limoncillo' "

515   *Z. thyrsiflora* Ecuador / Little, Jr. & Dixon 21165 / 1965 / "Frutos amarillos . . . comestibles" / " 'ébano; tillo' "

516   *Z. abyssinica* Tanganyika / R. E. S. Tanner 1260 / 1953 / "Sukuma tribe: Roots cooked and used mouth wash to treat sores on lips; used as eye wash" / "Kisukuma: Ngwata"
Tanganyika / R. E. S. Tanner 1193 / 1953 / "Sukuma tribe: Roots used to treat toothache — mouth rinsed in water in which roots were boiled" / "Kisukuma: Guguno"

517   *Z. mauritiana* Tanganyika / R. E. S. Tanner 3451 / 1957 / "Zigua: For gonorrhoea, leaves pounded up and squeezed and the fluid drunk" / "Zigua: Mkunazi"

518   *Z. mucronatus* Tanganyika / R. E. S. Tanner 3962 / 1958 / "Zigua: for infertility, roots and/or bark boiled and the water drunk" / "Zigua: Mguguni"

519   *Z. pubescens* Tanganyika / R. E. S. Tanner 3966 / 1958 / "Zigua: for blood in urine, bark boiled and the water drunk" / "Zigua: Mhungulu"

520   *Zizyphus* indet. China: Kiangsi / Y. Tsiang 9865 / 1932 / "frt. green edible"

521   *Z. rugosa* India / R. F. Hohenacker s.n. / 1847 / "Fructus edulis"
India / R. F. Hohenacker s.n. / 1847 / "Fructus acidulos comedunt"

522   *Zizyphus* indet. India / F. Kingdon Ward 18967 / no year / "Fruits . . . Said to be eaten, though there is almost no flesh, and what there is is tasteless"

VITACEAE

523   *Ampelocissus acapulcensis* Mexico / C. L. & A. A. Lundell 7097 / 1937 / "fruits edible, sour"
El Salvador / P. C. Standley 21982 / 1922 / "Fr. used for vinegar" / " 'Uva' "

524   *A. asarifolia* Tanganyika / R. E. S. Tanner 1248 / 1953 / "Sukuma tribe: Leaves used as hot poultice to treat sprains, cuts, as cold poultice for boils" / "Kisukuma: Ngombe ya hasi ngima"

525   *Ampelocissus* indet. Borneo / J. & M. S. Clemens 20387 / 1929 / "wintergreen odor"

526   *Ampelopsis cantoniensis* China: Kwangtung / W. T. Tsang 20629 / 1932 / "Edible" / "Teen Po Chai"

2527 *A. cantoniensis* var. *grossedentata* China: Kwangtung / W. T. Tsang 21352 / 1932 / "fruit, red, edible" / "Tin Poh Ch'a"

2528 *A. barbatus* Indochina / R. W. Squires 912 / 1932 / "fls. maroon, with odor like putrid meat"

2529 *Caryatia geniculata* Philippine Is. / R. S. Williams 2794 / 1905 / "Fruit edible"

2530 *Cissus rhombifolia* Mexico / Y. Mexia 1307 / 1926 / "Fr . . . bitter and acrid to taste"

2531 *C. sicyoides* Mexico / Y. Mexia 8740 / 1937 / "Leaves used on swellings as poultice" / " 'Bejuco de Alquilon' "

2532 *Cissus* indet. Brazil / M. Barbosa da Silva 14 / 1942 / "Plant used in tepid baths for weak or sore legs" / " 'parreira' "

2533 *C. buchananii* Tanganyika / R. E. S. Tanner 3086 / 1956 / "Fruits edible" / "Zigua: Tongo tongo"
Tanganyika / R. E. S. Tanner 2932 / 1956 / "Bondei: for stomach ache around navel. Roots boiled and the water drunk" / "Kibondei: Muuka"

2534 *C. cornifolia* Tanganyika / R. E. S. Tanner 1233 / 1953 / "Sukuma tribe: Pounded roots given in water to cattle to treat diarrhoea and to aid lactation; also in butter making to increase volume" / "Kisukuma: Ngwitankole; Ngombe ya hasi ngosha"

2535 *C. petiolata* Tanganyika / R. E. S. Tanner 1617 / 1953 / "Sukuma tribe: Roots and leaves pounded and used as poultice to treat discharging ears" / "Kisukuma: Matuga mpuli magosha"
Tanganyika / R. E. S. Tanner 1212 / 1953 / "Sukuma tribe: Sap from leaves used to treat discharging ears" / "Kisukuma: Matugampuli"

2536 *Rhoïcissus erythrodes* Tanganyika / R. E. S. Tanner 1556 / 1953 / "Sukuma tribe: Roots used to treat leprosy" / "Kisukuma: Imala"

2537 *R. revoilii* Tanganyika / R. E. S. Tanner 4355 / 1959 / "Roots used on swellings" / "Kizanaki: Irembero"

2538 *Tetrastigma planicaulis* China: Kwangsi / W. T. Tsang 24129 / 1934 / "fr. yellow, edible"

2539 *Vitis bicolor* U.S.: Vt. / H. S. B. 110 / '97 / "Wild Grapes"

2540 *V. coriacea* U.S.: Fla. / J. W. Harshberger s.n. / 1912 / "Grapes edible and sweet"

2541 *V. tiliaefolia* Guatemala / P. C. Standley 23934 / 1922 / "fr. used to make vinegar" / " 'Uva de vinaigre' "
Cuba / C. F. Baker 22 / 1907 / ". . . fruit . . . make first class jelly"

2542 *Vitis* indet. Paraguay / W. A. Archer 4636 / 1936 / "Taste sweet but insipid. Used for wine"

543 *V. flexuosa* China: Kwangtung / W. T. Tsang 21270 / 1932 / "fruit, egg yellow, edible" / "Ye Po Tai Tsz"

544 *V. obovata* China / A. Henry 12051A / no year / "edible"

545 *Vitis* indet. China: Kweichow / Y. Tsiang 7021 / 1930 / "Frt. green, edible"

LEEACEAE

546 *Leea aculeata* Philippine Is. / M. Oro 58 / 1929 / "young leaves eaten as vegetable & as medicine for worms (sp.?)"

547 *L. indica* Fiji / O. Degener 15385 / 1941 / "For lung trouble, rub chest with the leaf" / " 'Naikusanithovu'(Ra)"

ELAEOCARPACEAE

548 *Elaeocarpus lanceaefolius* China: Kwangsi / W. T. Tsang 24177 / 1934 / "fr. black, edible" / "Yeung Sze Nin Shue"
China: Kwangtung / W. T. Tsang 21246 / 1932 / "fr. black, edible" / "Tung to Tse Shue"

549 *E. sylvestris* China: Kwangtung / W. T. Tsang 21249 / 1932 / "fruit, black, edible" / "Sai Ip Tung To Tse Shue"

550 *E. curranii* Philippine Is. / M. Felix 30912 / 1929 / "Fruit edible . . ." / "Calumboya (Ilocano)"

551 *E. pedunculatus* Borneo / J. W. Purseglove P.5059 / 56 / "Blue edible fruits"

552 *Muntingia calabura* Sumatra / W. N. & C. M. Bangham 663 / 1932 / " 'Cherry' natives call it"

553 *Sloanea laurifolia* Venezuela / J. A. Steyermark 61354 / 1945 / "fresh fruit said to cause a loco type of feeling . . . ground up fruit cooked" / "arepa de maiz; 'taque' "

554 *S. pseudodentata* Venezuela / J. A. Steyermark 60188 / 1944 / "The bird 'wará-mi' eats the fruit" / " 'wará-mi-yú-yek' "

555 *S. terniflora* Peru / Y. Mexia 6510 / 1932 / "Seeds fall into water and are food for fish" / " 'Anallo-Caspi', (ant tree)"

556 *S. elegans* China: Kwangtung / no collector s.n. / 1924 / "Yeuk un (Drug origin); Ts'ing shaan t'o (Green mountain peach)"

557 *Vallea stipularis* Venezuela / J. A. Steyermark 55367 / 1944 / "fruit edible" / " 'membranito' "

TILIACEAE

2558 **Corchorus hirsutus** Dominican Republic / E. J. Valeur 132 / 1929 / "Caution! Powder dangerous to eyes"

2559 **C. hirtus** Colombia / F. R. Fosberg 19824 / 1943 / "leaves rubbed up and eaten for dysentery" / " 'escoba real' "

2560 **C. aestuans** Tanganyika / R. E. S. Tanner 3019 / 1956 / "Leaves used as vegetable" / "Bondei: 'Kibwando' "

2561 **Glyphaea laterifolia** Liberia / Zolu-Traub 324 / 1952 / "used medicinally" / "sĩĩ tulé"

2562 **Grewia bicolor** Tanganyika / R. E. S. Tanner 969 / 1952 / "Sukuma tribe: roots pounded, put in cold water, drunk at once twice daily for pain in stomach" / "Kisukama: 'Mkoma' "
Tanganyika / R. E. S. Tanner 2774 / 1956 / "Leaves used as vegetables" / "Zigua: 'Mnangu' "
Tanganyika / R. E. S. Tanner 995 / 1952 / "Sukuma tribe: roots used to treat diarrhoea, by pounding and mixing with water, this drunk" / "Kisukuma: 'Mkomamkuru, Mkomakoma' "

2563 **G. fallax** Tanganyika / R. E. S. Tanner 1032 / 1952 / "flowers and fruit aromatic" / "Kisukuma: 'Mkomahunga' "

2564 **G. forbesii** Tanganyika / R. E. S. Tanner 3372 / 1957 / "Zigua: For stomach pains, roots boiled and the water drunk" / "Sambaa: 'Mkole ngombe'; Zigua: 'Mkongodeka' "
Tanganyika / R. E. S. Tanner 1998 / 1955 / "to cause diarrhoea leaves are burned and the ash cooked in rice and eaten" / "Kibondei: 'Ngongoteka' "

2565 **G. microcarpa** Tanganyika / R. E. S. Tanner 2690 / 1956 / "Zigua: For severe menstrual flow roots boiled and drunk" / "Kisigua: 'Mkloe' "

2566 **G. mollis** Tanganyika / R. E. S. Tanner 4418 / 1959 / "Roots and bark used to treat persistent cough" / "Kizanki: 'Mukomaurheru' "

2567 **G. platyclada** Tanganyika / R. E. S. Tanner 1286 / 1953 / "Sukuma tribe: Bark pounded, used as poultice to treat sores" / "Kisukuma: 'Shingisha' "

2568 **G. stuhlmannii** Tanganyika / R. E. S. Tanner 2689 / 1956 / "Zigua: for vomiting; Bark stripped off, boiled and drunk" / "Kizigua: 'Mkongodeka' "
Tanganyika / R. E. S. Tanner 3369 / 1957 / "Zigua: for heavy cold, leaves boiled and patient sits over covered with blanket" / "Zigua: 'Mondeka mbago gosi' "

2569 **G. eriocarpa** Hainan / S. K. Lau 11 / 1932 / "Fruit not yet seen but natives say it is edible. Water in which the seeds and leaves have been boiled can be used as a remedy for stomach aches" / "(Lois) Tsi kow"

2570 **G. vitiensis** Fiji Is. / O. Degener 15242 / 1941 / "Fijians crush leaves in cold water and drink liquid as remedy for 'spit blood' " / "Nithi in Serua"

71 *Microcos chungii* Hainan / C. I. Lei 71 / 1932 / "fruit black, edible" / "Chong Tze Shu"

72 *M. antidesmifolia* N. Borneo / L. Madani 47165 / 1965 / "edible fruit sour taste"

73 *M. crassifolia* N. Borneo / B. Evangelista 761 / 1928 / "Fr. sour edible" / "Sandaken"

74 *Triumfetta* indet. Tanganyika / R. E. S. Tanner 1264 / 1953 / "Sukuma tribe: roots pounded, put in hot water, drunk for snake bite of waterdwelling snake. Person bitten must stay in water until after taking drink" / "Kisukuma: Nghole nghole"

75 *T. pilosa* China: Kwangsi / W. T. Tsang 24181 / 1934 / ". . . ill smelling"

76 *T. procumbens* Gilbert Is. / E. T. Moul 8044 / 1951 / "medicine" / " 'te kiaou' "

MALVACEAE

77 *Abelmoschus moschatus* Marianne Is. / P. Nelson 56 / 1918 / "used as cure for fevers" / " 'kahamong' "

78 *Abutilon trisulcatum* Mexico / Galar s.n. / 1971 / "Se usa contra la diabetes" / "tronadora"

79 *A. umbelliflorum* Argentina / J. E. Montes 14718 / 1955 / 'hojas y flores medicina popular"

80 *A. asiaticum* Tanganyika / R. E. S. Tanner 1394 / 1953 / "Sukuma tribe: Roots sliced, boiled, mixed with white mtama flour, taken with water to treat syphilis" / "Kisukuma: 'Iboshya lya mlugulu' "

81 *Althaea lavateraefolia* Afganistan / E. Bacon 68 / 1939 / "Infusion made of flower, drunk for coughs"

82 *Anoda cristata* El Salvador / P. C. Standley 19273 / 1922 / "Fls crushed and put in ears for deafness" / " 'Malva' "

83 *Hibiscus trionum* U.S.: N.J. / H. N. Moldenke 3092 / 1926 / "Bladder ketmia"

84 *H. rosa-sinensis* Cuba / C. F. Baker 25 / 1907 / "The leaves are prescribed by the native in small-pox, but are said to check the eruption too much"

85 *H. bifurcatus* Surinam / Lanjouw & Lindeman 3229 / 1949 / "Teas of leaves against cold" / "haimjaia mngkarare (Cor.)"

86 *H. aponeutus* Tanganyika / R. E. S. Tanner 1325 / 1953 / "Sukuma tribe: Roots pounded, dried, taken orally with water, used to treat fever" / "Kisukuma: Ng'wihunge"

87 *H. articulatus* Tanganyika / R. E. S. Tanner 1432 / 1953 / "Sukuma tribe: Roots chewed raw to treat coughs" / "Kiskuma: Ilaba lya hasi"

2588 **H. boranensis** Tanganyika / R. E. S. Tanner 3300 / 1956 / "Masai: Used as a vegetable" / "Masai: Ingarani"

2589 **H. garckeana** Kenya / Fosberg & Mwangangi 49926 / 1968 / "Fruit said to be eaten (pulp around seeds)"

2590 **H. lunariifolius** Tanganyika / R. E. S. Tanner 942 / 1952 / "eaten as a vegetable"

2591 **H. micranthus** Tanganyika / R. E. S. Tanner 3304 / 1956 / "Masai: for stomach ache, roots boiled and the water drunk" / "Masai: Alarrani" Tanganyika / R. E. S. Tanner 3174 / 1956 / "Zigua: For stomach trouble, leaves pounded up, soaked in water and drunk" / "Zigua: Mhasha Mbuzi" Tanganyika / R. E. S. Tanner 1263 / 1953 / "Sukuma tribe: Roots pounded, then chewed to treat cough; pounded roots used as poultice to treat sores" / "Kiskuma: Shingwa mindi; Mbagaswa jamamba"

2592 **H. physaloides** Tanganyika / R. E. S. Tanner 3151 / 1956 / "Bondei: For boils, leaves pounded up and applied as a paste" / "Bondei: Umaka"

2593 **H. syricus** China: Kwangtung / W. T. Tsang 21266 / 1912 / "Flower is edible" / "Kai Yak Fa"

2594 **H. tiliaceus** Mariana Is. / M. Evans 743 / 1965 / "Young leaves squeezed to obtain juice for babies" / " 'Pago' "
Mariana Is. / F. R. Fosberg 25357 / 1946 / "buds as medicine for boils" / " 'pago' "

2595 **Lavatera assurgentiflora** Ecuador / W. H. Camp 5028 / 1945 / ". . . root boiled in water and the extract used as a purgative" / " 'Malva pectoral' "

2596 **Malacothamnus howellii** U.S.: Calif. / J. T. Howell 48114 / 1917 / "very fragrant herbage"

2597 **Malva rotundifolia** U.S. & CAN. / Dr. Gregg s.n. / 47 / "flower white used for enemas, emolient, poultices" / "Malva de Castillo"
U.S.: Penn. / H. N. Moldenke 23881 / 1925 / " 'Cheeses' "

2598 **M. parviflora** Argentina / J. Semper 37 / 1944 / "como laxante"
Ecuador / W. H. Camp E-2466 / 1945 / "Roots used as a purgative" / " 'Cuchi malba echada' "

2599 **Malva** indet. Ecuador / J. A. Steyermark 53307 / 1943 / "After taking a purgative, this plant is boiled and the infusion drunk in order for the purgative to take effect" / " 'malva alta' "

2600 **M. rotundifolia** Afghanistan / E. Bacon 17 / 1939 / "seed makes one 'drunk' " / "panirac"

2601 **Malvaviscus populifolius** El Salvador / P. C. Standley 22300 / 1922 / "fr. eaten" / " 'Manzanillo' "
El Salvador / P. C. Standley 19339 / no year / "used to produce sweat"

602   *Pavonia rosea* Guatemala / J. A. Steyermark 45202 / 1942 / "Used as a remedy for certain sicknesses" / " 'mozote' "

603   *Pavonia* indet. Mexico / T. MacDougall s.n. / 1969 / "Regarded as cure for 'pinto' "

604   *Sida hoepfneri* Tanganyika / R. E. S. Tanner 654 / 1952 / "Roots chewed for dry cough" / "Kisukuma: Libumbo"

605   *S. fallax* Marshall Is. / F. R. Fosberg 26969 / 1946 / "plant used medicinally" / " 'kio' "

606   *Thespesia danis* Tanganyika / R. E. S. Tanner 3866 / 1957 / "Zigua: for swollen stomach, roots boiled and the water drunk" / "Zigua: Mkuluchemba"
Tanganyika / R. E. S. Tanner 2686 / 1956 / "Zigua: for sharp periodic stomach pains, roots boiled and drunk" / "Kizigua: Mkoko"
Tanganyika / R. E. S. Tanner 2957 / 1956 / "Fruits edible" / "Swaheli: Mkoko"

607   *T. garckeana* Tanganyika / R. E. S. Tanner 1412 / 1953 / "Sukuma tribe: Roots used to treat diarrhoea, syphilis" / "Kiskuma: Itobo"

608   *T. populnea* Mariana Is. / F. R. Fosberg 25361 / ". . . bark boiled down and used by women for lotion" / " 'banalo' or 'kiluk' "

BOMBACACEAE

2609   *Catostemma altsoni* Brit. Guiana / A. S. Pinkus 238 / 1939 / "fruit edible"

2610   *Ceiba pentandra* Tanganyika / R. E. S. Tanner 3968 / 1958 / "Zigua: for pain in heart, bark boiled and the water drunk" / "Zigua: Mnamba"

2611   *Durio testudinarum* Borneo / J. & M. S. Clemens 20146 / 1929 / "smells like limberger and tastes like ambrosia"

2612   *Durio* indet. N. Borneo / A. Cuadra A.1318 / 1948 / "fruit green, prickly, edible" / "Durian puteh (Brunei)"

2613   *Matisia alata* Ecuador / E. L. Little, Jr. 6200 / 1943 / "Fruits eaten" / " 'Sapote' 'sapotillo' "

2614   *Pseudobombax* indet. Brazil / B. A. Krukoff 2068 / 1933 / "Bark used for stomach" / " 'carrasco' "

2615   *Quararibea putumayensis* Brazil / J. M. Pires 236 / 1947 / "Frutos esreideados"
Colombia / G. Klug 1881 / 1930 / "fruit edible" / " 'Zapotillo' "

STERCULIACEAE

2616   *Helicteres angustifolia* Hainan / F. A. McClure 7829 / 1921 / "drug plant" / "Po-ma Ka yau ma"

2617   *Melochia odorata* Fiji Is. / O. Degener 15399 / 1941 / "Extract of leaves used medicinally. When sick from drinking too much yagoue, squash leaf and mix with water. Drink liquor" / " 'Seti (Ra)' "

2618   *Sterculia apetala* Honduras / C. & W. von Hagen 1066 / 1937 / "fruit edible; oil extracted from seeds" / " 'Castano' "
Honduras / C. & W. von Hagen 1279 / 1937 / "Edible fruit" / " 'Castaña' "
Mexico / E. Matuda 17464 / 1948 / "fr. edible" / "Castaña"
Colombia / H. H. Smith 1889 / '98 / "The leaves are used by the country people as an external remedy for inflammatory diseases"

2619   *Theobroma lemniscata* Venezuela / J. J. Wurdack 336 / 1955 / "Fruit said to be edible" / " 'Guayabori' "

2620   *T. speciosum* Bolivia / H. H. Rusby 654 / 1886 / "Pulp edible and equal to that of T. cacao, seeds white not used. (?)"

2621   *T. subincanum* Brazil / Prance, Maas, Atchley et al. 13933 / 1971 / "Bark used in snuff. The bark is stripped, burnt, and the ash mixed with tobacco (13929) to produce a narcotic snuff" / " 'Cowadimani' "
Brazil / Prance, Maas, Woolcott et al. 16515 / 1971 / "the bark ash used as an ingredient of Deni Indian snuff; the fruit eaten by the Deni Indians" / " 'Mapanahã' (Deni) 'Cupuī' (Port.)"
Brazil / Campbell, Nelson, Ramos et al. P21258 / 1974 / "Bark ash mixed with tobacco leaves as component of Jamamadi snuff" / " 'Shina' (Jamamadi)"

2622   *Waltheria americana* Brazil / A. Ducke 7889 / 1942 / "Stem decoction used for fevers" / " 'malva branca' "
Hawaii / D. L. Topping 2905 / 1924 / "A shrub whose flowers and roots are used as a remedy for the disease called ea" / " 'Uhaloa, Alaalapuloa' "

2623   *W. indica* Tanganyika / R. E. S. Tanner 3880 / 1957 / "Zigua: for pain in urinating, roots boiled and the water drunk" / "Zigua: Mwezuzi mwezuruzi"
Tanganyika / R. E. S. Tanner 2015 / 1955 / "Leaves used as poultice on suppurating sores"
Tanganyika / R. E. S. Tanner 583 / 1952 / "Leaves ground up, soaked in water with cattle urine, for bathing to treat erysipelas" / "Kisukuma: Matamagabagikuru"
Tanganyika / R. E. S. Tanner 1239 / 1953 / "Sukuma tribe: Roots pounded, soaked, used to wash patient with leprosy; root flour smoked for infected throat" / "Kisukuma: Ishuda"

DILLENIACEAE

2624   *Davilla rugosa* Brazil / M. B. da Silva 94 / 1942 / "Juice burns skin and for this reason called 'Fire vine' " / " 'cipó de fogo' "
Brazil / Y. Mexia 4703 / 1930 / " 'Cipo Carijóa' "

25   *Davilla* indet. Brazil / G. & E. Eiten 8737 / 1968 / " 'sambaibinha' "

26   *Dillenia indica* Honduras / A. Molina R. 21914 / 1968 / " 'Manzanote' "

27   *Doliocarpus coriacea* Brazil / Prance, Steward, Harter et al. 10538 / 1971 / "Stem yields potable water"

28   *D. coriaceous* Venezuela / J. A. Steyermark 93141 / 1964 / " 'bejuco de agua' " / " 'chparrillo' "
Brit. Guiana / A. C. Smith 2903 / 1938 / "stem yields potable water"

29   *D. major* Brazil / W. A. Archer 8350 / 1943 / "Juice of young leaves blisters skin" / " 'cipó de fogo' "

30   *Hibbertia scandens* Australia: Queensland / C. T. White 1368 / 1929 / "This flower has a very nauseating smell"
Australia: Queensland / L. Durrington s.n. / 1973 / "Flowers yellow with strong unpleasant odour"
Australia: Queensland / C. T. White 6350 / 1929 / "fls yellow with a disagreeable scent like excrement"

ACTINIDIACEAE

31   *Actinidia coriacea* China: Nanking / Y. Tsiang 5698 / 1930 / "Frt. green, edible"

32   *Saurauia tomentosa* Colombia / D. D. Soejarto 504 / 1963 / "Frts. edible, said to be sold in markets" / " 'Moquillo' "

33   *S. ursina* Colombia / J. M. Duque Jaramillo 2621 / 1946 / "Frutos comestibles" / " 'Dulumoco' Caldas, 'Moquillo' Valle"

34   *Saurauia* indet. Colombia / L. U. Uribe 5399 / 1965 / "Frutos comestibles, agradables"

35   *S. fasciculata* China: Kwangsi / R. C. Ching 7357 / 1928 / "Fruit . . . transparent rich in honey-like juice, sweet edible"

36   *S. tristyla* China: Canton / F. A. McClure 9129 / 1921 / "frs., edible" / "Nam Fung"
Hainan / F. A. McClure 20092 / 1932 / "Fruits edible"

37   *S. elegans* Philippine Is. / A. D. E. Elmer 8703 / 1907 / "the ripe fruits are very juicy and sweet. The native suck the juice" / " 'Oyoc' "

38   *S. polysperma* Philippine Is. / A. D. E. Elmer 8860 / 1907 / "the mature fruits called by the Igorrotes 'Oyok' is eaten; the fruits are juicy, sour and eaten by the natives green" / " 'Oyok' "

OCHNACEAE

39   *Brackenridgea zanguebarica* Tanganyika / R. E. S. Tanner 3929 / 1958 / "Bondei: for jaundice bark cooked and eaten with porridge" / "Bondei: Mkomazi"

2640 **B. nitida** Fiji Is. / A. C. Smith 6331 / 1947 / "leaves used for wrapping cigarettes" / " 'Mbelembele' "

2641 **Ochna ovata** Tanganyika / R. E. S. Tanner 954 / 1952 / "Sukuma tribe: Roots used to treat violent stomach pains and diarrhoea—cut roots soaked in hot water, this drunk" / "Kisukuma: Mfifi"

2642 **Ouratea cassinefolia** Brazil / R. Fróes 1830 / 1932 / "Fruits used for oil" / " 'Bate puta' "

2643 **Rhytidanthera magnifica** Colombia / E. L. Little, Jr. 9088 / 1944 / "Local man says that this is a kind of quina. Of interest as a possible false bark that might be mixed with cupres" / " 'calisaya', 'quina calisaya' "

2644 **Wallacea multiflora** Brazil / Ducke 257 / 1935 / " 'catinga' "

CARYOCARACEAE

2645 **Anthodiscus mazarunonsis** Venezuela / Steyermark & Wurdack 115 / 1955 / "Fls spicy-fragrant" / "Arekana: 'O-to-ye-mu-yek' "

2646 **Caryocar amygdaliforme** Peru / J. Schunke V. 3979 / 1970 / " 'Almendra Blanco' "

2647 **C. coriaceum** Brazil / Dr. v. Luetzelburg 26455 / 34 / "Pigui branca"

2648 **C. glabrum** Peru / A. Arostegui V. 11 / 1961 / " 'Almendra' "
Brazil / R. de L. Fróes 20764 / 1945 / " 'Piquiarana' "
Brazil / J. T. Baldwin 3183 / 1944 / "nut quite sweet and tasty" / " 'pequía' "
Venezuela / P. E. Berry 2138 / 1976 / "Nuez comestible. Se dice que tiene un sabor parecido a la Nuez de Pará o de Brasil" / " 'Jigua' "
Brazil / Prance, Maas, Woolcott et al. 15583 / 1971 / "Fruit ground up for potent fish poison" / " 'Pursh' (Maku)"
Peru / M. A. Soria S. 19 / 67 / "los frutos son comestibles" / " 'Almendra' "
Peru / A. Arostegui V. 72 / 1962 / "Semmollas comestibles aceitosas" / "Almendro"
Colombia / Schultes & Cabrera 14139 / 1951 / "seeds uncooked used as food" / "Puinave = how"

2649 **C. gracile** Brazil / Prance, Maas, Woolcott et al. 15963 / 1971 / "Flowers very fragrant"
Brazil / A. Ducke 1101 / 1942 / "floribus . . . odoratis" / " 'Piquia-rana' "
Brazil / Maguire, Wurdack & Maguire 41933 / 1957 / "Fruit said to be edible" / " 'Jigua' "

2650 **C. microcarpum** Brazil / Wurdack & Adderley 43577 / 1959 / "fruit used as fish poison" / " 'Jigua' "
Brit. Guiana / Jenman 3925 / 1887 / " 'Cula' "

Venezuela / P. E. Berry 2105 / 1976 / "Es usado como barbasco para matar peces, pilando los frutos u otras partes del arbol hasta producir espuma. Se describe como un planta muy venenosa. Tronco con resina roja" / " 'Cojón de verraco' "

Brazil / Ducke 1695 / 1945 / " 'Piquiá-rana do igapó' "

Venezuela / Maguire, Wurdack & Maguire 42640 / 1958 / "fruit is said to furnish a good fish poison" / " 'Cojon de verraco' or 'Barbasco' "

Surinam / J. C. Lindeman 1244 / 1950 / "Vern. names: (rough) sopohoedoe (Sur.)"

Venezuela / L. Williams 13846 / 1942 / "V. n. 'Jigua'. 'coco de Mono rebalsero' "

651 *C. pallidum* Brazil / Prance, Fidalgo, Nelson & Ramos 21357 / 1976 / "cotyledons eaten by Sanama & Mayongong Indians" / " 'Luamãs' (Sanama), 'Yeheedede' (Mayongong)"

652 *C. villosum* Brazil / W. A. Archer 8202 / 1943 / "Frt. edible" / " 'piquiá' "

Brazil / J. M. Pires 51720 / 1961 / "Nom. vulg.: 'piquia-verdad' "

MARCGRAVIACEAE

653 *Marcgravia membranacea* Panama / Maas & Dressler 746 / 1972 / "flowers . . . visited by hummingbirds"

654 *M. trianae* Venezuela / J. A. Steyermark 91956 / 1963 / "Bird seen visiting nectaries"

655 *Marcgravia* indet. Ecuador / J. A. Steyermark 54859 / 1943 / "flower . . . with odor of faeces"

656 *Norantea subsessilis* Panama / L. H. Durkee 71.147 / 1971 / "humming-bird pollinator"

657 *N. goyazensis* Brazil / Harley & Souza 10242 / 1968 / "Plant covered in small black ants"

658 *Ruyschia tremadena* Colombia / L. U. Uribe 3353 / 1959 / "Florecitas verdes muy perfumadas"

Venezuela / J. A. Steyermark 56977 / 1944 / "Inflorescences erect, fragarant"

659 *Souroubea exauriculata* Nicaragua / Wilbur & Almeda 16454 / 1972 / "Flowers with gardenia-like fragrance"

660 *S. sympetala* Panama Canal Zone / T. B. Croat 5294 / 1968 / "flowers w. sweet aroma"

Brit. Honduras / W. A. Schipp 514 / 1929 / "flowers . . . sweetly perfumed" / " 'Rare' "

Venezuela / J. d. Bruijn 1307 / 1966 / "Flowers smelling like peaches"

Venezuela / J. A. Steyermark 61000 / 1945 / "cut stem yields good water, first cut above and then below to get water to flow" / " 'bejuco de cuspa' "

2661   *S. guianensis* Fr. Guiana / F. Hallé 620 / 1962 / "petites fleurs . . . très odorantes"
Venezuela / F. J. Breteler 3723 / '64 / "Flowers with strong, very fine smell"

QUIINACEAE

2662   *Lacunaria crenata* Brit. Guiana / For. Dept. of Brit. Guiana F2711 / 47 / "fruit . . . edible"
Brazil / J. M. Pires 51799 / 1961 / "Nom. vulg. 'Muela de mutum' "

2663   *L. jenmanii* Surinam / J. G. W. Boer 946 / 1963 / "fruit . . . pulp juicy . . . tasting bitter"
Brazil / J. M. Pires 51881 / 1961 / " 'muela de mutum' "

2664   *Quiina schippii* Brit. Honduras / P. H. Gentle 2929 / 1939 / " 'pigeon plum' "
Brit. Honduras / W. A. Schipp 1220 / 1933 / "fruit of a fine tarty flavor"

THEACEAE

2665   *Adinandra acutifolia* China: Kwangsi / W. T. Tsang 24013 / 1934 / "fr. black, edible" / "Wong Pan Cha Shue"

2666   *A. millettii* Hong Kong / Y. W. Taam 1681 / 1940 / "fruit black, edible"
China: Kwangtung / W. T. Tsang 21076 / 1932 / "fruit black, edible" / "Wong Pan Ch'a"

2667   *Cleyera integrifolia* Mexico / G. B. Hinton et al. 7227 / 35 / "Fruit edible" / "Capulincillo"

2668   *Laplacea fruticosa* Venezuela / Curran & Haman 1019 / 1917 / "Bark used locally for tanning" / " 'Pedralejo' "

2669   *Laplacea* indet. Ecuador / J. A. Steyermark 54123 / 1943 / "Bark used often for Cinchona by novices" / " 'cascarilla dulce' "

2670   *Ternstroemia tepezapote* Guatemala / E. Contreras 7943 / 1968 / " 'Nance de Monte' "

2671   *T. hartii* Jamaica / W. Harris 8786 / 1904 / " 'Wild Mammee Sapota' "

2672   *Ternstroemia* indet. Brazil / M. Barbosa da Silva 156 / 1942 / "Latex. Frt. edible" / " 'marapajuba' "

2673   *T. gymnanthera* China: Kwangsi / W. T. Tsang 24148 / 1934 / "fr. edible" / "Sai Yeung Pat Kok Shu"

2674   *T. toquian* Philippine Is. / E. D. Merrill 499 / 1902 / "Fruit eaten by natives" / " 'Talipopo' "
Philippine Is. / P. T. Barnes 208 / 1904 / "According to the natives, the fruit is used to poison fish" / "T., Toquian, Garamansatay"

Philippine Is. / G. Edaño 2929 / 1930 / "The juice from the powdered bark is fish poison" / "Dagis Dialect Iloc."

75   *Ternstroemia* indet. New Guinea / D. Frodin NGF 26698 / 66 / "Bark has substance used as fishpoison" / "Local name Nasen"

76   *Thea caudata* China: Kwangtung / W. T. Tsang 21436 / 1932 / "Fr. yellow, edible" / "Sai Ip Shan Yau Ch'a Muk"
China: Kwangsi / W. T. Tsang 24813 / 1934 / "fr. edible" / "Sai Yeung Che Tez Shue"

577   *T. oleosa* China: Kweichow / Steward, Chiao & Cheo 610 / 1931 / "Seeds used for oil"
China: Kwangtung / T. M. Tsui 647 / 1932 / "Seeds used for oil. Oil used for food and light" / "Cha Yau Tsz"
China: Kwangtung / W. T. Tsang 21069 / 1932 / "fruit, cream, edible" / "Yau Ch'a Tse"
China: Kwangtung / W. T. Tsang 21478 / 1932 / "Fl. white, fragrant, edible. Used to make oil for food"
Hainan / F. A. McClure 7673 / 1931 / "oil pressed from seeds used as hair dressing by the Hainanese women . . . also boiled with water and poured on the ground to kill worms" / "Yau ch'a, Ch'a tsai"

578   *T. paucipunctata* Hainan / H. Fung 20336 / 1932 / "Oil from fruit for cooking or as hair tonic" / "Shan Yau"

579   *T. sinensis* China: Kwangtung / W. T. Tsang 21281 / 1932 / "flower white, edible" / "Shan Ch'a Ip'

580   *T. speciosa* var. *yunnanensis* China: Kwangsi / R. C. Ching 7967 / 1928 / "commonly cultivated by natives for its seed being exfered for oil for cooking purposes"

681   *Tutcheria symplocifolia* China: Kwangtung / W. T. Tsang 21196 / 1932 / "fruit yellow, edible" / "Shan Ch'a Shue"

BONNETIACEAE

682   *Bonnetia tepuiensis* Venezuela / J. A. Steyermark, G. G. K. & E. Dunsterville 92248 / 1963 / "Lilly Bark Sample and Leaf Sample"

CLUSIACEAE

683   *Allanblackia stuhlmannii* Tanganyika / R. E. S. Tanner 3000 / 1956 / "Sambaa: For rheumatism, seeds pounded up & boiled & rubbed into joints. Also used in cooking as flavouring" / "Kisambaa: Msambu Kiswaheli: Mkanyi"

2684 *Calophyllum antillanum* W.I.: Martinique / College of Pharmacy Herb.
s.n. / no year / "Balsam und Rinde werden in West-Indien als Arznei
verwendet" / "Westindisches Schönblatt"

2685 *C. inophyllum* Tanganyika / R. E. S. Tanner 2855 / 1956 / "Swaheli: For
person possessed by devil. Leaves pounded up, sap mixed with young
leaves and oil, cooked in fire and patient covered with blanket and breathes
it" / "Swaheli: Mtondoo"
Fiji Is. / A. C. Smith 8101 / 1953 / "fruit used to scent coconut oil" /
" 'Ndelo' "
Caroline Is. / F. R. Fosberg 25783 / 1946 / "used for medicine" / " 'Btaes' "

2686 *C. rubiginosum* Borneo / A. Kostermans 10584 / 1955 / "Latex used to stun
fishes"

2687 *Chrysochlamys guttifera* Venezuela / Killip & Lasser 37750 / 1943 / "Used
for malaria and rheumatism" / " 'Capuchino' "

2688 *Clusia rosea* Mexico / C. L. Lundell 1107 / 1931 / "Thick yellow latex used
to cure 'granos' "

2689 *C. alba* W.I.: Montserrat / J. A. Schaefer 200 / 1907 / " 'mountain
cherry' "

2690 *C. columnaris* Venezuela / J. A. Steyermark 57733 / 1944 / "resin of bark
used for putting over sprains and cuts" / " 'copey' "
Colombia / W. A. Archer 1907 / 1931 / " 'Guayabo agrio' "

2691 *C. grandiflora* Venezuela / J. A. Steyermark 58544 / 1944 / "Resin used in
treating leprosy and boils" / " 'uruyek' (copey)"

2692 *C. insignis* Brazil / A. Miranda Bastos 41 / 1955 / " 'Cebola brava' "

2693 *C. minor* Venezuela / J. A. Steyermark 61001 / 1945 / "Birds eat fruit" /
" 'cupey cillo' "

2694 *C. multiflora* Colombia / Fosberg & Drew 223191 / 1944 / ". . . oozing
aromatic yellow latex when broken"

2695 *C. nemorosa* Brazil / W. A. Rodrigues 640 / 1954 / "O leite usa-se para
colar corda de flecha" / "Manga brava (Manaus, apúi do branco)"

2696 *C. sellowiana* Brazil / Ratter et al. 1126 / 1968 / "Copious white latex with
a slight lemon tint" / "Cebola Braba"

2697 *Clusia* indet. Bolivia / O. Buchtien 1880 / 1907 / "Volksname 'Incensio' (=
Weihrauch)"

2698 *Garcinia oblongifolia* China: Kwangsi / W. T. Tsang 24268 / 1934 / "fr.
yellow, edible" / "Nam Na Kit Shue"
Hainan / Chun & Tso 43445 / 1932 / "fr. yellow, edible"
Hainan / S. K. Lau 73 / 1932 / "fr. yellow; edible" / "Sai Tai Kau (Lois)"
Hainan / C. I. Lei 873 / 1933 / "fruit yellow, edible" / "Shan Chuk Shu"

99   *G. kajewskii* Australia: Queensland / C. T. White 1418 / 1929 / "Fruit exceptionally acid"

00   *G. ovalifolia* E.I.: Canara / College of Pharmacy Herb. s.n. / no year / "Liefert ein Gummi-Guttae" / "Oval-blättriges Calysaccion"

01   *G. picrorhiza* Indonesia: Bogor / A. Kostermans 11096 / 1955 / "Pulp of fruit . . . sweet with bitter aftertaste. Bats are fond of the fruit"

'02   *G. pictorum* E.I.: Mangalor / College of Pharmacy Herb s.n. / no year / "Giebt ein zum Malen sehr brauchbares Gummigutt. Die Früchte welche im December reifen, werden gegessen"

'03   *G. pseudoguttifera* New Hebrides / S. F. Kajewski 389 / 1928 / "fr. red, when ripe, eaten by natives" / " 'Neyah-even' "
Fiji Is. / O. Degener 15392 / 1941 / "Fruit edible. For body pain squash leaf in water, strain & drink. Or for babies, squash leaf in coconut oil and rub body with oil" / " 'Mbuluwai' (Ra)"
Fiji Is. / A. C. Smith 1557 / 1934 / "Fruit used to scent coconut oil" / " 'Mbulu' "
Fiji Is. / A. C. Smith 997 / 1934 / "fruit oil used as a scent" / " 'Mbulu' "

'04   *G. rostrata* Java / N. Wirawan 229 / 1964 / "fruit-wall eaten by animal"

'05   *G. sessilis* Samoa / A. J. Eames 71 / 1921 / "Fls. to scent coconut oil; bark in medicine" / "Seilala"

'06   *G. vidalii* Philippine Is. / Ramos & Convocar 183 / 1931 / "The fruit of this plant is edible"

707   *Garcinia* indet. Philippine Is. / A. C. Mallonga 31211 / 1930 / "The pulp of the fruit is edible" / "Hanggos Dialect Visaya"

708   *Garcinia* indet. Sumatra / R. S. Toroes 5240 / 1933 / "fruit edible" / "Kajoe Kardis"

709   *Garcinia* indet. Borneo / A. Kostermans 10579 / 1955 / "Fr. sweet with much yellow latex"

710   *Haronga madagascarensis* Liberia / L. C. Okele 2 / 1950 / "with orange juice. Used for ringworm"

711   *Kielmeyera corymbosa* Brazil / E. P. Heringer 4074 / 55 / "The leaves are used as drink for stimulating instead coffee" / "Pan Santo, Boisinho"

712   *K. petiolaria* var. *longifolia* Brasil / E. P. Heringer 10782 / 1965 / ". . . do material quimico"

713   *Marila grandiflora* Venezuela / J. A. Steyermark 95062 / 1966 / " 'cacao de montaña' "

714   *Marila* indet. Ecuador / E. L. Little 6671 / 1943 / " 'Mamey de montana' "

715   *Rheedia edulis* Panama / Cooper & Slater 160 / 1927 / "The greenish yellow gum used for wounds, cuts, etc." / " 'Cero' "

716   *R. intermedia* Guatemala / E. Contraras 4354 / 1964 / " 'Limoncillo' "

2717   *Rheedia* indet. Mexico / J. N. Rovirosa 543 / 1889 / "vulgo *Limoncillo* dicitur"

2718   *Rheedia* indet. Brit. Honduras / P. H. Gentle 3482 / 1939 / " 'Waika plum' "

2719   *R. benthamiana* Guayana / C. Blanco 109 / 1965 / "Fruto comestible amarillo rosado" / "N.v.: Cozoiba rebalsera"

2720   *R. floribunda* Colombia / E. L. Little 8700 / 1944 / "Frts. eaten"
Brazil / Chagas 21207 / 1955 / "frutos amarelos, pôlpa comestível, azêda" / "Bacurí pau"

2721   *R. gardneriana* var. *glaziovii* Brazil / P. R. Reitz 4414 / 1952 / "Fruto amarelo, comestível" / "Bacoparí"

2722   *R. longifolia* Peru / J. M. Schunke 55 / 1935 / "fruit yellow, edible" / " 'Charichuelo' "

2723   *R. spruceana* Bolivia / R. S. Williams 598 / 1902 / "Fruit edible" / " 'Chichiru' "

2724   *Rheedia* indet. Venezuela / C. A. Blanco C. 1189 / 1971 / "n.v.: Naranjillo"

2725   *Rheedia* indet. Brazil / Pessoal do C. P. 1943 / 55 / "A polpa e comestivel e azeda" / "Bacuri-paú"

2726   *Rheedia* indet. Venezuela / A. L. Bernardi 7415 / 1959 / "N.V.: Naranjita"

2727   *Rheedia* indet. Venezuela / A. L. Bernardi 7068 / 1959 / "frutos comestibles" / "Cozoiba, cotoiba"

2728   *Rheedia* indet. Argentina / T. Meyer 8960 / 1945 / "fruta comestible" / "Pacuri"

2729   *Rheedia* indet. Bolivia / I. Steinbach 6434 / 1924 / "Fruta comible"

2730   *Rheedia* indet. Guayana / C. Blanco 88 / 1965 / "Fruta comestible" / "Cozoiba negra"

2731   *Rheedia* indet. Venezuela / A. L. Bernardi 7415 / 1959 / "Naranjita"

2732   *Rheedia* indet. Colombia / Idrobo & Schultes 752 / 1950 / "Fruto dulce, pulpa comestible"

2733   *Symphonia globulifera* Brit. Honduras / S. J. Record BH 25 / 26 / "Waika Chewstick"
Peru / J. J. Wurdack 2364 / 1962 / "latex used for curare matrix" / " 'Pengüe-nunú' (Aguaruna)"
Venezuela / J. A. Steyermark 87422 / 1960 / "branches covered with stinging ants in abundance"

2734   *Tovomita weddelliana* Venezuela / J. A. Steyermark 94744 / 1966 / "Eli Lilly sample"

2735   *Tovomita* indet. Colombia / Idrobo & Schultes 849 / 1951 / "Fruit sweet"

36   *Tovomita* indet. Venezuela / J. A. Steyermark 86497 / 1960 / " 'naranjillo negro' "

37   *Tovomita* indet. Peru / C. Schunke 302 / 1929 / " 'Naranja silvestre' "

38   *Tovomita* indet. Venezuela / Killip & Lasser 37740 / 1943 / "Used to purify blood" / " 'Quina' "

39   *Tovomita* indet. Brazil / A. Silva 236 / 1944 / "Frt. edible" / " 'abricó rana' "

40   *Tovomitopsis nicaraguensis* Brit. Honduras / P. H. Gentle 6436 / 1948 / " 'Waika Plum' "

HYPERICACEAE

41   *Hypericum perforatum* U.S.: Colo. / W. A. Weber 8517 / 1953 / "ca. 17,000 acres now dominated by this species . . . beetles introduced this year to help combat the spread"

42   *H. laricoides* Venezuela / J. A. Steyermark 55501 / 1944 / "Good forage for cattle; also make a liquor out of crude plant putting in a bottle; also dried leaves pulverized for putting over sores and cuts" / " 'dictamo matico' "

43   *H. mutilum* Brazil / G. & L. T. Eiten 2408 / 1960 / "Lightly grazed by horses"

44   *H. strictum* Ecuador / R. Espinosa 319 / 1946 / " 'romerillo' "

45   *H. thesiifolium* Ecuador / M. Ownbey 2611 / 1943 / "said to be used in the treatment of blood poisoning"
Ecuador / Fosberg & Giler 22900 / 1945 / "plant not even eaten by goats"

46   *H. adenocladum* Turkey / T. Goell 4 / 1956 / "flower used to put on animals' wounds" / " 'Saridarman' "

47   *H. hyssopifolium* Turkey / T. Goell 6 / 1956 / "In June, animal fodder is gathered for winter"

48   *H. leschenaultii* Sumatra / W. N. & C. N. Bangham 921 / 1932 / "Fragrant, exquisite"

49   *H. moserianum* U.S.: Hawaii / O. & I. Degener 31346 / 1967 / "visited by honeybees . . . cultivated"

50   *Vismia camparaguey* Guatemala / J. A. Steyermark 44178 / 1942 / "leaves cooked for use in headache" / " 'Camparaguay' "

51   *V. baccifera* subsp. *dealbata* Colombia / Killip & Smith 16335 / 1926 / "sap used as coloring matter" / " 'Sangrito' "
Brazil / Jaccoud M G n. 21.195 / 1955 / "látex vermelho" / " 'Lacre vermelho' "

2752   *V. buchtienii* Bolivia / Steinbach 6527 / 1924 / "Los tallos heridos filtran una recina colerada que coajula en el aire y que lloman lacre" / "Lacre"

2753   *V. cayennensis* Surinam / W. A. Archer 2706 / 1934 / "Bark is scraped in water as poison for fish" / " 'Swinani' (Carib)"

2754   *V. falcata* W.I.: Tobago / W. E. Broadway 4139 / 1910 / "sweetly scented" W.I.: Trinidad / R. E. Schultes 18637 / 1953 / "Latex orange-brown"

2755   *V. guianensis* Panama Canal Zone / Stern & Chambers 36 / 1957 / "fruits . . . with skunk-like aroma"
Brazil / Gouvêa M G 21210 / 1955 / "resina alaranjada. Casca tónica e febrífugo" / " 'Lacre' "
Brazil / Chagas M G 21215 / 1955 / "casca verde-amarelada . . . aromática"
Venezuela / Aristeguieta & Agostini 4789 / 1961 / "Latex rojizo, poco abundante"

2756   *V. japurensis* Colombia / E. L. & R. R. Little, Jr. 8252 / 1944 / "Orange latex—dye" / " 'pepa de lacre' "

2757   *V. latifolia* Panama / G. P. Cooper 659 / 1928 / "resin . . . is aromatic and stains hands yellow"
Brazil / Harley & Souza 11092 / 1968 / "with aromatic cinnamon like scent with orange sap"

2758   *Vismia* indet. Colombia / Soejarto & Cardozo 667 / 1963 / "Cut of branch gives clear sap with aromatic smell"

DIPTEROCARPACEAE

2759   *Anisoptera marginata* N. E. Borneo / A. Kostermans 8896 / 1953 / "Heart-wood . . . with sour smell"

2760   *Dipterocarpus dyeri* Indo-China / L. Pierre 1592a / 1866 / "Exploité pour son bois et son oléo résine" / "Aun: Dzao xam nau. Moi: rat soul"

2761   *D. tuberculatus* Burma / K. Ithil 38 / 1931 / "The oil is used in lighting fire" / "In"

2762   *D. crinitus* N. Borneo / A. Cuadra A.3296 / 51 / "When cut white sap exudes from the bark" / "Kerring buluh (Malay)"

2763   *D. warburgii* Philippine Is. / A. D. E. Elmer 13373 / 1912 / "Wood . . . odorless, slightly bitter, exuding a sticky clear sap . . . flower odorous" / " 'Balaou' in Manobo"

2764   *Dipterocarpus* indet. E. Borneo / A. Kostermans 12664 / 1956 / "Seed edible. Dispersed by water" / " 'Laran' "

2765   *Dryobalanops lanceolata* N. Borneo / Harvey A108 / 47 / "Fruit pale greenish color 5 wings of exuding fragrant oily substance of camphor" / "Kapur (Malay) Keladan (Malay) Kapur bulist (Malay) Kapur barus (Malay)"

2766 *Monotes engleri* S. Rhodesia / G. M. McGregor 23/37 / 37 / "fairly durable timber . . . is used for fencing. The tree harbours a black caterpillar which is eaten by women only" / "Native name 'Munete' "

2767 *Shorea thorelii* Cochinchinae / L. Pierre 1688 / 1866 / "copiosé resinosa . . . Bois putrescible" / "Kmer: Ktym ou Ktyon. Aun: Xên; — vin vin; — ehây"

2768 *S. leprosula* S. Sumatra / L. J. W. Dorst 35E3PT 322 / 1924 / "pistil white, smell delicious, taste bitter . . . seed inside light yellow, taste harsh" / "meranti samak"

2769 *S. polita* Philippine Is. / L. Aguilar 31042 / 1929 / "Flower odorous . . . smells like honey . . . Economic uses for banea" / "Common name Bantahon Dialect Mand"

2770 *S. teysmanniana* Philippine Is. / M. Oro 31055 / 1929 / "Economic uses for banea particularly" / "Common name — Malagmat. Dialect — Tag"

2771 *Shorea* indet. Brit. N. Borneo / Puasa 2247 / 1926 / "edible fruit" / "Cang Cawang"

2772 *Vatica mangachapoi* Philippine Is. / A. D. E. Elmer 13398 / 1912 / "wood with sticky resin . . . flowers sweetly fragrant" / " 'Binuñgo' in Manobo" Philippine Is. / A. D. E. Elmer 13680 / 1912 / "wood . . . odorless but bitter . . . the middle portion contains some viscid juice" / " 'Liloan' in Manobo" N. Borneo / Harvey A122 / 1947 / "Black damar exceeding" / "selangan kuning (Malay) selangan kacha (Malay) selo-selo (Bajau E. C.)"

FRANKENIACEAE

2773 *Frankenia salina* S. AM.: Prov. de Santiago / E. Bernath 4107 / 1955 / " 'Hierba del salitre' incol."

TAMARICACEAE

2774 *Myricaria prostrata* Punjab / W. Koelz 2073 / 1931 / "Flower . . . interesting odor" / "Common name — Ombu. Dialect — Tibet"

BIXACEAE

2775 *Bixa* indet. Brazil / R. de L. Fróes 20868 / 1945 / " 'Urucurana' "

COCHLOSPERMACEAE

2776 *Cochlospermum orinocense* Venezuela / E. Foldats 260-A / 1971 / "Frutos rojos, comestibles" / "Carnestolendo"

CANELLACEAE

2777 **Canella winterana** U.S.: Fla. / Brizicky, Stern & Nagle 426 / 1956 / "aromatic leaves, flowers velvety, deep red"
U.S.: Fla. / D. H. Caldwell 8744 / 1952 / "Common name Cinnamon Bark"
Puerto Rico / E. L. Little, Jr. 13157 / 1952 / "barbasco"

2778 **Capsicodendron dinisii** Brazil / G. Hatschbach 16703 / 1967 / "casca com . . . forte sabor de pimenta, fruto . . . com igual sabor" / "Pimenteira"
Brazil / Reitz & Klein 6507 / 1958 / "Pau paratudo"

2779 **Cinnamodendron axilare** Brazil / Pabst, Hatschbach & Pereira 9138 / 64 / " 'Casca de Cotia' "
Brazil / P. Dusén 7615 / 1909 / " 'Pimenta brava' incol."

2780 **Pleodendron macranthum** Puerto Rico / L. R. Holdridge 103 / 1940 / "Aceitillo"
Puerto Rico / L. R. Holdridge 222 / 1940 / " 'Chupa callo' "

2781 **Warburgia stuhlmannii** S. Africa / L. E. Codd 8733 / 54 / "Used medicinally by natives"

VIOLACEAE

2782 **Amphirrhox surinamensis** Brazil / Silva & Souza 2450 / 1969 / "flor branca, perfumada"

2783 **Corynostylis arborea** Brit. Honduras / P. H. Gentle 3725 / 1941 / " 'monkey melon' "

2784 **C. carthagenensis** Colombia / Garcia y Barriga & Lozano-C. 18238 / 1965 / " 'Granadilla' "

2785 **Hybanthus attenuatus** El Salvador / P. C. Standley 22064 / 1922 / " 'Hierba del rosario' "
Honduras / A. Molina R. 27477 / 1972 / " 'Ipecacuanilla' 'hierba del rosario' "

2786 **H. calceolaria** Brazil / R. Fróes 11655 / 1939 / " 'Ipecoaconha branca' "

2787 **H. enneaspermus** Tanganyika / R. E. S. Tanner 2611 / 1956 / "flowers blue, aromatic"

2788 **Rinorea hummelii** Brit. Honduras / P. H. Gentle 8701 / 1955 / " 'Wild Coffee' "

2789 **R. falcata** Peru / J. Schunke V. 5419 / 1971 / " 'Canilla de Vieja' "

2790 **R. lindeniana** Peru / J. Schunke V. 2677 / 1968 / " 'Canilla de vieja' "
Venezuela / J. A. Steyermark 61014 / 1945 / "Birds eat the fruit" / " 'gaspalilla' "

2791 **R. pubiflora** Brazil / R. Fróes 1908 / 1932 / " 'Canela de Jacamin' "

792   *R. elliptica* Tanganyika / R. E. S. Tanner 3963 / 1958 / "Zigua: For
threatened abortion, roots boiled and the water drunk" / "Zigua: Muyama"

793   *Viola fimbriatula* U.S.: N.Y. / Dr. O. R. Willis s.n. / ca. 1880 / "Abounds
in mucilage. Used in medicine as a substitute for Slippery Elm" / "Arrow-
leaved or Fever Violet"

794   *V. glabella* Alaska / M. W. Gorman 10 / 1895 / "Roots are boiled & the
decoction drunk as a purgative by the natives"

795   *V. rafinesquii* U.S.: Ga. / A. Cronquist 4226 / 1947 / "Roots pleasantly
aromatic"

796   *Viola* indet. U.S.: Fla. / P. P. Sheehan s.n. / 1919 / "Medicinal Plants of
the Seminole Indians. Used as a remedy for kidney trouble. The leaves are
used to make a tea" / "Blue-violets"

797   *V. diffusa* China: Kwangtung / W. T. Tsang 20214 / 1932 / "Medicinal
value"

FLACOURTIACEAE

798   *Abatia verbasciflora* Colombia / J. M. Duque Jaramillo 3113 / 1946 /
" 'Duraznillo' "

799   *Banara guianensis* var. *guianensis* Brit. Guiana / For. Dept. of Brit.
Guiana F 2979 / 50 / "fr. has an unpleasant smell"

800   *B. regia* Ecuador / J. A. Steyermark 54217 / 1943 / "Ants of biting stinging
type found among flower"

801   *Barteria fistulosa* Cameroons / E. Bates 1279 / no year / "the hollow stems
. . . are inhabited by a kind of large fiercely black ant, so that the tree is
carefully avoided" / " 'engoom' or 'mebongo' "

802   *Carpotroche amazonica* Brazil / Prance et al. 15584 / 1971 / "the bark
scraped and put in armadillo holes as a poison to catch them by Makú Indians" / " 'Warapash' "

803   *C. longifolia* Ecuador / Cazalet & Pennington 7531 / 62 / "Frt . . . with
HCN scent"
Peru / G. Klug 1121 / 1930 / "Fruit edible" / " 'Huira guayo' (Inca: huira-
lard, guayo-fruit)"

804   *C. pacifica* Colombia / J. M. Idrobo 1887 / 1955 / "Semillas granate . . .
muy perseguido por los niños de la región" / " 'Chocolate de indio' "

805   *Casearia aculeata* Brit. Honduras / P. H. Gentle 3081 / 1939 / " 'wild
lime' "
Santo Domingo / P. Fuertes 259 / 1910 / " 'Limoncillo' "
Brazil / G. & L. T. Eiten 10285 / 1970 / " 'cafe brabo' "

806   *C. arguta* Mexico / G. B. Hinton 10270 / 37 / "fruit edible" / "Vernac.
name: Bonetillo"

Colombia / J. A. Duke 9817 / 67 / ". . . the pulp around the seed is edible" / " 'raspa lengua' "

2807   *C. corymbosa* Mexico / J. N. Rovirosa 515 / 1889 / "Cafeillo"

2808   *C. nitida* Costa Rica / A. Gentry 809 / 1971 / "Cafecillo' "

2809   *Casearia* indet. Mexico / P. Moreno 327 / 1975 / "fruto . . . lo comen las aves"

2810   *C. decandra* Puerto Rico / E. L. Little 13026 / 1950 / "Yellowish fruit, edible" / " 'cerezo' "

2811   *C. sylvestris* var. *sylvestris* Jamaica / A. Moore 10646 / 1908 / "Leaves used for feeding stock"
Haiti / G. V. Nash 605 / 1903 / "Fls. yellowish, smelling something like old vinegar" /
Brazil / A. Silva 58 / 1944 / "Frt. yellow, edible"

2812   *C. tremula* Netherlands Antilles / Fr. M. Arnoldo-Broeders 3793 / 1969 / " 'Guayaba baster' 'Palu de veneno' "
Venezuela / A. Bernardi 2499 / 1955 / "Manzano"

2813   *C. combaymensis* Brazil / R. S. Cowan B8432 / 1954 / ". . . seeds covered with sweet, odorific pulp"
Brazil / Prance et al. 16344 / 1971 / "the fruit eaten by Deni Indians" / " 'Mukuvava' (Deni)"

2814   *C. grandiflora* Fr. Guiana / Fr. Guiana For. Service 7042 / 1955 / "Bois de feu—Cendre résidu employé en mélange avec le tabac prise par iles primitifs" / "BITA TIKI (Idiome Paramaka)"
Brazil / G. & L. T. Eiten 10702 / 1970 / " 'farinha sêca' "

2815   *C. guianensis* Brazil / A. Ducke 2407 / 1955 / " 'Café bravo' "

2816   *C. javitensis* Venezuela / A. Bernardi 1925 / 1955 / "Cafecillo"

2817   *C. macrophylla* Fr. Guiana / R. A. A. Oldeman 2654 / 1964 / ". . . comestible"

2818   *C. negrensis* Brazil / Berg et al. P18646 / 1973 / " 'Canella de cutia' "

2819   *C. pitumba* Brazil / Berg et al. P18600 / 1973 / " 'Caferana da terra firme' "

2820   *C. prunifolia* Peru / T. B. Croat 18754 / 1972 / "fruits orange; seeds surrounded by very tasty sweet, gelatinous pale orange substance"
Peru / R. Kayap 1320 / 1974 / " 'uci nahahaip' "

2821   *C. zizyphoides* Venezuela / Saer 265 / 1925 / "Guyabito"
Venezuela / H. Pittier 10520 / 1922 / "fruit black, sweet"

2822   *C. cinera* Philippine Is. / F. Paraiso 248 / 1929 / "Leaves are boiled & water taken to cure culebra" / "Lamlamuyot Dialect Ilocano"

2823   *Casearia* indet. New Guinea / D. Frodin NGF 26596 / 66 / "Leaves are chewed and the extract spat onto wounds to relieve soreness"

2824   *Doryalis zeyheri* S. Africa / B. Kellogg 864 / 1939 / "Fruit orange. Edible"

325 **D. hebecarpa** Hawaii Is. / P. C. Hutchison 2794 / 1967 / " 'Ceylon Gooseberry' 'Kelambilla' "

326 **Flacourtia indica** Tanganyika / R. E. S. Tanner 1502 / 1953 / "Sukuma tribe: Roots pounded up, eaten to treat leprosy" / "Kisukuma: Mbuguswa" Hainan / C. I. Lei 586 / 1933 / "fruit red, edible" / "Lah Tz"

327 **F. euphlebia** Philippine Is. / Mallonga & Burnea 31167 / 1929 / "Used for the manufacture of pestles for pounding rice" / "Banauo Dialect Visayan"

328 **Homalium racemosum** Panama / N. Bristan 1180 / 1967 / "bark with odor" Honduras / A. & A. R. Molina 24785 / 1969 / " 'Quina' "

329 **H. longifolium** Cochinchina / L. Pierre 30 / 1867 / "Charbon excellent pour la fabrication de la poudre" / "tnnam: chá ran. Kmer: dónstéo"

330 **H. zeylanicum** Ceylon / Jayasuriya & Sumithraarachchi 1556 / 1974 / "flowers smell rotten dried fish" / " 'Eta-Heraliya' "

331 **Hydnocarpus octandrus** Ceylon / Jayasuriya & Sumithraarachchi 1559 / 1974 / "bark smell almond"

332 **Kiggelaria africana** S. Rhodesia / N. C. Chase 572 / 1947 / "Seeds much sought after by green pigeons"

333 **Lindackeria latifolia** Brazil / E. Oliveira 4324 / 1968 / " 'Farinha sêca' "

334 **L. paraensis** Brazil / J. M. Pires 51862 / 1961 / " 'canela de velha' "

335 **Ludia sessiflora** Tanganyika / R. E. S. Tanner 3704 / 1957 / "Zigua: Roots boiled and the water drunk as a male aphrodisiac" / "Zigua: mzuchi zuchi"

336 **Neosprucea grandiflora** Brazil / Krukoff's 4th Exped. to Braz. Amaz. 5812 / 1933 / "Bark is scraped off and placed in water and then is ready for cozsu. Drink resembles beer" / "Sazacuza cozsu"
Brazil / Prance et al. 16486 / 1971 / "the stem bark scraped & mixed with water & stirred, then drunk as a cure for stomach complaints" / " 'Mado' "

337 **Paropsia laevigata** Sierra Leone / J. M. Dalziel s.n. / 1912 / "The very sticky white latex is used as a bird lime. Young leaves ground with flour & dried are taken as a laxative" / "Awan (Temne); Bonjie (Mende)"

338 **Ryania mansoana** Brazil / D. R. Gifford G. 119 / 68 / "Roots pounded make a strong poison as food. The smoke from burning it said to be fatal" Brazil / Ratter et al. 1306 / 1968 / "Very poisonous, hence vernac. name (= it kills silently)" / "Vernacular name: Mata Calado"

339 **R. speciosa** var. **bicolor** Brasil / R. de Lemos Fróes 21157 / 1945 / " 'Dveahochequê' (Mata calado)"

340 **R. speciosa** var. **minor** Brazil / Prance et al. 15587 / 1971 / "Bark formerly used as human poison to end lives of elderly" / " 'Caramã' (Makú)" Brazil / R. Fróes 182 / 1942 / "Poisonous" / "Uairú mirá"

341 **R. speciosa** var. **stipularis** Venezuela / F. Bianco 74 / 1976 / "fruto rosado el cual, al ser ingerido por los animales, produce ciertos efectos de ebriedad"

Venezuela / J. A. Steyermark 60982 / 1945 / "fruit poisonous to all cattle, goats, mules, burros, and horses, but doesn't poison lapa or acure" / " 'aguacero' "
Venezuela / F. Flamayo 628 / 1938 / "Arbol venenoso del oriente de Venezuela" / " 'Ciezo' "

2842 **R. speciosa** var. **subuliflora** Brasil / R. de Lemos Fróes 20849 / 1945 / " 'Mata calado' "
Brazil / W. A. Archer 8451 / 1943 / "Root poisonous, used to kill dogs" / " 'mata calado' "

2843 **Ryparosa** indet. New Guinea / P. van Royen NGF 20121 / 64 / "Flowers greenish white, with sickly sweetish smell"

2844 **Scolopia brownii** Australia: Queensland / M. S. Clemens s.n. / 1945 / "Birds eat abundant fruit"

2845 **Xylosma tessmannii** Peru / J. Schunke V. 4879 / 1971 / " 'Limón casha' "

2846 **Xylosma** indet. Venezuela / E. Foldats 155A / 71 / "Limoncillo"

2847 **X. suaveolens** Marquesas Is. / F. B. H. Brown 429 / 1921 / "Berries used for making ink" / " 'piapian' "

PERIDISCACEAE

2848 **Peridiscus** indet. Brazil / A. Ducke 113 / 1932 / " 'Pao santo' "

TURNERACEAE

2849 **Turnera ulmifolia** Bahamas / O. Degener 18816 / 1946 / "tea from leaves medicinal" / " 'Buttercup' "
Bahamas / A. E. Wight 16 / 1905 / "Leaves strong scented, slightly aromatic"
W.I.: Grand Cayman Is. / W. Kings GC 391 / 38 / "Leaves used as a medicine for liver & kidney trouble" / "Catbush"

PASSIFLORACEAE

2850 **Adenia** indet. Tanganyika / R. E. S. Tanner 1284 / 1953 / "Sukuma tribe: Roots and leaves used as poultice to treat headache, leaves pounded, soaked, taken orally for snake bite" / "Kisukuma: Ibombolwa lya muluguru"

2851 **Adenia** indet. Liberia / R. M. Warner 28 / 1942–43 / "Fish poison"

2852 **Dilkea acuminata** Colombia / J. A. Duke 11254 / 1967 / "seeds with a juicy tasteless hyaline pulp"

2853 **Passiflora adenopoda** Nicaragua / A. Molina R. 20400 / 1967 / " 'grenadilla de monte' "

Colombia / Idrobo & Jaramillo Mejía 2252 / 1956 / "arilos anaranjados de sabor dulce, comestibles"

2854 *P. ambigua* Panama / T. B. Croat 13010 / 1971 / "seeds at first sweet, becoming sour after sucking for a while"

2855 *P. biflora* Brit. Honduras / P. H. Gentle 3823 / 1941 / " 'melon de raton' "
Colombia / H. H. Smith 1597 / 1898–99 / "The fruit has a pleasant flavor but is not eaten by the natives"
Philippine Is. / E. Canicosa 909 / 1952 / "Frt. edible"

2856 *P. bryoniodes* Mexico / G. B. Hinton 4187 / 33 / "Granada cimarrona"

2857 *P. ciliata* Mexico / J. N. Rovirosa 560 / 1889 / "Jujito colorado"

2858 *P. foetida* El Salvador / P. C. Standley 21633 / 1922 / "Fr. eaten" /
" 'Grenadilla montés' "

2859 *P. foetida* var. *gossypiifolia* Mexico / Y. Mexia 8747 / 1937 / "edible fruit"
El Salvador / P. C. Standley 19727 / 1922 / "Plant with strong offensive odor. Fr. eaten" / " 'Granadilla silvestre' "
Mexico / G. B. Hinton 8045 / 35 / "Granada de zorro"

2860 *P. foetida* var. *hibiscifolia* Mexico / G. Aguirre B. 12 / 1945 /
" 'chichicamole' "

2861 *P. jorullensis* Mexico / G. B. Hinton 7834 / 35 / "Granada de Zorro"
Mexico / G. B. Hinton 6558 / 34 / "Bad smelling"

2862 *P. nitida* Panama / T. B. Croat 16505 / 1971 / "seeds eaten"

2863 *P. palmeri* Mexico / I. L. Wiggins 5451 / 1931 / "Fruit with not unpleasant, slightly acid taste"

2864 *P. platyloba* Costa Rica / J. L. Whitmore 32 / 1967 / "Crushed leaf has very sweet odor"
El Salvador / P. C. Standley 19487 / 1921 / " 'Granadilla montés' "

2865 *P. pulchella* El Salvador / S. Calderon 1659 / 1923 / "Has diuretic properties" / " 'Calzoncillo' "

2866 *P. cuprea* Bahamas / R. A. & E. S. Howard 9993 / 1948 / " 'Devils pumpkin' "

2867 *P. foetida* var. *ciliata* Jamaica / W. Harris 11816 / 1914 / "Pulp edible"

2868 *P. foetida* var. *hispida* W.I.: Martinique / P. Duss 8856 / 1877 / "fruits muis jaunes, bon à manger" / "Vulgo: Marie Caujat"
Ecuador / W. H. Camp E-3742 / 1945 / "Fr. said to be eaten"
Surinam / E. A. Mennega 1000 / 1949 / "Strong scent of tomato plants" / "sosapóro (Car.)"

2869 *P. penduliflora* Cuba / E. L. Ekman 2747 / 1914 / " 'Bejuco manteca' cub."

2870 *P. rubra* W.I.: Dominica / W. H. Hodge 3284 / 1940 / " 'pomme de liane zombie' "

Ecuador / F. Prieto CP-36 / 1944 / "Fruits said to containe a narcotic; used by adding to chicha"

2871 **P. serrato-digitata** W.I.: Dominica / W. H. & B. T. Hodge 2846 / 1940 / "according to H. L. Shillingsford fruit tastes like that of P. edulis but is not generally eaten"

2872 **P. suberosa** Cuba / A. Luna 395 / 1920 / " 'Meloncillo' "

2873 **P. aristulata** Peru / Y. Mexia 6424 / 1932 / "edible fruit" / " 'Granadilla' "

2874 **P. auriculata** Colombia / J. M. Idrobo 2196 / 1956 / "Frutos morado-oscuros, comestibles, de sabor dulce"

2875 **P. caerula** Argentina / W. A. Archer 4614 / 1936 / "pulp around seed is red, edible"

2876 **P. coccinea** Peru / J. Schunke V. 1165 / 1966 / "Dice Schunke que la raíz es abortiva" / " 'Granadilla colorada' "
Bolivia / H. H. Rusby 475 / 1921 / "Fruit yellow, edible"

2877 **P. coriacea** Peru / J. Schunke V. 3823 / 1970 / "Las hojas en infusión las toman los nativos para curarse de infeccion en el bazo" / " 'Bazo Sacha' "

2878 **P. costata** Brazil / R. S. Cowan 38543 / 1954 / "Fruit edible"

2879 **P. cuadrangularis** Colombia / J. M. Duque Jaramillo 4644 / 1947 / "Frutos grandes, verdosos y comestibles. Cultivada" / " 'Badea' 'Granadillo grande' "

2880 **P. cumbalensis** Colombia / J. M. Duque Jaramillo 3074 / 1946 / ". . . frutos rosados, éstos apenas comestibles" / " 'Curuba' "

2881 **P. foetida** var. **foetida** Peru / A. Sagástegui A. 7877 / 1973 / "Sus bayas dulceinas son comestibles" / " 'granadilla de culebra' "
Peru / Simpson & Schunke 556 / 1968 / "Fruit edible; leaves as medicine for 'Carache'—translation unknown, may mean a skin fungus—(economic uses fide Pablo Leon)"
Hawaiian Is. / J. A. Harris C242288 / 1924 / " 'Lemon apple' "
Caroline Is. / F. R. Fosberg 24426 / 1946 / "fruit orange, sweet and of fair flavor"

2882 **P. foetida** var. **hirsuta** Peru / Killip & Smith 27828 / 1929 / "fruit edible" / " 'Puru-puru' "

2883 **P. foetida** var. **sanctae-martae** Colombia / H. H. Smith 1532 / 1898–99 / " 'Gallinago' 'Maiz tostada' "

2884 **P. foetida** var. **strigosa** Brazil / B. A. Krukoff 1208 / 1931 / "Seeds are poisonous" / "Maracuja de rato"

2885 **P. manicata** Ecuador / E. Heinrichs 181 / 1933 / "Früchte essbar" / " 'Taseo' "
Ecuador / F. Prieto E-2535 / 1945 / "the juice of the fruit used to bathe 'nervous' babies. (probably this treatment 'gums up' the lice and fleas, cutting down their perambulatory and other activities, and thus soothes the child by getting at the source)"

886 *P. moriflora* Bolivia / J. Steinbach 5349 / 1921 / "Es comible pero poco apetecible"

887 *P. serrulata* Venezuela / L. Aristeguieta 4723 / 1961 / "Comestible" / " 'Parachita' "

888 *P. triloba* Bolivia / H. H. Rusby 739 / 1921 / "Fruit edible"

889 *P. vitifolia* Colombia / J. A. Ducke 9864 / 1967 / "fruits green, eaten by the natives"

890 *Passifloracea* indet. Bolivia / Herb. Cardenasium 571 / 1933 / "Edible fruits"

CARICACEAE

2891 *Carica cauliflora* Venezuela / A. L. Bernardi 1911 / 1955 / "Frutos comestibles" / "Tapaculo"

2892 *C. goudotiana* Colombia / H. St. John 20827 / 1944 / "Flesh with taste of mild acid apple, pulp around seeds milder and more pleasant. Natives use flesh of fr. with sugar as a dessert" / " 'papayuelo' "

2893 *C. horovitziana* Ecuador / E. André 4229 / 1876 / "fructus edulis"

2894 *C. microcarpa* var. *microcarpa* Venezuela / E. Ijjász 353 / 1964 / " 'Lechoza de monte' "

2895 *C. parviflora* Peru / R. Espinosa 1056 / 46 / "Flores rojos carmín. Se la cree venenosa"

2896 *C. sphaerocarpa* Colombia / Fassett & St. John 25242 / 1944 / "Fruit . . . edible when ripe" / " 'Papaya del monte' "

2897 *C. papaya* Caroline Is. / F. R. Fosberg 32106 / 1950 / "said to be used as medicine, leaves and roots pounded, placed in bag, soaked in water, drunk for tuberculosis"

2898 *Jacaratia digitata* Brazil / Prance et al. 7816 / 1968 / "Latex used as a vermifuge" / " 'Javacatia' "
Peru / G. Klug 1222 / 1930 / "The natives cut this tree and make holes in it, so that in the soft part are formed maggots of the size of a fat finger (Coleoptera Caterpillers) for the purpose of eating them" / " 'Shamb 'uru' "

LOASACEAE

2899 *Cevallia sinuata* U.S.: Tex. / Dr. Gregg s.n. / 47 / "nettle—very severe sting" / "Ortega, or ortiguilla"

2900 *Gronovia longiflora* Mexico / G. B. Hinton 13131 / 38 / "Stings"
El Salvador / P. C. Standley 19210 / 1922 / "Hairs sting flesh badly" / " 'Pega-pega, Juancaliente' "

2901   *G. scandens* Mexico / Y. Mexia 8734 / 1937 / "stem with pricking hairs" /
        " 'Quemador' "

2902   *Mentzelia gracilenta* var. *veatchiana* U.S.: Calif. / C. C. Parry s.n. / 1850 /
        "The seeds are used for shortening, in making Piñolé"

2903   *M. arborescens* Mexico / Y. Mexia 9116 / 1938 / "Twigs made into infusion
        for renal disorders" / " 'Zagaduche' (Zapotecan)"

DATISCACEAE

2904   *Datisca cannabina* Persien / College of Pharmacy Herb. s.n. / no year /
        "Das Decoct der Wurzel und des Krautes ist gegen Fieber in Italien und
        England in Gebrauch" / "Hanfartiges Strickkraut"

2905   *Octomeles* indet. New Guinea / D. Frodin NGF 26715 / 66 / "Odour bitter"

BEGONIACEAE

2906   *Begonia gracilis* var. *mariana* Mexico / C. V. Hartman 744 / 1891 / "mex.
        'Pito moreal' "

2907   *B. gracilis* var. *martiana* Mexico / Martius Herb. s.n. / 1845 / " 'sangre de
        doncella' "

2908   *B. semperflorens* W.I.: Guadeloupe / P. Duss 3567 / 1894 / " . . . on el
        pousse a poison"

2909   *B. juntasensis* Bolivia / J. Steinbach 5012 / 1920 / "Los indigenas ocupan
        las hojas para hacer una bebida fresca" / "Binu-binú"

2910   *B. rossmaniae* Ecuador / C. C. Fuller 103 / 53 / "Infusion of leaves used to
        cure itch" / "Sinjimbu: Quichua"

2911   *Begonia* indet. Argentina / J. E. Montes 14728 / 1955 / "Uso: medicinal" /
        " 'Agrial' "

2912   *B. fimbristipula* China: Kwangtung / W. T. Tsang 21542 / 1932 / "Used
        to make medicine" / "Tin Kwai"
        China: Kwangtung / W. T. Tsang 20502 / 1932 / "for medicine"

2913   *B. dipetala* Ceylon / A. H. M. Jayasuriya 2058 / 1975 / "lvs. & fl. taste
        sour and then very bitter"

2914   *B. oxysperma* Philippine Is. / A. D. E. Elmer 7533 / 1906 / "the stalks
        possess an acrid taste and are eaten by the natives"

CACTACEAE

2915   *Cactus maxoni* Guatemala / J. A. Steyermark 43867 / 1942 / "fruit deep
        rose-red, sweet, edible. Pulpy inside when boiled, used for inflammation of
        stomach" / " 'chile' "

2916 *C. acunai* Cuba / Fre. Leon 12410 / 1924 / " 'Melón de costa' "

2917 *C. intortus* W.I.: Virgin Is. / Little & Woodbury 2373 / 1969 / "Fruits pink, edible" / "turkshead cactus"

2918 *Cephalocereus millspaughii* Cuba / J. A. Schafer 725 / 1909 / "odor of garlic"

2919 *Cereus robinii* var. *keyensis* U.S.: Fla. / Mr. Bennett s.n. / no year / " 'Flowers smell of Garlic' "

2920 *C. hexagonus* W.I.: Guadeloupe / P. Duss 3075 / 1895 / ". . . fruits mangeables" / "Ciergé pascal"

2921 *Hylocereus undatus* W.I.: Martinique / P. Duss 904 / 1884 / "fleurs . . . d'une odeur forte et d'agréable" / "Chardon liane"

2922 *H. lemairei* Trinidad / W. E. Broadway 3154 / 1907 / "Fruits . . . pleasantly tasting"

2923 *Lemaireocereus* indet. Mexico / Y. Mexia 893 / 1926 / "Fruit edible" / " 'Pitahaya' "

2924 *Opuntia scheeri* Mexico / A. Berger Succulent Herb. s.n. / no year / "Fleisch rot, saft rot, süss und wohlschmeckend"

2925 *O. tuna* W.I. / A. Berger Succulent Herb. s.n. / no year / "Fleisch rot, Saft blutrot, säuerlich süss"

2926 *Opuntia* indet. Cuba / Herb. Collegii Pharmaciae Neo-Eboracensis s.n. / no year / "The fruit gives a crimson dye" / "Tuna colorado"

2927 *O. soehrensis* Bolivia / Mr. & Mrs. J. N. Rose 1892 / 1914 / "From the seed confined within the round and spiny fruit is derived a color of a clear violet, brilliant and extremely agreeable to the eye, but very superficial and very light"

2928 *O. tomentosa* Sicily / College of Pharmacy Herb. s.n. / no year / "Auch auf dieser Art wird Cochenille gezogen" / "Filzige Nopalpflanze"

2929 *O. dillenii* Tanganyika / R. E. S. Tanner 1148 / 1952 / "Fruits edible" / "Kisukuma: Sanvenge"

2930 *O. monacantha* India / W. Koelz 3219 / 1931 / "fruit edible"

2931 *Pereskia nicoyana* Costa Rica / A. Berger Succulent Herb. s.n. / no year / "The fruit very aromatic. The leaves for cataplasms. Used in the preparation of Ung-Altea" / "Mateáre"

2932 *P. weberiana* Bolivia / H. H. Rusby 116 / 1921 / "Fruit edible"

THYMELAEACEAE

2933 *Aquilaria sinensis* Hainan / F. A. McClure 20072 / 1932 / "The injured wood develops or deposits a fragrant substance and when isolated by decay the pieces of this wood have a high commercial value. This is the incense wood of Hainan chips secured" / "Hio Ch'a (H) Loi: t'oon"

2934 **Daphne gnidioides** Turkey / Lambert & Thorp 556 / 1968 / "Constituent of maquis"

2935 **D. gnidium** Italy / College of Pharmacy Herb. s.n. / no year / "Im sudlichen und westlichen Europa wird die Rinde, Cortex Gnidii S. Thymelaeae, wie der Seidelbast angewendet. Die früher gebrauchlichen Kellerhalskörner, Semen Coccognidii s. Grana Gnidii, kommen von diesem Strauche" / "Italienischer Seidelbast"

2936 **Daphnopsis americana** var. **guatemalensis** Costa Rica / W. C. & M. Burger 7696 / 1971 / "Stems . . . have a sharp odor"

2937 **D. americana** var. **caribaea** W.I.: Dominica / F. E. Lloyd 666 / 1903 / "bark has pepper like taste" / " 'Mahaut piment' "
W. I.: Tobago / W. E. Broadway 3413 / 1909 / "Fruits white, eaten by the Blue Pigeon" / " 'Burn nose' "
W.I.: Dominica / W. H. Hodge 1534 / 1940 / "bark aromatic and used for rope"

2938 **D. macrophylla** Ecuador / W. H. Camp E-3336 / 1945 / "The ripe fruits a potent purgative, and often used by the local gente who seriously caution that, in its use one must take it as follows: 1–½ fruits, or 2–½ fruits; but must never be taken as 1 or 2 fruits, otherwise it is quite poisonous (i.e. there must always be ½ of a cut fruit in the dose)" / " 'Zapan' "

2939 **Dirca palustris** U.S.: Ill. / V. H. Chase 5025 / 1934 / "Probably a pre-historic planting"

2940 **Funifera** indet. Venezuela / J. A. Steyermark 58569 / 1944 / " 'imuyek' (mata de almedón)"

2941 **Gnidia capitata** S. Africa / J. Pentz 15977 / 1933 / "Plants listed in connection with suspected poisoning in cattle by Director, Vet. Research Lab."

2942 **G. polycephala** S.W. Africa / R. Seydel 4431 / 66 / "Blueten mit Honigduft. Giftig?" / "Tollkraut"
S. W. Africa / R. Seydel 3578 / 1963 / ". . . soll sehr giftig sein"

2943 **G. spicata** AFR. / F. M. Leighton 1675 / 46 / "Plants eaten down by animals"

2944 **Gonystylus** indet. Borneo / A. Kostermans 13901 / 1957 / "Fr. pale brown, used as fish poison" / " 'Tempe-eng' (Bassap-Mapulu)"

2945 **Lagetta wrightiana** Cuba / J. A. Schafer 8860 / 1911 / " 'Guava' "

2946 **Phaleria perrottetiana** Philippine Is. / A. D. E. Elmer 9952 / 1908 / "the scant whitish meat juicy and bitterish to taste" / " 'Malacopa' "

2947 **Phaleria** indet. Philippine Is. / M. Oro 221 / 1929 / "Fruit edible when fresh 2 or 3 times as larger as the dry condition"

2948 **Synaptolepis kirkii** Tanganyika / R. E. S. Tanner 2597 / 1956 / "flowers white, aromatic" / "Kizigua: Mchole"

949   *Thymelea tartonraira* France / College of Pharmacy Herb. s.n. / no year / "Die Blätter in Sardinien als Brech—und Purgiermittel officinell"

950   *Wikstroemia stenantha* China: Kwangtung / Lingnan (To & Ts'ang) 12423 / 1924 / "Ye min t'o (Wild downy peach)"

951   *W. indica* Hainan / C. I. Lei 689 / 1933 / "Fish may be poisoned by the roots / "Por Mun"

952   *W. uva-ursi* Hawaii Is. / O. & I. Degener 27510 / 1961 / "avoided by cattle"

953   *W. viridiflora* New Hebrides / S. F. Kajewski 302 / 1928 / "Bark taken and thrown into pools to stupefy fish so they are easily caught by hand" / " 'Too-wop' "

954   *Wikstroemia* indet. Hawaii Is. / L. M. Andrews 336 / 1968 / "made into deadly drink for suicide or execution. Fish poison" / "Akia"

ELAEAGNACEAE

2955   *Elaeagnus angustifolia* U.S.: Penn. / T. C. Porter s.n. / 1889 / "Fruit fleshy, edible, pleasant to the taste"
EUR. / Herb. of A. Woods s.n. / no year / "Dieser Strauch soll das wahre Elemi der früheren Griechen und Araber liefern" / "Schmalblättriger Oleaster"
Kashmir / R. R. Stewart 20566 / 1940 / "Fruit used . . . usually cultivated"

2956   *E. pungens* U.S.: Calif. / A. Griffiths 3967 / 60 / "strongly fragrant, spicy"

2957   *E. bockii* China: Yunnan / G. Bonati 383P / 1909 / "drupes roses comestibles"

2958   *E. cuprea* China: Kwangsi / W. T. Tsang 24513 / 1934 / "fr. red, edible" / "Tiu Chung Tsz Tang"

2959   *E. lanceolata* China: Kweichow / Steward, Chiao & Cheo 666 / 1931 / "Fr. edible in March when mature"

2960   *E. gonyanthes* Indochina / F. A. McClure 90738 / 32 / "fruits eaten" / "Lai Ngoht"

2961   *Elaeagnus* indet. India & Burma / F. Kingdon Ward 18529 / 1949 / "As they rot on the ground, the fruits emit a disgusting odour"

2962   *Hippophaë rhamnoides* Himalaya / W. Koelz 1292 / 1932 / "Fruit used for silver ornaments. Boiled with silver ornaments. Medicine for unsettled stomach" / "Tairua"

LYTHRACEAE

2963   *Adenaria floribunda* Colombia / J. M. Duque Jaramillo 2616 / 1946 / " 'Guayabillo' "

2964   *Cuphea aequipetala* var. *hispida* Mexico / Hinton et al. 8246 / 36 / "Hierba del cancer"

2965   *C. sanguinea* Guatemala / J. A. Steyermark 51618 / 1942 / " 'Chokshan' "

2966   *C. wrightii* var. *wrightii* Honduras / A. Molina R. 26241 / 1971 / ". . . between Acropolis and Jaguar Temple of Copán Ruins"

2967   *C. linarioides* Argentina / J. E. Montes 27570 / 58 / "pegajosa al tacto" / " 'Sietesangrias' (Variedad)"

2968   *C. racemosa* Colombia / J. M. Duque Jaramillo 3647 / 1946 / " 'Yerba-buenilla' "

2969   *C. strigulosa* Peru / Hutchison & Wright 3385 / 1964 / "Tops of plants used in tea for upset stomach" / " 'Sangre Toro' "

2970   *Lawsonia inermis* Colombia / H. H. Smith 2739 / 1898–99 / "Flowers whitish, with a strong odour of raspberries. An infusion of the flowers is used as a remedy for intestinal complaints" / "Called *resada*" Sumatra / W. N. & C. M. Bangham 614 / 1931 / "Natives make red dye from leaves, to use on fingernails" / " 'Partja' "

2971   *Nesaea cordata* Tanganyika / R. E. S. Tanner 3666 / 1957 / "Whole plant used as a vegetable" / "Swa: Kidevu cha Mbuzi"

2972   *Pehria compacta* Venezuela / J. A. Steyermark 61080 / 1945 / "Birds eat fruit" / " 'indiacita' or 'coralita' "

2973   *Pemphis acidulis* Caroline Is. / F. R. Fosberg 32214 / 1950 / "leaves slightly acid and astringent when chewed"

2974   *Lythracea* indet. Rhodesia / P. J. Tyrer 163 / 62 / ". . . leaves and stem all smelling of cat"

SONNERATIACEAE

2975   *Duabanga borneensis* Philippine Is. / M. Oro 115 / 1929 / "sulphur like substance filling the pores of the heartwood"

2976   *Sonneratia alba* Marshall Is. / F. R. Fosberg 41391 / 1960 / "sweet nectar from flowers eaten by Marshallese" / " 'pulabl' "

LECYTHIDACEAE

2977   *Allantoma lineata* Venezuela / Wurdack & Adderley 43035 / 1959 / "bark used as cigarette wrapper" / " 'Tabari' "

2978   *Barringtonia asiatica* Fiji Is. / O. Degener 15104 / 1941 / "fruit scraped on stones and mashed, then used as fish-poison" / " 'Vutu' (Serua)" / Marshall Is. / F. R. Fosberg 34028 / 1952 / "Plant used medicinally" / " 'wuj', 'woip' "

Marshall Is. / F. R. Fosberg 26809 / 1946 / "fruits used to poison fish" / " 'wop' "

Caroline Is. / F. R. Fosberg 26328 / 1946 / "said to be used for fish poison in Kity" / " 'wi' "

979   **B. edulis** Fiji Is. / Degener & Ordonez 13993 / 1943 / "Kernel edible" / " 'Vutu' "

980   **B. excelsa** New Hebrides / S. F. Kajewski 35 / 1928 / "It is very pleasant to eat" / " 'Nevingen' "
New Hebrides / S. F. Kajewski 247 / 1928 / "nuts greatly prized & eaten by natives" / " 'Velingeh' "

981   *Cariniana micrantha* Brazil / B. A. Krukoff's 4th Exped. to Braz. Amaz. 5095 / 1933 / "Wood is occasionally sold in mixture with 'Switenia' as bakorin" / "Cornimboque"

982   *Couratari guianensis* Venezuela / J. A. Steyermark 87065 / 1960 / "The inner living wood under the bark is tapped, separated into strands, and wrapped around as cigars for smoking" / " 'Tapa tabaco' "
Colombia / Nee & Mori 3770 / 1971 / "Wood smells like rotten eggs"
Fr. Guiana / Fr. Guiana For. Service 7727 / 1957 / "cigare gris verdâtre" / "Nom commercial: Mahot cigare"

983   *C. multiflora* Surinam / I. Zanderij 44 / '44 / "used for cigarette paper" / "OELEMARI, ingipipa"

984   *C. oblongifolia* Fr. Guiana / Fr. Guiana For. Service 7728 / 1957 / "cigares brun rougeâtre" / "Nom commercial: Mahot cigare"

985   *Couroupita idolica* Panama / J. D. Dwyer 1311 / 61 / "this is the 'Idol Tree' for years worshipped by natives during Holy Week"

986   *C. amazonica* Peru / A. Gutierrez Ruiz 209 / 1965 / "Los frutos maduros son comestibles para las aves de corral" / " 'Ayahuma' "

987   *C. guianensis* var. *guianensis* Venezuela / J. A. Steyermark 61484 / 45 / " 'coco de mono' "

988   *Eschweilera longipes* Guayana / L. M. Berti 99 / 1964 / "Cacao grande"

989   *E. subglandulosa* Brit. Guiana / Jenman 2403 / 1886 / "bark yields a valuable bast" / " 'Kakeralli' "
Venezuela / J. A. Steyermark 86775 / 1968 / " 'tapa de tabaca' "
Venezuela / Wurdack & Monachino 39410 / 1955 / " 'Cacao' "

990   *E. tenax* Venezuela / J. A. Steyermark 61112 / 1945 / "Wood useful for pests and beams . . . fruit eaten by animals" / " 'coco de mono' or 'cacao' "

991   *Eschweilera* indet. Venezuela / J. A. Steyermark 90815 / 1962 / "flowers with fragrance of *Magnolia* . . . loaded with stinging ants"

992   *Eschweilera* indet. Peru / Killip & Smith 27865 / 1929 / "Fruit edible"

2993   *Lecythidacea* indet. Venezuela / M. M. Suárez 35 / 1968 / "Uso: Gripe. Fiebre" / "Nombre indigena: Osibuakua-Rusori"

2994   *Lecythidacea* indet. Venezuela / Maguire, Steyermark & Maguire 53512 / 1962 / ". . . said to be remedy for colitis and dysentery—boil bark until water is discolored, then drink"

RHIZOPHORACEAE

2995   *Bruguiera cylindrica* Maldive Is. / F. R. Fosberg 36903 / 1956 / "embryo radicle boiled and eaten" / " 'Kandu' "

2996   *Carallia brachiata* China: Kwangtung / F. A. McClure 00525 / 1931 / "The fruits . . . are eaten by children and are said to be mildly subacid in flavor" / "Ngeh Sun Muk"

2997   *Ceriops tagal* Philippine Is. / A. D. E. Elmer 11986 / 1909 / "Tongal = the commercial bark for use as a ferment and a dye" / " 'Tongog' "

2998   *Rhizophora mangle* W.I.: Martinique / P. Duss 1836 / 1879 / "Uit en societe dans les maricages d'eau salei" / "Mangle rouge"
Colombia / H. H. Smith 1953 / 1898–99 / "The fruit . . . is used to preserve fish lines and color them red; the lines are rubbed with the fruit on a section of it"

2999   *R. stylosa* Mariana Is. / F. R. Fosberg 25362 / 1946 / "bark used to stain fish nets" / " 'white mangle' "

MYRTACEAE

3000   *Amomis caryophyllata* Jamaica / W. Harris 8523 / 1903 / " 'Bay Rum Tree' "
Cuba / E. L. Ekman 14839 / 1922 / " 'Pimiento' "
W.I.: Guadeloupe / H. Stehlé 115 / 1935 / ". . . exploité aboutement pour le bay-oil et les bay rhun"

3001   *A. grisea* W.I.: Virgin Is. / J. S. Beard 336 / 1944 / "leaves with aromatic essential oil" / " 'Wild cilliment' "

3002   *Angophora lanceolata* Australia: Queensland / C. J. Trist 4 / 1931–32 / " 'Brown Applegum' "
Australia: N.S.W. / C. T. White 5561A / 1927 / ". . . these were the most valued 'gum tips' of Sidney florists; the average value of them in the market being L6 per cartload"

3003   *Aulomyrcia hostmanniana* Brazil / W. A. Archer 8330 / 1943 / "Lvs. used as a cure for diabetes" / " 'pedra-hume-caá' "

3004   *A. minutiflora* Surinam / J. C. Lindeman 3559 / 1955 / "seed sticky with resinous odour"

3005   *A. tenuifolia* Brazil / R. Fróes 1993 / 1932 / "Bark used in making paint" / " 'Cumaty' "

006   *A. tomentosa* Colombia / E. L. Little 8987 / 1944 / " 'sanquemula', 'guayaro agrio' "

007   *Calycolpus glaber* var. *glaber* Brit. Guiana / J. S. de la Cruz 952 / 1921 / "seed edible" / " 'Kakurio' "

008   *Calyptranthes* indet. Mexico / B. P. Reko s.n. / 1946 / "The leaves and the red fruit . . . have a strong smell of lemon" / " 'thé de limon silvestre' "

009   *Calyptranthes* indet. Mexico / G. B. Hinton 7952 / 35 / "Fruit edible" / "Arrayan prieto"

010   *Calyptranthes* indet. Mexico / G. Aguirre B. 22 / 1945 / " 'té limón del monte' "

011   *C. syzygium* W.I.: Grand Cayman / W. Kings GC 413 / 38 / "Red Strawberry"

012   *Psidiopsis moritziana* Venezuela / J. A. Steyermark 55956 / 1944 / "Fruit good to eat, sweet" / " 'guayabo' "

013   *Psidium* indet. Costa Rica / A. M. Brenes s.n. / 1902 / ". . . fruto comestible" / "Cas"

014   *P. amplexicaule* W.I.: Brit. Virgin Is. / E. L. Little 26062 / 1972 / "Fruit edible"

015   *P. rotundatum* Cuba / Bro. Leon 12692 / 1926 / "fruit yellow, sweet"

016   *P. salutare* W.I.: Dominican Republic / J. de Js. Jimenez 2433 / 1952 / "the fruits are used for aromatizing alcoholic beverages" / " 'MANAGUA' "
Cuba / Schafer & Leon 13570 / 1912 / "fruit edible . . . aromatic"
W.I.: Isla de Piños / A. H. Curtiss 350 / 1904 / "It bears a large soft berry which is gathered in summer for use with aguardiente"

017   *P. acutangulum* Brit. Guiana / For. Dept. of Brit. Guiana F 2890 / 49 / "fruit . . . when ripe fleshy, edible / "Arisa"

018   *P. densicomum* Peru / F. R. Fosberg 29174 / 1947 / " 'guayaba', 'guayabilla' "
Peru / F. Woytkowski 5167 / 58 / "fruit used as bait for fishing" / "Yacu-guayabillo"

019   *P. galapageium* Galapagos Is. / I. L. Wiggins 18399 / 1964 / "Leaves with distinct, pleasant fragrance when crushed"

020   *P. rufum* Brazil / Y. Mexia 5714 / 1931 / "green, edible fruit" / " 'Araça' "

021   *P. sprucei* Venezuela / Aristeguieta & Zabala 7066 / 1969 / "Frutos verdosos, comestibles cuando maduros" / " 'Guayabo de agua' "

022   *Psidium* indet. Brit. Guiana / A. A. Abraham 254 / 1919 / "Guava"

023   *Psidium* indet. Argentina / J. E. Montes 14803 / 1955 / "Frutos comestibles" / " 'Guayabo' "

024   *Psidium* indet. Argentina / J. E. Montes 27604 / 58 / "Frutos comestibles" / " 'Guayabo' 'Güayabita' 'Arara' "

3025 **P. littorale** Fiji Is. / A. C. Smith 9509 / 1953 / "fruit used for jellies" / " 'Ngguava' "

3026 **Rubachia glomerata** Brazil / J. G. Kuhlmann 71948 / 1950 / "frutos amarelos, comestiveis" / "Cambucá"

3027 **Syzygium aromaticum** W.I. / Herb. of A. Wood s.n. / no year / "Liefert die Gewürznelken, Caryophylli und die Mutternelken, Anthophylli" / "Gewürz nelkenbaum"
W.I.: Grenada / W. E. Broadway s.n. / 1905 / " 'Clove' "

3028 **S. cumini** Venezuela / E. Tejera s.n. / 1974 / "fruto morado-negro, comestible"
Hainan / S. K. Lau 8 / 1932 / "Used as food" / "Tsopun (Lois)"

3029 **S. malacense** Venezuela / J. A. Steyermark 87485 / 1960 / "Fruit edible" / " 'pomarosa' "

3030 **S. hancei** China: Kwangtung / F. A. McClure 229 / 1921 / "Bark used in dying and make cloth receptive and retentive of dye" / "Tsz Ling Shue"

3031 **S. tetragonum** China: Yunnan / A. Henry 12650C / no year / "fr. edible" / "tien kan kuo"

3032 **S. bullockii** Hainan / C. I. Lei 1022 / 1932 / "fr. edible" / "Huck Hau Tsi"

3033 **S. claviflorum** Hainan / F. A. McClure 20128 / 1932 / "fruits edible"

3034 **S. caryophyllata** Ceylon / C. F. Baker 108 / 1907 / "The clove of commerce"

3035 **S. fibrosa** Australia: Queensland / S. F. Kajewski 1457 / 29 / "seeds abortive"

3036 **S. malacensis** Fiji Is. / A. C. Smith 6597 / 1947 / "fruit . . . edible" / " 'Kavika ndamu' "

3037 **S. multipetala** New Caledonia / I. Franc 2476 / 1930 / "fruits rouge violacée, comestibles"

3038 **S. neurocalyx** Fiji Is. / A. C. Smith 1008 / 1934 / "fruit oil used as a skin lotion" / " 'Lemba' "

3039 **S. rariflora** Fiji Is. / Degener & Ordonez 14217 / 1941 / "Fresh crushed leaves used for thrush ni babies" / "Tomitomi Dialect Vanua Levu"

3040 **S. samarangense** Borneo / J. & M. S. Clemens 6511 / 1929 / "Edible"

3041 **Syzygium** indet. Sumatra / W. N. & G. M. Bangham 875 / 1932 / "Frt. edible" / " 'Tinkeri' "

3042 **Syzygium** indet. Papua / C. E. Carr 15358 / 36 / "Fruit red. Eaten by natives in Gulf Div. but not by Biogi people" / "MAITA (Motuan) and MASISA (Orikolo)"

3043 **Syzygium** indet. New Hebrides / J. P. Wilson 1003 / 1929 / "Fr. eaten" / " 'Inyhueg' "

3044 **Syzygium** indet. New Hebrides / S. F. Kajewski 255 / 1928 / "fr. . . . eaten by natives" / " 'Wer-veh' "

45 *Ugni oerstedii* Costa Rica / Holm & Iltis 570 / 1949 / "Fruit . . . aromatic, bitter"

46 *U. molinae* Chile / W. J. Eyerdam 10681 / 1958 / ". . . the berries edible"

47 *Myrtacea* indet. Argentina / J. E. Montes 27613 / 58 / "Frutos comestibles" / " 'Güamiriu negro' "

48 *Myrtacea* indet. Argentina / J. E. Montes 195 / 44 / "frutos comestibles" / " 'Iguahay' 'Ubahay' "

49 *Myrtacea* indet. Argentina / J. E. Montes 14782 / 1955 / "Frutos comestibles?" / " 'Arazá-guarii' "

50 *Myrtacea* indet. Argentina / J. E. Montes 14798 / 1955 / "Frutos comestibles" / " 'Guayabo colorado' "

51 *Myrtacea* indet. Venezuela / L. Williams 14205 / 1942 / "los indíos Bamba preparan una infusion del liber, con la cual pintan las totumas" / "Curamo"

52 *Myrtacea* indet. Brazil / M. Silva 867 / "com odor de hervacidreira"

53 *Myrtacea* indet. Surinam / I. Zanderij 274 / '44 / "the red latex used (alin) for staining kölē(hē)" / "KÖLE(HE) ALIN"

54 *Myrtacea* indet. Venezuela / A. L. Bernardi 2405 / 1955 / "Frutos comestibles" / "Panijil"

55 *Myrtacea* indet. Argentina / J. E. Montes 27597 / 58 / "Frutos comestibles" / " 'Güamiriu negro' "

56 *Myrtacea* indet. Argentina / J. E. Montes 14801 / 1955 / "Frutos comestibles" / " 'Iguajay-mi' "

57 *Myrtacea* indet. Argentina / J. E. Montes 14802 / 1955 / "Frutos comestibles" / " 'Sete-capotes' 'Nandu-piusa' "

58 *Myrtacea* indet. Argentina / J. E. Montes 14764 / 1955 / "Frutos comestibles" / " 'Igua-jay' 'Ubajay-guazu' "

59 *Myrtacea* indet. Guayana / L. M. Berti 558 / 1965 / "N.v.: Carne asada"

60 *Myrtacea* indet. Colombia / H. Garcia y Barriga 18050 / 1964 / " 'MANZANO' "

61 *Myrtacea* indet. Brazil / G. Hatschbach 18654 / 1968 / "Fruto quase preto . . . sabor adstringente"

62 *Myrtacea* indet. Fr. Guiana / Fr. Guiana For. Service 7004 / 55 / "comestibles" / "PÉ PÉ CHI idiome (Caramake)"

63 *Myrtacea* indet. Fr. Guiana / M. Lemoine 7911 / 1961 / "Fruits comestibles" / "(Taki-Taki): Liba-Banda"

64 *Myrtacea* indet. Fr. Guiana / M. Lemoine 7801 / 1961 / "Nom. Vernac. (Taki-Taki): Bouchi Mango"

065 *Myrtacea* indet. Brazil / E. P. Heringer 14789 / 1975 / "Produz fruto alimento para passarinhos"

3066   *Myrtacea* indet. Brazil / E. P. Heringer 11916 / 1970 / "Frutos comestíveis" / "Sinhazinha"

3067   *Myrtacea* indet. Indonesia / Kuswata & Soepadmo 205 / 1959 / "Fr. . . . astringent"

COMBRETACEAE

3068   **Buchenavia capitata** Fr. Guiana / W. E. Broadway 536 / 1921 / "Fruits yellow, succulent, acid"

3069   **B. grandis** Venezuela / T. Lasser 1801 / 1946 / ". . . frutos amarillos, comestibles, sabor dulce"

3070   **B. parvifolia** Brit. Guiana / A. C. Smith 2833 / 1937 / "leaves boiled with *Trichomanes* (no. 2824) to make a liquid; said to be effective by Waiwais to check internal bleeding"

3071   **B. tomentosa** Brazil / Ratter & Ramos 564 / 1967 / "Tree infested by ants"

3072   **Combretum cacoucia** Brazil / W. A. Archer 8226 / 1943 / "Frt. poisonous to pigs"
Brit. Guiana / J. S. de la Cruz 1488 / 1922 / "Poisonous tree"
Venezuela / J. J. Wurdack 323 / 1955 / "fruit yellow, said to be poisonous" / " 'Sanajoro' "
Brit. Guiana / C. W. Anderson 79 / 1908 / "Fruit is poisonous & used for killing bats" / "YARRIMANNI"

3073   **Combretum** indet. Brazil / A. Silva 226 / 1944 / "Oil from seed much used in local medicine to cure skin diseases" / " 'andiroba-rana' "

3074   **C. apiculatum** S.W. Africa / R. Seydel 4110 / 65 / "Sehr geschaetzter Futterbusch von Kuddu u. Grossvieh"

3075   **C. brosigianum** Tanganyika / R. E. S. Tanner 2966 / 1956 / "Zigua: for diarrhoea, bark boiled in water and drunk when cool" / "Zigua: Mgunku"

3076   **C. constrictum** Tanganyika / R. E. S. Tanner 2367 / 1955 / "flowers red, aromatic" / "Kizigua: Mkwikwi"

3077   **C. gueinzii** Tanganyika / R. E. S. Tanner 1242 / 1953 / "Sukuma tribe: Roots used in solution to wash skin to prevent scabies; ground-up roots used as poultice to treat sores; flour of roots taken orally in water for constipation" / "Kisukuma: Nama"
Tanganyika / R. E. S. Tanner 3227 / 1956 / "Masai: For gonorrhoea, roots boiled and drunk" / "Masai: Ol Maroroi"

3078   **C. mwanzense** Tanganyika / R. E. S. Tanner 1218 / 1953 / "Sukuma tribe: Roots used to treat boils — boiled in water which is drunk" / "Kisukuma: Namatha"

079 *C. verticillatum* Tanganyika / R. E. S. Tanner 657 / 1952 / "Used for building animist shrines; pounded leaves to heal boils; leaves used to sift milk from butter" / "Kisukuma: Mnama"

080 *C. zeyheri* Tanganyika / R. E. S. Tanner 1409 / 1953 / "Sukuma tribe: Roots pounded, cooked in porridge to treat ankylostomiasis; roots chewed for Bilharzia" / "Kisukuma: Sanna"

081 *Pteleopsis myrtifolia* Tanzania / H. Faulkner 4329 / 70 / "Flowers cream, sweet scented"

082 *Terminalia amazonica* Guatemala / P. C. Standley 73131 / 1939 / " 'Naranjo' "

083 *T. pittieri* Panama / G. P. Cooper 55 / 1928 / " 'Guayabe de montaña' "

084 *Terminalia* indet. Guatemala / Jones & Facey 3224 / 1966 / " 'Naranjo' "

085 *T. arbuscula* Jamaica / G. R. Proctor 26436 / 1965 / " 'Olive' "

086 *T. dichotoma* Brit. Guiana / C. W. Anderson s.n. / 1908 / " 'Coffee-mortar' or 'Naharu' "

087 *T. sericea* S.W. Africa / R. Seydel 3771 / 1963 / "Die Blätter werden als Futter geschätzt" / " 'Gelbkolz' "

088 *Terminalia* indet. Tanganyika / R. E. S. Tanner 1244 / 1953 / "Sukuma tribe: Pounded roots used as poultice to treat syphilis; cooked roots for Bilharzia (water drunk)" / "Kisukuma: Njimya"

089 *T. microcarpa* Philippine Is. / Ahern's collector 39 / 1904 / "Fruit edible" / "T., Calumpit"
Philippine Is. / J. G. Pacis 42 / 1929 / "Children eat pulp of fruit when ripe. Taste sweet"

090 *T. nitens* Philippine Is. / A. D. E. Elmer 11873 / 1909 / "The nuts are relished by the Bogobos when they are ripe" / " 'Samulocco' "

091 *T. samoensis* Caroline Is. / F. R. Fosberg 25777 / 1946 / ". . . used medicinally" / " 'esemi' "
Caroline Is. / F. R. Fosberg 47663 / 1965 / "mature fruit red, eaten avidly by Palauans, but flesh rather bitterish or astringently sweet"
Society Is. / Fosberg & Sachet 54632 / 1973 / "fruits . . . said to be eaten; flesh neither sweet nor bitter, very slightly astringent" / "Tahitian name 'autara' "

092 *Terminalia* indet. Hawaiian Is. / O. Degener 11286 / 1935 / ". . . chopped down this date because of signed petition of neighbors complaining about its sewer-like odor"

093 *Terminalia* indet. New Guinea / L. J. Brass 31062 / 1959 / "flowers cream, honey-scented"

094 *Combretacea* indet. Mexico / G. B. Hinton et al. 7608 / 35 / "Uses: medicinal" / "Suelda con suelda"

MELASTOMATACEAE

3095   *Aciotis rostellata* Brit. Honduras / W. A. Schipp 46 / 1929 / ". . . seems to be avoided by insects"

3096   *A. polystachya* Venezuela / Aristeguieta & Lizot 7384 / 1970 / "Las pájaros comen los frutos"

3097   *Arthrostemma volubile* Colombia / F. R. Fosberg 20510 / 1943 / ". . . used to make a refreshing drink for fever patients" / " 'cañagra' "

3098   *Axinaea grandifolia* Venezuela / J. A. Steyermark 55258 / 1944 / " 'canilla de venado' "

3099   *A. lepidota* Ecuador / J. A. Steyermark 53689 / 1943 / " 'cebolleta' "

3100   *A. macrophylla* Colombia / Idrobo, Pinto & Bischler 3501 / 1958 / " 'Manzanillo' "

3101   *Bellucia costaricensis* Guatemala / H. Pittier 8572 / 1919 / " 'Manzano de montaña' "
Panama / T. B. Croat 12485 / 1970 / "covered with ants"
Panama / S. Hayes 394 / 1860 / ". . . very fragrant flowers"
Costa Rica / A. Jimenez M. 3810 / 1966 / "Frutos . . . comestibles y algo astringentes"

3102   *B. axinanthera* Venezuela / J. Saer d'Heuguert 867 / 1942 / " 'Manzana de Corona' "
Brazil / J. M. Pires 3705 / 1951 / ". . . servir para pintar madeira (envernizar)" / " 'Muuba'. "

3103   *B. grossularioides* Venezuela / C. A. Blanco C. 1151 / 1971 / "Guayabo de danto"
Venezuela / J. A. Steyermark 60588 / 1944 / "fruit edible, sweet; flowers very fragrant, like rich perfume" / " 'karare-yek' "
Surinam / W. A. Archer 2799 / 1934 / "Frt. edible" / " 'Nispa' "
Brit. Guiana / Jenman 4549 / 1888 / "Useful fruit"
Brazil / J. M. Pires 51955 / 1961 / " 'goiaba de anta' "
Colombia / J. Cuatrecasas 6848 / 1939 / "fruto comestible indios y animales"

3104   *B. imperialis* Bolivia / R. S. Williams 576 / 1901 / "Fruit edible"

3105   *B. weberbaueri* S. AM.: Napo-Pastaza / E. Asplund 19446 / 1956 / ". . . flowers white, aromatic, fruits edible"

3106   *Bellucia* indet. Colombia / Schultes & Cabrera 13696 / 1951 / "Flowers white, with odour of citronella"

3107   *Blakea gracilis* Costa Rica / C. Schnell 750 / 1966 / "Fruits eaten by birds"

3108   *B. tuberculata* Costa Rica / R. W. Lent 789 / 1967 / "Strong sweet odor"

3109   *Blakea* indet. Costa Rica / P. H. Raven 21807 / 1967 / "flowers with fetid odor and conformation suggestive of bat pollination"

10   *B. grisebachii* Venezuela / J. A. & C. Steyermark 95339 / 1966 / " 'manzano' "

11   *B. subconnata* Ecuador / J. A. Steyermark 52809 / 1943 / "watery solution in inflated calyx buds used for kidney trouble" / " 'agua de mono' "
Ecuador / Y. Mexia 6707 / 1934 / "Buds & fls. filled with mucilaginous juice"

12   *Brachyotum weberbaueri* Peru / J. J. Wurdack 1635 / 1962 / "petal infusion drunk for pneumonia" / " 'Sarcilleja' "

13   *Centronia brachycera* Colombia / E. L. Little 7035 / 1944 / "Used for tanning leather" / " 'tuna roja' 'siete cueros' "

14   *Chaetolepis alpestris* Venezuela / J. A. Steyermark 55729 / 1944 / "Used medicinally" / " 'dictamo romero' "
Venezuela / J. A. Steyermark 55483 / 1944 / "Used as pasture for cattle" / " 'romerito' "

15   *Clidemia setosa* Guatemala / P. C. Standley 91889 / 1941 / "Remedy for sterility in women" / " 'Ixquil quen' "
Guatemala / J. A. Steyermark 49392a / 1942 / "This is used if the woman is considered to be sterile, but the man has to cook the potage and the woman shouldn't see the man do it" / " 'yerba de mico' (hembra)"

16   *C. erythropogon* Jamaica / G. R. Proctor 22592 / 1962 / " 'Soap-bush' "

17   *C. hirta* Venezuela / J. A. Steyermark 61239 / 1945 / "Fruit edible" / " 'uva' "
Brazil / Y. Mexia 5412 / 1930 / "Fruit edible" / " 'Quaresminha' "

18   *C. rubra* Peru / W. J. Dennis 29213 / 1929 / "fruit deep red, used to make ink" / " 'Mullaca' "
Colombia / W. A. Archer 1816 / 1931 / "frt. dark blue, edible" / " 'Uvita' "

19   *C. testiculata* Colombia / L. Uribe U. 3719 / 1961 / "En los domacios foliares habitan hormigas pequeñas, negras"

20   *C. umbonata* Peru / R. Scolnik 1174 / 1948 / ". . . fruto morado, comestible" / " 'mullaca' "

21   *Clidemia* indet. Colombia / J. A. Duke 11050 / 1967 / "fruits blue, edible"

22   *Conostegia speciosa* Panama / G. White 137 / 1938 / "fruit purple, sweet to taste"

23   *C. polyandra* Colombia / J. A. Duke 11567 / 1967 / "fruits . . . quite EDIBLE"
Colombia / Killip & Garcia 33112 / 1939 / "fruit purple, sweet, pulp inky"

24   *Dinophora spenneroides* Liberia / P. M. Daniel 424 / 1952 / "Local use: medicine for sore mouth"

25   *Henriettea succosa* Brit. Honduras / P. H. Gentle 2971 / 1939 / " 'wild guava' "

26   *H. ramiflora* Venezuela / J. A. Steyermark 60405 / 1944 / "fruit edible" / " 'yará-yek' "

3127　*Henriettella duckeana* Brazil / W. R. Anderson 10575 / 1974 / "mature fruit white and succulent"

3128　*Heterocentron undulatum* Mexico / Y. Mexia 1329 / 1927 / "used medicinally" / " 'Cana agra' 'Agrio' "

3129　*Leandra strigosa* Colombia / W. A. Archer 2073 / 1931 / "Stem used as 'cabos' or stems for clay or tin smoking pipes used by the Negros"

3130　*Loreya acutifolia* Brit. Guiana / Brit. Guiana For. Dept. 874 / 1926 / "The sap from the bark is used by the Indians as a varnish" / "ITARRA Arawak"

3131　*L. mespiliodes* Brazil / P. Cavalcante 2527 / 1970 / "frutos comestiveis, segundo os indios"
　　　Venezuela / J. A. Steyermark 75591 / 1953 / "fruit fleshy, edible"

3132　*Macairea parvifolia* Venezuela / J. A. Steyermark 60322 / 1944 / "Iguanas eat flowers" / " 'quita-pán-yek' "

3133　*Maieta guianensis* Peru / Y. Mexia 6357 / 1931 / "Bases of leaf blades inhabited by small reddish ants"
　　　Colombia / G. Klug 1882 / 1930 / ". . . fruit edible" / " 'Morillo' "
　　　Venezuela / J. A. Steyermark 60439 / 1944 / " 'mi-ke-yek' (ant-tree)"
　　　Ecuador / C. C. Fuller 101 / 53 / "Leaves chewed and masticated ball applied to wounds to check bleeding. Is excellent hemostatic" / "Situli panga; Quichua"

3134　*Marcetia andicola* Venezuela / A. Jahn 708 / 1921 / "N.v. Romero"

3135　*Marumia pachygyna* Brit. N. Borneo / Sator 814 / 1928 / "Fruit can be eaten" / "Totopo (Sungei)"

3136　*Medinilla pendula* Philippine Is. / A. D. E. Elmer 8441 / 1907 / "Ripe fruit are eaten by Igorote children" / "Balangbañg"

3137　*Melastoma polyanthum* Hainan / C. I. Lei 127 / 1932 / "Leaves for feeding pigs" / "Foo Tan"
　　　Hainan / C. I. Lei 477 / 1933 / "flower red, fragrant" / "Shan Tim Neung"
　　　Philippine Is. / A. D. E. Elmer 7590 / 1907 / "the fleshy part of the seeds strawberry red and edible, but without a flavor"

3138　*M. denticulatum* Melanesia / J. H. L. Waterhouse 63 / 33 / "Used in native medicine" / "Tupainaraku"

3139　*M. malabathricum* Fiji Is. / O. Degener 15065 / 1941 / "Leaves chewed as remedy for thrush. Flying foxes eat the fruit. Serua people do not eat fruit" / " 'Kaunisinga' (Serua)"
　　　Fiji Is. / O. Degener 14080 / 1941 / "fruit red-purple, edible"

3140　*Memecylon umbellatum* Ceylon / N. Wirawan 718 / 68 / "flower violet blue, sweet scented / " 'Korakaha' (S)"

3141　*M. densiflorum* Philippine Is. / M. Oro 310 / 1929 / ". . . bark to prevent fading of colors on breri water"

42 *Meriania splendens* Colombia / J. A. Ewan 16545 / 1944 / ". . . used as remedy for dandruff . . . the fls. are placed in cold water, allowed to steep for three days, the water is then poured off and used directly, the bark is also used but is somewhat less efficacious" / " 'garbunquillo' of native"

43 *Miconia aeruginosa* Costa Rica / R. W. Lent 869 / 1967 / " 'Uña de gata' "

44 *M. glaberrima* Guatemala / J. A. Steyermark 50073 / 1942 / " 'uva' "

45 *M. hyperprasina* Brit. Honduras / P. H. Gentle 4006 / 1942 / "fls. white, fragrant" / " 'sirin' "

46 *M. ibaguensis* Honduras / C. & W. von Hagen 1062 / 1937 / "Fruit eaten" / " 'Uva' "

47 *M. impetiolaris* Honduras / C. & W. von Hagen 1013 / 1937 / " 'Uva' "

48 *M. insularis* Panama / I. M. Johnston 448 / 1944 / ". . . flowers white, with offensive odor"

49 *M. lateriflora* Costa Rica / Burger & Matta U. 4304 / 1967 / ". . . lvs. with a sasparilla odor when crushed" Brazil / A. Silva 239 / 1944 / " 'canela-rana' "

50 *M. mathaei* Brit. Honduras / W. A. Schipp 1118 / 1933 / "Flowers white, sweetly perfumed"

51 *M. punctata* Costa Rica / A. F. Skutch 4640 / 1940 / " 'Canilla de mula' "

52 *M. rubens* Panama / C. & W. von Hagen 2042 / 1940 / " 'Uva' "

53 *Miconia* indet. Guatemala / J. A. Steyermark 50479 / 1942 / "Fruits used in dyeing leather" / " 'alumbre' "

54 *M. sintenisii* Puerto Rico / E. L. Little 21612 / 1966 / "Bark gray . . . as-tringent"

55 *M. alypifolia* Ecuador / J. A. Steyermark 53705 / 1943 / " 'duraznilla' "

56 *M. barbinervis* Colombia / O. Haight 1887 / 1936 / "Flowers white, very fragrant — odor like that of 'clove pinks' "

57 *M. candolleana* Brazil / Y. Mexia 5715 / 1931 / "Tea made of leaves" / " 'Marceiro' " Brazil / R. Fróes 1775 / 1932 / "Leaves boiled to make an ink dye" / " 'Tinteiro' "

58 *M. caudata* Colombia / Archer & Ballou 1301 / 31 / "Sweet-scented" Colombia / W. A. Archer 1636 / 31 / " 'Carate' "

59 *M. chaetodon* Venezuela / W. Gehriger 404 / 1930 / "fruto . . . comestible" / "N.v. Mortiño"

60 *M. cuspidata* Brazil / E. Oliveira 4527 / 1968 / " 'Tinteiro branco' "

61 *M. diaphanea* Surinam / Daniëls & Jonker 735 / 1959 / "soewa-wiri (sour leaf)"

3162   *M. lepidota* Brazil / E. Oliveira 4598 / 1968 / " 'Tinteiro preto' "

3163   *M. longifolia* Brazil / E. Oliveira 4611 / 1968 / " 'Tinteiro' "

3164   *M. minutiflora* Peru / Killip & Smith 27346 / "Leaves used for ink" / " 'Riparillo' "

3165   *M. mirabilis* Venezuela / J. A. & C. Steyermark 95216 / 1966 / " 'manzano' "

3166   *M. molybdea* Bolivia / J. Steinbach 6120 / 1924 / "En las hojas hornean el pan (bizcocho) = por eso el nombre" / "Bizcochero"
Argentina / Schulz & Varela 5298 / 1944 / "flor rosada, con mucho perfume"
Bolivia / J. Steinbach 5675 / no year / "Fruta negra, comible. Las hojas orupan las horneadoras para moldas para biscochos por eso el nombre" / "Biscochéro"

3167   *M. nervosa* Ecuador / C. C. Fuller 102 / 53 / "Claimed that chewed leaves stuffed up nostrils will stop nose bleed" / "Payas situli: Quichua"

3168   *M. petropolitana* Brasil / Lindeman & de Haas 2547 / 1966 / "flowers white, honey-scented"

3169   *M. poeppigii* Brazil / E. de Oliveira 3690 / 1967 / " 'Tinteiro branco' "

3170   *M. prasina* Brazil / E. de Oliveira 3719 / 1967 / " 'Tinteiro' "

3171   *M. pyrifolia* Brazil / E. Oliveira 4046 / 1968 / " 'Tinteiro branco' "

3172   *M. regelii* Brazil / E. de Oliveira / 3918 / 1968 / " 'Tinteiro' "

3173   *M. rubiginosa* Brazil / P. Cavalcante 2544 / 1970 / "flor branca, perfumada"

3174   *M. scorpioides* Ecuador / E. L. Little 6401 / 1943 / "Berries used in medicine"

3175   *M. sellowiana* Brasil / Irwin & Soderstrom 6235 / 1964 / "Flowers white, with disagreeable odor"

3176   *M. serrulata* Peru / Killip & Smith 29157 / 1929 / "lvs. used as a dye" / " 'Rifari' "
Venezuela / J. A. Steyermark 81094 / 1945 / "Fruit edible" / " 'camburito' "

3177   *M. staminea* Brazil / Y. Mexia 5961 / 1931 / "Black dye made from bark" / " 'Tinteira' "

3178   *M. surinamensis* Brazil / J. M. Pires 51956 / 1961 / " 'tinteiro' "

3179   *M. theaezans* Colombia / J. A. Ewan 15786 / 1944 / "Fls. cream-white, very fragrant as to fill the air in the vicinity with its sweetness"

3180   *M. tomentosa* Surinam / Wood Herb. 331 / '45 / "wood decoct emetic!" / "SAREROKONA"
Venezuela / L. Williams 14816 / 1942 / "Panápe (Kuripako)"

1 *M. triplinervis* Peru / Y. Mexia 6109 / 1931 / "fruit eaten by birds" / " 'Chinchaque ubiambo' "

2 *M. violacea* Colombia / Lehmann 3842 / 1884 / "Ein Dekokt bei Typhus angewandt"

3 *Miconia* indet. Venezuela / J. A. Steyermark 55383 / 1944 / " 'canilla de venado' "

4 *Miconia* indet. Brazil / E. de Oliveira 3516 / 1967 / " 'Tinteiro preto' "

5 *Miconia* indet. Colombia / Espina & Giacometto A176 / 30 / "Com. Name Árnica (árbol)"

6 *Microlicia goyazensis* Brazil / Fróes 2070 / 1933 / "Used for fever" / "Cha de vereda"

7 *Mouriri parvifolia* Panama / T. B. Croat 15151 / 1971 / "fruits plum-brown, sweet tasting"

8 *M. elliptica* Brazil / Harley, de Castro & Ferreira 10900 / 1968 / "Fruit edible"

9 *M. grandiflora* Venezuela / G. Agostini 1543 / 1973 / ". . . la pulpa mucilaginosa de sabor dulce"

0 *Mouriri* indet. Venezuela / Steyermark, Bunting & Blanco 101802 / 1968 / "fruit turning deep red, edible"

1 *Pachycentria glauca* Sarawak / J. W. Purseglove P.4894 / 56 / ". . . often ant inhabited"

2 *Pterolepis glomerata* W.I.: Guadeloupe / P. Duss 2266 / 1894 / "Herbe á vache mále"

3 *Rhexia nashii* U.S.: N.C. / J. F. McRee s.n. / no year / "Deer Grass"

4 *Rhynchanthera mexicana* Panama / Wilbur, Weaver & Correa 11162 / 1970 / ". . . with sticky glandular resinous smelling hairs"

5 *Tococa coriacea* Nicaragua / Davidse & Pohl 2310 / 1970 / "with active ant colonies"

6 *T. guianensis* Colombia / E. P. Killip 34242 / 1939 / ". . . ant-infested"

7 *T. nitens* Venezuela / Maguire, Maguire & Steyermark 53538 / 1962 / ". . . fruit blue-black at maturity, edible"

8 *Topobea maurofernandeziana* Costa Rica / A. Smith 62 / 1937 / "Flowers pulpy, basal parts very sweet (edible)"

9 *T. floribunda* Colombia / W. A. Archer 1888 / 1931 / ". . . frt. red, edible" / " 'Cojón de toro' "

00 *Melastomatacea* indet. Ecuador / W. von Hagen 250 / 1935 / "Sometimes applied externally for skin irritation" / "Shashake Jiv."

01 *Melastomatacea* indet. Papua / C. E. Carr 15452 / 36 / "The ripe fruit yields the natives a dull purple dye" / "Native name PAIT"

ONAGRACEAE

3202 *Circaea alpina* var. *alpina* U.S.: Wis. / A. M. Keefe s.n. / 38 / "Enchanter's Nightshade"

3203 *C. lutetiana* U.S.: N.C. / L. M. Stewart s.n. / 1939 / " 'Puffed rice' "
U.S.: N.D. / Lee s.n. / 1891 / "Wild Rice"
U.S.: N.J. / J. Oehler s.n. / 1877 / "Enchanter's Nightshade Pariser Hexenkraut"
U.S. & CAN. / Meisner Herb. s.n. / no year / "*Officin. Deutsch.* Hexenkraut. *Franz.* La circée. Sorcière. l'herbe enchanteresse"

3204 *Fuchsia arborescens* Mexico / G. B. Hinton 3520 / "Fruit edible" / "Vernac. name: Pipilito"

3205 *F. fulgens* Mexico / Hinton et al. 15533 / 40 / "Fruit eaten"

3206 *F. microphylla* Mexico / G. B. Hinton 3687 / 33 / "Fruit edible"

3207 *F. microphylla* var. *aprica* El Salvador / P. C. Standley 21549 / 1922 / "fr. black, sweet"

3208 *F. apetala* Venezuela / J. A. Steyermark 56293 / 1944 / " 'pimiento' "

3209 *F. canescens* Ecuador / W. H. Camp E-1988 / 1945 / "The ripe fruits are eaten" / " 'Zarcillo' "

3210 *F. petiolaris* Colombia / H. St. John 20850 / 1944 / " 'pepino de monte' "

3211 *Gaura coccinea* var. *glabra* U.S.: Ariz. / E. A. Mearns 347 / 1888 / "This is 'a great bee plant'. It is valuable to the apiarists"

3212 *G. coccinea* Mexico / Aguirre & Reko 72 / 1946 / " 'hierba del golpa' "

3213 *Ludwigia palustris* var. *americana* U.S.: N.J. / E. H. Day s.n. / 82 / " 'Water-Purslane' "

3214 *L. peruviana* Peru / D. McCarroll 6 / 1942 / "Used as a purgative"
Colombia / J. M. Duque Jaramillo 26281 / 1946 / "Medicinal" / " 'Clavo de pantano' "

3215 *L. sedioides* Bolivia / J. Steinbach 5342 / 1921 / "Se usa la planta para ensaladas y jarabe. Dicen que es provechosa para higados enfermos" / "n.v. Berro"

3216 *L. erecta* Tanganyika / R. E. S. Tanner 3604 / 1957 / "Leaves used as a vegetable" / "Zigua: Mrashi"

3217 *L. octovalvis* Fiji Is. / Degener & Ordonez 13545 / 1940 / "Used by Fijians as black-hair dye but not lasting"
Fiji Is. / O. Degener 14963 / 1941 / "Boil leaf with hala leaf to stain it black for plaiting designs in mat" / " 'Naingisa' (Sabatu)"
Caroline Is. / M. Evans 504 / 1965 / "Used as medicine. Informant Francisco Steven" / " 'SOW' "

8 *Oenothera glandulosa* U.S.: Nev. / P. Train 2119 / 1938 / "Flr. yellow, fragrant"

9 *O. nuttallii* U.S.: Mont. / F. W. Anderson 2683 / 1887 / "At night the fls. emit foetid puffs of warm breath at regular intervals"

0 *O. kunthiana* Mexico / Dr. Gregg s.n. / 47 / "Used as poultice for bruises, etc." / "Yerba del golpe"

1 *O. rosea* Mexico / Dr. Gregg s.n. / 47 / "Said to make good poultice for bruises" / "Yerba del golpe"
Mexico / Stanford, Retherford & Northcraft 144 / 1941 / "heavily grazed by goats"
Mexico / G. B. Hinton 3727 / 1933 / "concoction used for stomache-ache" / " 'Yerba del golpe' "
Ecuador / W. H. Camp E-2512 / 1945 / "used to refresh the intestines" / " 'Zchullo' or 'Zchungir' "

2 *O. epilobifolia* Ecuador / W. H. Camp E-2122 / 1945 / "an infusion of the boiled plant used to wash and disinfect wounds" / " 'Platanillo' "

3 *O. laciniata* var. *pubescens* Ecuador / W. H. Camp E-2637 / 1945 / "made into plasters for external application and also an infusion used internally for wounds and contusions" / " 'Platanillo' "

4 *O. magellanica* Chile / F. Meigen 539 / 1892 / "Yerba loca"

5 *O. tetraptera* Ecuador / W. H. Camp E-2537 / 1945 / "the lvs. are ground and placed in wounds" / " 'Platanillo' "

6 *Onagracea* indet. Mexico / Dr. J. Gregg 313 / 1848–49 / " 'Used as a culinary spice' "

HALORAGACEAE

7 *Serpicula brasiliensis* Venezuela / J. A. Steyermark 62704 / 1945 / "Boil in water to serve for snake bite" / " 'balsamo' "

ARALIACEAE

8 *Aralia hispida* U.S.: N.Y. / W. T. Davis s.n. / 1893 / "Bristly Sarsaparilla"

9 *A. nudicaulis* var. *nudicaulis* U.S.: N.J. / H. L. Fisher s.n. / 1893 / "False Sarsaparilla"
U.S.: N.Y. / H. N. Moldenke 11131 / 1939 / " 'Wild-sarsaparilla' "

30 *A. racemosa* var. *racemosa* U.S. & CAN. / Meisner Herb. s.n. / no year / "Wild liquorice"
Canada: Québec / Fr. Fabius 356 / 46 / "Indian root"

31 *A. dasyphylla* China: Kwangtung / W. T. Tsang 21777 / 1932 / "head as medicine" / "Niu Pat Kei"

3232   *A. spinifolia* China: Kwangtung / F. A. McClure 3533 / 28 / "For washing boils" / "Kai Na Lak"

3233   *Boerlagiodendron novoguineense* New Guinea / J. H. L. Waterhouse 319 / 1934 / "Sap is injurious to skin, but bark is scraped and given with coconut to pigs apparently as a tonic" / "Pulaka"

3234   *Cussonia paniculata* S. Africa / A. O. D. Mogg 14758 / 1935 / "Good fodder"

3235   *C. spicata* S. Africa / R. D. A. Bayliss BS5622 / 1973 / " 'Cabbage tree' "

3236   *Dendropanax arborea* Puerto Rico / W. R. Stimson 3644 / 1966 / " 'palo de pollo' "
        Venezuela / J. A. Steyermark 92138 / 1963 / "leaves spicy fragrant"

3237   *D. caucana* Ecuador / Little & Dixon 21149 / 1965 / "Poca resina en la corteza interior color canela o café, sabor oleaginoso" / " 'algodoncillo' "

3238   *D. colombianum* Colombia / J. Cuatrecasas / 14907 / 1943 / "Hoja . . . con olor a trementina"

3239   *D. cuneatum* Brazil / Ratter & Ramos 438 / 1967 / "Foliage has a peppery odor when crushed"

3240   *D. fendleri* Venezuela / J. A. Steyermark 62026 / 1945 / "fruit spicy fragrant" / " 'mangle' "

3241   *D. macrocarpum* Colombia / J. Cuatrecasas 19458 / 1945 / "Fruto . . . resinoso aromático"

3242   *D. macrophyllum* Colombia / J. Cuatrecasas 21626 / 1946 / "Fruto . . . trementinoso resinoso"

3243   *D. tessmannii* Peru / G. Klug 2016 / 1931 / "Leaves chewed to strengthen teeth, making them black" / " 'Chirez' "

3244   *Dendropanax* indet. Bolivia / J. Steinbach 7155 / 1925 / ". . . madera blanda y blanca, que enquemada de una ceniza rica su tejia para fabricar jabon" / "Yerba"

3245   *Didymopanax glabratum* Venezuela / J. A. Steyermark 94975 / 1966 / " 'mango de montaña' "

3246   *Nothopanax fruticosum* Fiji Is. / O. Degener 15446 / 1941 / "if testicles are swollen, mix leaves of this & of Acalypha, squash in water & drink" / " 'Dangidangi' in Ra dialect"
        Fiji Is. / O. Degener 15036 / 1941 / "Cultivated by Fijians. For stomach trouble Fijians mash leaves, put in water and then drink the water" / " 'Ndalindali' in Serua dialect"

3247   *Oreopanax capitatus* Venezuela / J. A. Steyermark 61038 / 1945 / "Fruit eaten by birds; wood used for making into bateas (a rectangular tub for washing clothes, etc.)" / " 'higuerote' "

48 *Panax trifolium* U.S.: Vt. / E. T. & H. N. Moldenke 9588 / 1937 / "Roots edible" / " 'Dwarf Ginseng' "
U.S.: N.Y. / G. T. Hastings s.n. / 1894 / "Ground Nut"

49 *Pentapanax angelicifolius* Brazil / Fróes 16 / 1942 / "People use the leaves to kill fish" / "Tinguy"

50 *Plerandra pickeringii* Fiji Is. / O. Degener 15440 / 1941 / "When Devil takes possession of a Fijian, they bring trunk into house and there cut away bark, mix with water and drink to drive him away" / " 'Sole' in Ra"

51 *Polyscias elegans* Australia: Queensland / Smith & Webb 3681 / 1948 / "inner bark . . . with a somewhat celery-like odour" / " 'Celery Wood' "

52 *P. fruticosa* New Hebrides Is. / S. F. Kajewski 931 / 1929 / "Leaves boiled and eaten by natives"

53 *P. grandifolia* Mariana Is. / F. R. Fosberg 24887 / 1946 / "plant with slight celery odor"

54 *Schefflera fragrans* Colombia / J. Cuatrecasas 14960 / 1943 / "Olor a violetas"

55 *S. leucantha* China: Kwangtung / F. A. McClure Y-124 / 1930 / "Drug plant" / "Tsát Ip Muk"

56 *S. octophyllum* Hainan / C. I. Lei 221 / 1932 / "fl. fragrant" / "Aap Mo Kwat"

57 *Schefflera* indet. Ceylon / Nawicke & Jayasuriya 378 / 1973 / "foliage with manganiferous odor"

58 *S. blancoi* Philippine Is. / A. D. E. Elmer 8421 / 1906 / "the chopped up leaves are put in water for fish poison" / " 'Abchal' "

59 *S. oblongifolia* Philippine Is. / A. D. E. Elmer 8693 / 1907 / "Bees were found swarming about the flowers . . . inflorescence and fruit have a strong 'carrot odor' . . . the leaves are pounded in small pieces and thrown in the streams for poisoning fish. It is sour" / " 'Abcal' "

60 *S. odorata* Caroline Is. / F. R. Fosberg 32383 / 1950 / "leaves with strong oily odor when crushed. Brought in by informant Alkong to substantiate ethnobotanic information" / " 'Bungaruau' or 'pungaruau' "

61 *S. vitiensis* Fiji Is. / O. Degener 15147 / 1941 / "Fijians mash leaf and add cup water, drinking strained liquid as remedy for lung trouble" / " 'Sole lewa' in Serua"

62 *Sciadodendron excelsum* El Salvador / P. C. Standley 20829 / 1922 / " 'Carroncha de lagarto' "
Venezuela / L. Aristeguieta 6157 / 1966 / "Frutos . . . cuando maduros, con pulpa azucarada, comida por los pájaros"

APIACEAE

3263   **Ammi majus** Bermuda Is. / F. S. Collins 241 / 1913 / ". . . reputed poisonous if handled when moist. Said to cramp feet of chickens and turkeys so they cannot open their claws" / "locally called 'May Weed' "

3264   **Angelica lucida** Canada: Quebec / Fernald & Smith 25924 / 1923 / "Badly eaten by moose"

3265   **Apium leptophyllum** Honduras / A. Molina R. 27347 / 1972 / " 'Perejil de montés' "
Mexico / Dr. Gregg s.n. / 47 / "Application for pain in side" / "Culantro cimarron"
Colombia / F. W. Pennell 23333 / 1917 / "Herb, odor of celery"
Ecuador / J. N. & G. Rose 22218 / 1918 / " 'Colantrillo del monte' "
Argentina / J. E. Montes 14579 / 55 / "Uso: toda la planta: med. pop."
Australia: Queensland / L. S. Smith 3062 / 1947 / "Wild Carrott"

3266   **A. australe** Falkland Is. / B. F. s.n. / 38 / "Celery"

3267   **Arracacia longipedunculata** Mexico / G. B. Hinton 1142 / 32 / "Root juicy. Juice smells like turpentine"

3268   **Bifora** indet. Mexico / Rose, Standley & Russell 13993 / 1910 / "Bought in market. Used for seasoning"

3269   **Bunium flexuosum** Ireland / G. Macloskie s.n. / 1878 / "Earthnut or Pignut"

3270   **Centella asiatica** Hainan / C. I. Lei 474 / 1933 / "Used as medicine to help digestion" / "Ma Tei Tso"

3271   **Cicuta douglasii** U.S.: Wash. / F. W. Johnson 1635 / 1913 / "Roots very poisonous" / " 'Western Water Hemlock' "

3272   **C. virosa** EUR. / no collector s.n. / no year / "Feuilles peu employées en médecine & seulement á l'extérieur contre le squirrhe &c.;—odeur forte, piquante, saveur approchant de celle du persil; trés vénéneuse" / "giftiger Wasserschierling, Wüterich"

3273   **Daucus montanus** Ecuador / W. H. Camp E-2665 / 1945 / "Said to kill guinea pigs if fed to them. (The guinea pig is commonly kept in the house—running around loose—and used as food" / " 'Cuy-guañuna' (cuy-guinea pig, or cavia; guañuna—to kill)"

3274   **Eryngium yuccaefolium** Nord-Amerika / Herb. of A. Wood s.n. / no year / "In Nordamerika gegen Wassersucht und Schlangenbiss gebraucht"

3275   **E. bonplandii** Mexico / Bell & Duke 16759 / 1959 / "Plants have a strong, sweet scent"

3276   **E. deppeanum** Mexico / R. Endluh 1013 / 1905 / "n.v.: Yerba del sapo"

7   *E. phyteumae* Mexico / H. S. Gentry 6297 / 1941 / ". . . with pungent odor"

8   *E. foetidum* W.I.: Martinique / P. Duss 1752 / 1892 / ". . . racines et feuilles alexitéres et employées contre les morsures du serpent" / "Chardon beni"
Dominican Republic / E. J. Valeur 241 / 1929 / "Aromatic fragrance" / " 'Silandro sabanero' "
W.I.: Martinique / Herb. of A. Wood s.n. / no year / "Wird in Amerika wie contrayerva bei bösartigen Fiebern gebraucht"
Peru / G. Tessmann 3435 / 1923 / "Used in medicine"

'9   *Ferula jaeschkeana* var. *parkiana* India / W. Koelz 1792 / 1931 / "Seeds used as poultice for infections" / "Kyett—Dialect Rampur"
India / W. Koelz 1135 / 1930 / "Root crushed and used as poultice to bring the boils to head" / "Bakhyud—Dialect Lahuli"

30   *Ferula* indet. Afghanistan / E. Bacon 20 / 1939 / "Sap used medicinally—put on open wounds"

31   *Glehnia littoralis* Ryukyu Is. / F. R. Fosberg 37036 / 1956 / ". . . said to be used for a tea to treat colds or coughs" / " 'shi'i boku' or 'shibo' "

32   *Heracleum sphondylium* EUR. / Herb. of A. Wood / no year / "Wurzel u. Blätter als R. u. Fol. Brancae ursinae spuriae bekannt"

33   *H. pinnatum* W. Himalayas / W. Koelz 245 / 1931 / "Plant scented queer interesting, enduring odor"

84   *Hydrocotyle bonplandii* Ecuador / Grubb, Pennington & Whitmore 1001 / 1960 / "Mixed in paste with salt and water and taken for boils" / "Chupana"

85   *H. globiflora* Peru / Mathias & Taylor 516 / 1960 / "Strong carrot odor"

86   *H. leucocephala* Venezuela / Steyermark & Rabe 96137 / " 'orega de mono' "

87   *H. asiatica* Abyssinien / W. Schimper 13 / 1852 / "Die Pflanze enthält ein scharfes Gift"

88   *Ligusticum canadensis* U.S.: Ky. / C. W. Short s.n. / 1837 / "Root strongly scented with the angelicous odour" / "Angelica"

89   *Lomatium dissectum* var. *eatonii* U.S.: Nev. / J. Heinrichs 551 / 1940 / "Medicinal plant of Washoe & Paiute Indians"

90   *Niphogeton angustisecta* Colombia / L. U. Uribe 1618 / 1947 / "Usada por el pueblo para endurecer las encías. También para combatir el gorgojo" / " 'ahumaria negra' "

91   *N. dissecta* Ecuador / G. W. Prescott 146 / 53 / "Toma-toma, a medicinal plant used for stomach disorders"

92   *Osmorhiza longistylis* U.S.: S.D. / L. J. Harms 2425 / 1965 / "Smell of anice oil when crushed"

Canada: Québec / A. Dubois s.n. / 1929 / "Anise-root"

3293   *Ottoa* indet. Venezuela / L. Aristeguieta 2622 / 1956 / "Aromàtica, medicinal para el parto" / " 'Cebolleto' "

3294   *Perideridia oregana* U.S.: Calif. / Fremont's Exped. to Calif. 487 / 1846 / "The Indians use the roots as food and collect them in considerable quantities"

3295   *Peucedanum* indet. China: Hopei / J. C. Liu L1856 / 1928 / "Rt. very fragrant"

3296   *Pimpinella flava* Georg. cauc. / R. F. Hohenacker s.n. / 1836 / "Tinctoribus praebet colorem flavum"

3297   *Pleurospermum brunonis* India / W. Koelz 22017 / 1948 / "used to scent butter, for incense, against moths" / " 'Lesur' "

3298   *Pseudotaenidia montana* U.S.: Va. / A. Cronquist 10941 / 1972 / "Plants with a heavy, rather unpleasant, somewhat anise-like odor"

3299   *Sanicula canadensis* U.S.: N.Y. / E. Springfield s.n. / 76 / " 'Black Snakeroot' "
U.S.: Wyo. / T. A. Williams 190 / 1903 / "Medicine Root"

3300   *S. gregaria* U.S. & CAN. / E. G. Knight s.n. / 76 / " 'Black Snakeroot' "

3301   *S. graveolens* Chile / W. J. Eyerdam 10076 / 1957 / "the roots with unpleasant odor"

3302   *Selinum papyraceum* W. Himalayas / W. Koelz 2538 / 1931 / "Plant pleasantly scented. Eaten with sattu when small" / "Ussu"

3303   *Seseli tortuosum* France / Columbia College of Pharmacy Herb. s.n. / no year / "Davon Sem. Seseleos massiliens"

3304   *Sium suave* U.S.: N.J. / E. G. Knight s.n. / 79 / " 'Water-Parsnip' "

3305   *Smyrnium dioscorides* Greece / Columbia College of Pharmacy Herb. s.n. / no year / "Von dieser Pflanze wurde Semen Smyrnii cretici gesammelt"

3306   *Torilis japonica* China: Kwantung / Lingnan (To & Ts'ang) 12123 / 1924 / "Ngai ts'oi (Ant vegetable)"

3307   *Zizia aurea* U.S.: N.J. / J. Oehler s.n. / 1878 / "Meadow-Parsnip"

3308   *Apiacea* indet. China: Shangtung / C. Y. Chiao 2207 / 1929 / "Drug plant"

CORNACEAE

3309   *Cornus florida* U.S.: N.Y. / Columbia College of Pharmacy Herb. s.n. / no year / "Die Rinde ist als Surrogat der China empfohlen worden und das chocoladefarbige Holz wird zu feineren Arbeiten verwendet" / "Schönblühender Hornstrauch"

3310   *C. capitata* China: Kwangsi / W. T. Tsang 238941 / 1934 / "fr. black, edible"
China: Kwangtung / W. T. Tsang 20685 / 1932 / "Fr. red, edible" / "Shan Li Tsz Shu"

3311   *C. capitata* var. *khasiana* W. Himalayas / W. Koelz 31711 / 1931 / "pinkish, edible but not particularly pleasant. Suggests banana flavor" / "Barnor sherli 'Monkey Apple' Rampur Dialect"

3312   *C. hongkongensis* China: Kwangsi / R. C. Ching 6000 / 1928 / ". . . edible for its sweet taste" / "locally known as young mai"

GARRYACEAE

3313   *Garrya veatchii* U.S.: Calif. / W. M. R. s.n. / 1927 / "Quinine Bush"

3314   *G. laurifolia* Mexico / Y. Mexia 1560 / 1927 / " 'Zapotillo' "

ALANGIACEAE

3315   *Alangium kurzii* China: Kwangsi / Steward & Cheo 867 / 1933 / "Juice of fruit poisonous to the man's skin — causing itching"

NYSSACEAE

3316   *Nyssa sinensis* China: Kwangtung / Lingnan (To & Ts'ang) 12508 / 1924 / "Tai shui li (Large water pear); Shui ts'it li (Water joint pear)"

CLETHRACEAE

3317   *Clethra alnifolia* U.S.: N.J. / A. Niederer s.n. / 1883 / "Sweet Pepperbush"

3318   *C. hartwegi* Mexico / G. B. Hinton 13536 / 38 / "flower white and very sweet smelling" / "Vernac. name: Cucharo"

3319   *C. lanata* El Salvador / S. Calderon 331 / 1922 / " 'Zapotillo de montaña' "
Mexico / Herb. of the Phila. Mus. A. 12 / no year / " 'Aguacatillo' or 'Mamey cimaron' "
Panama / Kirkbride & Duke 872 / 1968 / "flowers very aromatic"
Panama / P. H. Allen 2438 / 1941 / " 'Mame cillo' "
Venezuela / J. A. Steyermark 61374 / 1945 / "odor of lily-of-valley perfume in flowers" / " 'carbonero' "

3320   *C. occidentalis* Honduras / C. & W. von Hagen 1230 / 1937 / " 'Nance de cerro' "
Mexico / Schultes & Reko 938 / 1939 / "Zapotec name = yĕ-ta-wey-dĕ = flor de encinal"

Brit. Honduras / W. A. Schipp 1171 / 1933 / "Flowers white, sweetly perfumed"

Jamaica / W. Harris 9001 / 1905 / " 'Blood Wood' or 'Soap Wood' "

3321   *Clethracea* indet. Peru / C. Ochoa 162 / 47 / "Usos: endulzar la 'jora' y hacer chicha"

PYROLACEAE

3322   *Chimaphila maculata* U.S.: Va. / E. S. Burgess s.n. / 1887 / "Wild arsenic and Toothache plant of the vernacular in the mountains of Virginia, Miss Edmonston"

3323   *Pyrola virens* U.S.: N.Y. / Torrey Bot. Club Herb. s.n. / no year / "Green-flowered Wintergreen"

MONOTROPACEAE

3324   *Monotropa hypopitys* U.S.: Va. / E. J. Grimes 3559 / 1921 / "Very fragrant"

ERICACEAE

3325   *Andromeda mariana* U.S.: N.J. / E. H. Day s.n. / 82 / " 'Stagger-bush' "

3326   *Arctostaphylos alpina* U.S.: N.H. / Morong Herb. s.n. / no year / "Black Bear-berry"

3327   *A. uva-ursi* U.S.: Colo. / J. S. Brandegee 207 / 1871 / "The 'Kinnikinnick' of the Western Indians"

U.S.: Ore. / M. W. Gorman 962 / 1899 / ". . . much eaten by grouse in early spring"

U.S.: Ore. / F. E. Lloyd s.n. / 1894 / "used for smoking" / " 'Kinnikinick' "

3328   *A. viscida* U.S.: Calif. / G. W. Hulse s.n. / ca. 1848 / " 'Mansinita' "

3329   *Befaria mexicana* Honduras / Davidse & Pohl 2252 / 1970 / ". . . very strong fragrance"

3330   *B. cinnamomes* Peru / Hutchison & Wright 6919 / 1964 / "Insects are trapped on the viscid parts of the inflorescence. Hummingbirds collect nectar and insects"

3331   *B. congesta* Colombia / J. M. Duque Jaramillo 3488 / 1946 / "Fls. rosado intenso, sin resina pegajosa" / "N.v. 'Pegamosco' "

3332   *Cavendishia callista* Costa Rica / H. E. Stork 1789 / 28 / " 'Colmillo de perro' "

3333   *C. melastomoides* var. *albiflora* Costa Rica / A. Smith NY 1383 / 1938 / "Frt. of this shrub edible and resembles in taste blueberries"

3334   *C. quereme* Costa Rica / Luteyn & Wilbur 4294 / 1974 / "Odor of winter-green strong. Visited by hummingbird"

3335   *C. cordifolia* Colombia / Bro. Thomas 220 / 1938 / " 'Uva de anis' "

3336   *C. guateapeensis* Colombia / W. A. Archer 1289a / 31 / " 'Uvito' "

3337   *C. pubescens* Colombia / W. A. Archer 1592 / 31 / "fruit edible" / "Uva de monte"

3338   *Diplycosia luzonica* Philippine Is. / A. D. E. Elmer 10184 / 1908 / ". . . very aromatic, similar to sweet chewing gum"

3339   *Disterigma empetrifolium* Ecuador / Fosberg & Prieto 22810 / 1945 / "fruit . . . at first pleasantly acid, then bitter to taste"

3340   *Epigaea repens* U.S.: Mass. / F. C. MacKeever MV 181 / 1959 / "The fruits of these plants in this area, quite a taste treat"

3341   *Gaultheria hispidula* Canada: Brit. Columbia / E. C. Marquand s.n. / 1934 / "Moxie plum"
Canada: Prince Edward Is. / D. Erskine 1628 / 1952 / "Berries . . . somewhat acid and dry, with 'wintergreen' aroma"
U.S.: N.Y. / E. G. K. s.n. / 1876 / " 'Teaberry' "

3342   *G. shallon* U.S.: Penn. / Columbia College of Pharmacy Herb. s.n. / no year / "Die Blätter sind in ihrem Vaterlande officinell, werden dort aber auch zu einem beliebten Thee verwendet, auch wird davon das Oleum Gaultheriae bereitet, das als Parfum gebraucht wird" / "Liegende Gaultheria"

3343   *G. acuminata* var. *acuminata* Mexico / C. Conzatti 3534 / 1919 / "Planta sagrada con olor a Salicilato de melilo"

3344   *G. odorata* Mexico / R. E. Schultes 516 / 1939 / "Fruits edible" / "Mije name — an-dzits"

3345   *G. ovata* Mexico / V. V. Torres W-309 / 1976 / "Usos: en baños para la fiebre" / "Nombre Loc.: Achocopa"

3346   *G. domingensis* Dominican Republic / Gastony, Jones & Norris 295 / 1967 / "fruits . . . edible, but insipid"

3347   *G. anastomosans* Venezuela / J. A. Steyermark 10486 / 1971 / " 'borra-chero' "

3348   *G. rigida* Colombia / J. M. Duque Jaramillo 3423 / 1946 / "N.v. 'Uvito' '

3349   *G. tomentosa* Ecuador / J. A. Steyermark 52597 / 1943 / " 'duraznillo' "

3350   *G. rudis* Ceylon / M. Jayasuriya 188 / 1971 / "smells oil of wintergreen"

3351   *Gaylussacia brasilensis* Brasil / P. R. Reitz 1996 / 1945 / "comestivel"

3352   *G. descipiens* Brasil / A. P. Duarte 2288 / 1949 / "frutos comestiveis muito gostosos"

3353   *Gaylussacia* indet. Brazil / Y. Mexia 5717 / 931 / "green edible fruit"

3354 **Kalmia angustifolia** var. **angustifolia** U.S.: Mass. / E. C. Marquand s.n. / 1929 / "Sheep-laurel—Lambkill"
U.S. & CAN. / Columbia College of Pharmacy Herb. s.n. / no year / "Die Pflanze hat giftig narkotische Eigenschaften"

3355 **Ledum glandulosum** var. **glandulosum** U.S. & CAN. / M. Rusby s.n. / 1915 / " 'Labrador Tea' "
U.S.: Mont. / J. G. Witt 1697 / 1949 / "Lemon fragrance"

3356 **L. groenlandicum** Alaska / M. W. Gorman 47 / 1895 / "The leaves of this plant are still used as a tea by the natives"

3357 **L. latifolium** U.S.: N.Y. / E. H. Day s.n. / 77 / " 'Labrador Tea' "

3358 **Leucothoë elongata** U.S.: N.C. / H. D. House 5114 / 1913 / "fls. with odor of fresh white clover honey" / "Called 'Honey-caps' by natives"

3359 **Lyonia ferruginea** U.S.: Ga. / J. S. Harper 17 / 1930 / " 'Poor grub' "

3360 **L. fruticosa** U.S.: Fla. / L. M. Andrews 843 / 1966 / "Staggerbush"

3361 **L. mariana** U.S.: N.Y. / C. F. Austin s.n. / no year / "Staggerbush"
U.S.: N.Y. / Columbia College of Pharmacy Herb. s.n. / no year / "Kill-lamb. Stagger-bush"

3362 **Macleania hirtiflora** Ecuador / W. H. Camp E-3939 / 1945 / "Sweetish but flat in comparison to real blueberries" / " 'Guayapa' "

3363 **M. rupestris** Colombia / J. Triana s.n. / 1851–57 / "Vulgo *Uva camarona*"

3364 **Menziesia ferruginea** var. **ferruginea** U.S. & CAN. / M. W. Gorman 205 / ca. 1902 / "Fungus not uncommon on the leaves . . . much relished by the natives who eat them raw" / "Called *Skluk-wud-dish* by the Haidas, from Skuk-wun or Stl-kwin a finger-nail"

3365 **Oxydendron arboreum** U.S.: S.C. / W. M. Canby s.n. / 1876 / "Leaves acid and used as a refrigerant"

3366 **Pernettya mexicana** Mexico / W. C. Leavenworth 281 / 1940 / "Said to poison mules"

3367 **P. mucronata** var. **mucronata** Chile / W. J. Eyerdam 10677 / 1958 / ". . . black, edible berries in large clusters"

3368 **P. prostrata** var. **purpurea** Colombia / F. R. Fosberg 20812 / 1943 / ". . . planted to potatoes, grazed . . . fruit black, fleshy"
Venezuela / J. A. Steyermark 105055 / 1971 / "Berries globose, fleshy, black, edible"

3369 **Pernettya** indet. Venezuela / H. Pittier 13214 / 1929 / " 'Borrachero' "

3370 **Psammisia williamsii** Costa Rica / Burger & Liesner 6388 / 1969 / "With a tart taste and said to give energy when eaten"
Venezuela / J. A. Steyermark 105055 / 1971 / "berries globose, fleshy, black, edible"

3371 *P. falcata* Colombia / H. St. John 20670 / 1944 / " 'uva' "

3372 *P. grandiflora* Colombia / H. St. John 20880 / 1944 / " 'uvitas' "

3373 *Psammisia* indet. Venezuela / Gehriger 405 / 1930 / "comestibles" / "N.v. Coral"

3374 *Rhododendron occidentale* U.S.: Calif. / W. A. Dayton 234 / 1913 / "Reputed to be poisonous"

3375 *R. arboreum* India / W. Koelz 1740 / 1931 / "Flowers acid, eaten with onion, mints, etc. as salad" / "Dialect Kulu 'Barass' "

3376 *Satyria* indet. Colombia / W. A. Archer 1178 / 30 / "fruit edible" / " 'Uvito' "

3377 *Thibaudia floribunda* Colombia / Idrobo et al. 3617 / 1958 / "Frutos inmaturos blancos de sabor ácido agradable, azuloso, al madurar menos ácido" / " 'Chaquilulo' "

3378 *Vaccinium constablaei* U.S.: Tenn. / W. H. Camp 1659 / 1936 / "mature fruit . . . delicously tart"

3379 *V. crassifolium* U.S.: Va. / Fernald & Long 11604 / 1939 / "Fruit . . . juicy; flavor sweet and bland"

3380 *V. globulare* U.S.: Idaho / Hitchcock & Muhlick 21725 / 1958 / ". . . of most excellent flavor"

3381 *V. lamarckii* U.S.: Vt. / Torrey Herb. s.n. / 1833 / "The berries are sweet" / "Huckleberry"

3382 *V. microcarpum* Alaska / Y. Mexia 2197 / 1928 / "Fruit a red berry, edible"

3383 *V. myrtilloides* Canada: Ontario / F. J. Hermann 7075 / 1935 / "fruit larger, with . . . much better flavor"

3384 *V. neglectum* U.S.: Penn. / H. N. Moldenke 2847 / 1926 / " 'Deerberry' " U.S.: Ala. / C. Mohr 7 / 1892 / "Edible" / "Wild Gooseberry"

3385 *V. occidentale* U.S.: Calif. / E. P. Copeland 438 / 1930 / "Berries delicious"

3386 *V. ovalifolium* U.S.: Mich. / H. Gillman s.n. / 1875 / "Called by the Indians Rabbit-berry"

3387 *V. ovatum* U.S.: Calif. / C. C. Parry s.n. / 1850 / "Berries good"

3388 *V. pallidum* U.S.: Tenn. / Smith & Jennison 2675 / 1936 / "Fruits dark blue and sweet"

3389 *V. parvifolium* Alaska / Y. Mexia 2320 / 1928 / "Fruit bright red berry. Edible" / " 'Huckleberry' "

3390 *V. stamineum* U.S.: Ark. / D. Demarée 27027IA / 48 / "Fruits somewhat edible"
U.S.: N.J. / no collector s.n. / 1861 / "Deerberry Squaw Huckleberry"
U.S.: Ind. / C. C. Dean 27823 / 1919 / " 'Wild Gooseberry' "

3391   *V. uliginosum* var. *glabrum* Alaska / Y. Mexia 2209 / 1928 / "Important article of food for Indians & Whites" / " 'Blueberry' "

3392   *V. uliginosum* var. *uliginosum* Greenland / H. E. Wetherill 61 / 1894 / "Partridgeberry"

3393   *V. vitus-idaea* var. *minus* Canada: Québec / B. Boivin 503 / 1938 / " 'Pommes de terre' "

3394   *V. stenophyllum* Mexico / Y. Mexia 1340 / 1927 / "Fr . . . edible" / " 'Capulin' "

3395   *V. floribundum* var. *floribundum* Venezuela / J. A. Steyermark 104845 / 1971 / "Reputed that goats are sick ('loco') after having eaten this plant" / " 'borrachero' (n.v.)"

3396   *V. symplocifolium* Ceylon / Jayasuriya & Sumithraanachchi 1568 / 1974 / "berry reddish, edible; pulp sweet"

3397   *V. macgillivrayi* New Hebrides / S. F. Kajewski 301 / 1928 / "Fruit eaten by natives" / " 'Autarm-tell' "

3398   *V. microphyllum* Philippine Is. / A. D. E. Elmer 11394 / 1909 / ". . . very juicy and possess a fine sweet flavor" / " 'Manalali' "

3399   *V. rapae* Rapa / F. R. Fosberg 11478 / 1934 / "Fruit . . . quite edible, good flavor"

3400   *Ericacea* indet. Canada: Northwest Terr. / H. T. Schacklette 3354 / 1948 / "Fruit edible"

3401   *Ericacea* indet. Canada: Northwest Terr. / H. T. Schacklette 2978 / 1948 / "Fruit . . . edible"

EPACRIDACEAE

3402   *Brachyloma daphnoides* Australia: Queensland / C. T. White 7098 / 1930 / "fls. white. very sweetly honey-scented"

3403   *Lysinema ciliatum* W. Australia / J. C. Anway 219 / 1965 / ". . . flowers . . . smell like curry" / " 'curry flower' "

MYRSINACEAE

3404   *Aegiceras corniculatum* China: Tungking / McClure & Fung Hom 00678 / 1932 / "Fruits cooked & eaten by the natives. The plants are up rooted and used for full" / "Hoi Laam"

3405   *Ardisia compressa* Brit. Honduras / P. H. Gentle 2990 / 1939 / " 'pigeon berries' "
Brit. Honduras / P. H. Gentle 3993 / 1942 / " 'blossom berry grape' "
El Salvador / S. Calderon 683 / 1922 / "in market" / " 'Cerezo' "
Honduras / C. & W. von Hagen 1181 / 1937 / " 'Uvilla de montana' "

Venezuela / J. A. Steyermark 62837 / 1945 / " 'mantequilla' "
Venezuela / J. A. Steyermark 61299 / 1945 / "Fruit edible" / " 'martinica' "

3406 *A. glanduloso-marginata* Costa Rica / A. Smith 4142 / 1937 / "Very strongly perfumed resembling lilacs"

3407 *A. revoluta* El Salvador / P. C. Standley 20934 / 1922 / " 'Uva' "
Mexico / Y. Mexia 1251 / 1926 / "Occasionally eaten" / " 'Capulin' "

3408 *Ardisia* indet. Costa Rica / H. E. Stork 3308 / 1932 / "Fruits edible"

3409 *A. guadalupensis* Virgin Is. / C. L. Horn s.n. / 1939 / "Said to be of insecticidal value"

3410 *Ardisia* indet. Ecuador / J. A. Steyermark 53772 / 1943 / "Fruit edible" / " 'uva' "

3411 *A. crenata* China: Kwangtung / W. T. Tsang 21009 / 1932 / "fl. white, as medicine" / "Lung Shan Che"
China: Kwangtung / W. T. Tsang 20862 / 1932 / "The roots are used for medicine" / "Lon San Shu"
China: Canton / F. A. McClure 7897 / 1921 / "It is reported to be used with five or six other herbs, made into a tea with wine and drunk for broken bones or sprain. It is also used externally" / "Siu lo san"
China: Kwangtung / Fung Hom 6 / 1929 / "Drug name Lok Tei Kam Ich"

3412 *A. henryi* var. *dielsii* China: Kweichow / Steward, Chiao & Cheo 463 / 1931 / "Roots medicinal"

3413 *A. pachyphylla* China: Canton / F. A. McClure 8882 / 1922 / "fls. edible" / "Lo san"

3414 *A. punctata* China: Kwangtung / W. T. Tsang 21165 / 1932 / "use the root to heal wounds" / "Sai Loh Shaan Shue"

3415 *A. quinquegona* var. *quinquegona* China: Canton / F. A. McClure 7930 / 1921 / "drug plant" / "Lo san Shu"

3416 *A. obovata* Indochine / M. Poilane 420 / 1919 / "Écorce grise feuilles lisses fruits roufatre fruits commestible"
Brit. N. Borneo / Matahis 746 / 1928 / "Fruit can be eaten" / "Surusup (Malay)"

3417 *A. copelandia* Philippine Is. / A. D. E. Elmer 10490 / 1909 / The Bogobos eat the berries which have a pleasant flavor of sour" / " 'Catigpo-tigpo' "

3418 *Cybianthus rostratus* W.I.: St. Lucia / J. S. Beard 480 / 1945 / " 'Cafe marron' "

3419 *C. parvifolius* Brazil / Harley & Souza 11183 / 1968 / "Flowers . . . with nauseous smell"

3420 *C. silvestris* Brazil / B. A. Krukoff's 6th Exped. to Braz. Amaz. 7663 / 1935 / "Used in Curare"

3421   *Cybianthus* indet. Colombia / J. M. Idrobo 2467 / 1957 / "Frutos . . . sabor astringente"

3422   *Cybianthus* indet. Venezuela / J. A. Steyermark 56612 / 1944 / " 'manteco' "

3423   *Embelia laeta* China: Kwangtung / To & Ts'ang 12154 / 1924 / "Kai t'ih sheung (Chicken kick injury)"

3424   *E. ribes* Hainan / C. I. Lei 372 / 1933 / "fr. poisonous" / "Siu Chung Nam Tang"

3425   *E. tsjeriam-cottam* India / C. J. Saldanha 13699 / 1969 / "Flowers . . . with a stench"

3426   *E. philippinensis* Philippine Is. / A. D. E. Elmer 8871 / 1907 / "fruits wine color and taste; called by the Igorotes 'Besodak', who eat the leaves as well as the berries; rough! The foliage is used in cooking with meat and which gives it a vinegar flavor"

3427   *Embelia* indet. Borneo / A. Kostermans 490 / 1951 / "Bark rubbed on the skin against leaches"

3428   *Geissanthus lineatus* Venezuela / A. Fendler 759 / 1855 / "fragrant"

3429   *Maesa subdentata* Cochinchina / M. Poilane 711 / 1919 / "Fruits comestibles"

3430   *M. corylifolia* Fiji Is. / O. Degener 13824 / 1940 / "fruit . . . sweetish"
        Fiji Is. / A. C. Smith 349 / 1935 / " 'Chambu di wawa' "

3431   *Maesa* indet. New Guinea / D. Frodin NGF 26766 / 66 / "Leaves chewed as relief for colds and bronchitis"

3432   *Parathesis villosa* Mexico / Y. Mexia 1325 / 1927 / "Fr. edible" / " 'Manzanita' "

3433   *P. vulgata* Honduras / C. & W. von Hagen 1015 / 1937 / " 'Uva' "

3434   *P. reticulata* Colombia / H. S. Kernan 136 / 1944 / " 'fruta de pava' "

3435   *Rapanea ferruginea* Honduras / C. & W. von Hagen 1041 / 1937 / " 'Uva' "
        W.I.: Dominica / G. P. Cooper 1841 / 1933 / " 'Petite plum' "
        Venezuela / J. A. Steyermark 864381 / 1960 / " 'cafecito' "

3436   *R. guyanensis* Brit. Honduras / P. H. Gentle 3613 / 1941 / " 'Bastard fig' "

3437   *R. andina* Ecuador / W. H. Camp E 2707 / 1945 / "The bark used to tan a form of white 'rawhide' " / " 'Yubar' "

3438   *R. latifolia* Ecuador / J. A. Steyermark 52651 / 1943 / "Bark used as substitute for Cinchona in treating malaria" / " 'casca' "

3439   *R. paulensis* Brazil / Y. Mexia 5068 / 1930 / "eaten by birds" / " 'Pau de azéite' 'Porroróco' "

3440   *Rapanea* indet. Ecuador / J. A. Steyermark 52339 / 1943 / "bark used for curing hides" / " 'samil' "

41　**Rapanea** indet. Ecuador / J. A. Steyermark 53428 / 1943 / "Bark used for tanning" / " 'giripe' or 'yobar' "

42　**R. lessertiana** Hawaii Is. / O. & I. Degener 30791 / 1963 / "Fruit starchy & mildly sour"

43　**Stylogyne laterifolia** W.I.: Guadeloupe / P. Duss 2283 / 1892 / "fleurs blanches d'une odeur exquise"

44　**S. amplifolia** Peru / G. Klug 2148 / 1931 / "Jipina coca. Coca silvestre" / "Huitoto Indian name: Taïfe jipina, Taïfe diablo"

45　**Wallenia venosa** Jamaica / W. Harris 8768 / 1904 / " 'Wild Mammee' "

46　**Myrsinacea** indet. Panama / G. P. Cooper 464 / 1928 / "Black berries eaten by birds" / "Black cherry"

THEOPHRASTACEAE

447　**Jacquinia barbasco** Brit. W.I. / W. Kings GC 334 / 1938 / "Bark is removed and soft cortex is scraped off and used as soap to wash clothes"

PRIMULACEAE

448　**Anagallis arvensis** Ecuador / F. Prieto E2452 / 1945 / "A plaster is made by grinding this plant and one is placed on the forehead and a similar one on the back and then an infusion is drunk; supposed to get rid of intestinal worms" / "Sache-false"
Ecuador / J. A. Steyermark 53260 / 1945 / "Used as a worm medicine; mash up green plant, squeeze out the liquid and drink 3 spoonfuls when you have worms (escrobota) in the stomach (symptoms of these are itching of the nose); after drinking the worms are reputed to leave" / " 'pitisacha' "

449　**A. repens** Algeria / A. Faure s.n. / 1923 / "Endroits humectes par les eaux"

450　**Lysimachia congestiflora** China: Kiangsi / Chung & Sun 512 / 1933 / "Flower-incense"

SAPOTACEAE

451　**Bumelia eriocarpa** Mexico / T. MacDougall s.n. / 1970 / "Flowers fragrant, attracting many bees" / " 'cinco negritos' "

452　**Bumelia** indet. Guatemala / P. S. Odell 12276 / 1941 / "Was first type of tree to be tapped for chewing gum" / " 'Abalo' "

453　**B. retusa** Bahama Is. / O. Degener 18818 / 1946 / "fruit edible" / "Milkwood"
Bahama Is. / O. Degener 18955 / 1946 / "Fruit chewed for chewing gum"

3454   *Chrysophyllum medicanum* Honduras / C. & W. von Hagen 1145 / 1937 / "fruit edible" / " 'Guayava de danto' "

3455   *C. oliviforme* Dominican Republic / H. A. Allard 13370 / 1945 / "Tastes like and astringent as choke cherries . . . Eaten by children"

3456   *Chrysophyllum* indet. Dominican Republic / M. G. Whiting 408a / 1964 / "Leaves used for fodder" / " 'Calmito' "

3457   *C. balata* var. *saunanyek* Venezuela / J. A. Steyermark 60197 / 1944 / "fruit edible" / " 'sau-nan-yek' "
Venezuela / J. A. Steyermark 60672 / 1944 / "fruit edible" / " 'saunan-yek' "

3458   *C. beardii* Venezuela / J. A. Steyermark 60040 / 1944 / "fruit edible" / " 'cacak-orai-yek' "

3459   *C. caracasanum* Venezuela / J. A. Steyermark 61492 / 1945 / "Fruits sweet, edible" / " 'Paramino' "
Venezuela / H. Pittier 12312 / 1927 / "fr. edible when ripe" / " 'Guarracuco' "

3460   *C. gonocarpum* Paraguay / A. L. Woolston 755 / 1956 / "fr. orange-yellow, milky, sweet, edible" / "agual"

3461   *Madhuca obovata* New Hebrides / S. F. Kajewski 464 / 1928 / "fr. eaten by natives"

3462   *Manilkara striata* Guatemala / P. S. Odell 12270/10 / 1941 / "Used for chewing gum" / " 'Zapote blanco' "

3463   *M. zapotilla* Guatemala / P. S. Odell 12269/9 / 1941 / "Used for chewing gum"
Guatemala / P. S. Odell 12777/17 / 1941 / "Used for chewing gum" / " 'Chico Zapote' "
India / E. W. Erlanson 5138 / 1933 / "Fruit eaten"

3464   *M. bidentata* W.I.: St. Lucia / J. S. Beard 511 / 1945 / "fr. yellow, plumlike, edible" / " 'Balata' "

3465   *M. inundata* Colombia / R. E. Schultes 6860 / 1945 / "Latex white. Wood used to smoke rubber" / " 'Massarandabi blanca' "

3466   *M. longiciliata* Peru / R. L. Fróes 22688 / 1947 / "(Good chewing gum)"

3467   *M. surinamensis* Venezuela / Wurdack & Adderley 42737 / 1959 / "Voucher for Warner Lambert drug sample" / " 'Massaranduba' "

3468   *M. mochisia* Tanganyika / R. E. S. Tanner 2777 / 1956 / "Zigua: leaves dried and pounded up and used as snuff in exorcism" / "Zigua: Kungo"

3469   *M. zanzibarensis* Tanganyika / R. E. S. Tanner 3658 / 1957 / "Fruits edible" / "Doe: Sinzilugulu"

3470   *M. hexandra* Ceylon / P. L. Comanor 401 / 1967 / "Fruit orange, edible . . . Tastes like a grape"

71 *Mimusops caffra* Port. E. Africa / G. Barbosa 2633 / 1949 / "Edible wild fruit" / " 'Tinzol' "

72 *M. fruticosa* Tanganyika / R. E. S. Tanner 3715 / 1957 / "Zigua: For pain in stomach, inner bark soaked in water and drunk" / "Zigua: Mhozohozo"

73 *M. zeyheri* Transvaal / J. M. Lanham 41 / 1940 / "yellow edible fruits" / " 'Muppel' "

74 *Mimusops* indet. Tanganyika / R. E. S. Tanner 1332 / 1953 / "Sukuma tribe: Roots used to aid lactation by adding powdered root to porridge; root flour put in hot water and drunk as secret poison antidote" / "Kisukuma: Melemele gwa mulugulu"

75 *M. hexandra* Hainan / S. K. Lau 501 / 1932 / "Fr. edible" / "Tit Sintsz"

76 *Palaquium bataanense* Philippine Is. / A. L. Zwiekey 296 / 1938 / "useful wood for bankas" / " 'Nato' "

77 *Pouteria hypoglauca* Mexico / C. L. & A. A. Lundell 7128 / 1937 / "Fruits edible"

78 *P. domingensis* Cuba / R. A. Howard 5182 / 1941 / "Frts. brown and edible when mature" / " 'canedodo' "

79 *P. cayennensis* Venezuela / J. A. Steyermark 60664 / 1944 / "fruit edible" / " 'amarun-yek' "

480 *P. egregia* Bolivia / J. A. Steyermark 86962 / 1960 / "fruit edible, favored by monkeys" / " 'Purguillo' "

481 *P. guyanensis* Venezuela / J. A. Steyermark 60991 / 1945 / "fruit orange-yellow . . . edible" / " 'purvio amarillo' "

482 *P. lucuma* Peru / Killip & Smith 22818 / 1929 / "Fruit eaten" / " 'Lucmo' "

483 *P. macrophylla* Fr. Guiana / W. E. Broadway 791 / 1921 / "Ripe fruits said to be edible"
Venezuela / A. L. Bernardi 5813 / 1956 / "frutos comestibles"
Peru / W. A. Archer 7745 / 1942 / "frt. edible"

484 *P. migonei* Paraguay / A. L. Woolston 1335 / 1953 / "Tree cultivated for edible fruit" / "aguai-guasu"

485 *P. sapota* Ecuador / Little & Dixon 21210 / 1965 / "Sembrado para los frutos comestibles" / " 'mamey' "

486 *P. venosa* Brazil / A. Ducke 1697 / 1945 / "Fructus edulis" / " 'Cutitiriba grande' "

487 *Pouteria* indet. Ecuador / J. A. Steyermark 54348 / 1943 / "Fruit edible" / " 'lucma' " or " 'luma' "

488 *Pouteria* indet. Venezuela / Steyermark & Rabe 96112 / 1966 / "fruit pear-shaped, edible ripe" / " 'buyero' "

3489   *Pouteria* indet. Brit. Guiana / For. Dept. of Brit. Guiana F809 / 1942 / "sweet edible pulp"

3490   *Pouteria* indet. Brit. Guiana / For. Dept. of Brit. Guiana WB213 / 1948 / "seeds embedded in edible sweet pulp"

3491   *Pouteria* indet. Peru / W. A. Archer 7732 / 1942 / "Frt. edible, yellow when ripe" / " 'Abiú' "

3492   *Pouteria* indet. Ecuador / E. L. Little, Jr. 6552 / 1943 / "Fruit edible" / " 'Logma' "; " 'lucma' " or " 'lugma' "

3493   *Pouteria* indet. Peru / J. M. Schunke 56 / 1935 / "Fruit yellow; edible" / " 'Sachacaimito' "

3494   *Pouteria* indet. Colombia / A. Dugand G. 393 / 1933 / ". . . somewhat perfumed latex. Fruit edible"

3495   *Pouteria* indet. Brit. Guiana / A. C. Smith 3247 / 1938 / "seed edible"

3496   *Pouteria* indet. Ecuador / Fosberg & Giler 23076 / 1945 / "fruit edible" / " 'lugma' "

3497   *Vitellariopsis kirkii* Tanganyika / R. E. S. Tanner 3176 / 1956 / "Zigua: for increased fertility in women, roots boiled and the water drunk"

3498   *Sapotacea* indet. Peru / R. L. Fróes 24055 / 1949 / "yellow fruit edible"

3499   *Sapotacea* indet. Peru / R. L. Fróes 23440 / 1948 / "yellow fruit when ripe, latex eadable fruit"

3500   *Sapotacea* indet. Peru / A. Silva 122 / 1944 / "Frt yellow, edible" / " 'Guajara' "

EBENACEAE

3501   *Diospyros ferrea* var. *buxifolia* Ryukyu Is. / F. R. Fosberg 38158 / 1956 / "fruits orange, edible"
Ryukyu Is. / F. R. Fosberg 38224 / 1956 / "fruit orange when immature, dark brownish red when ripe, edible, with sweetish thin flesh"

3502   *D. diversilimba* Hainan / S. K. Lau 1457 / 1933 / "use the seed for dyeing purposes" / "Kwo Po Tsz"
Hainan / C. I. Lei 859 / 1933 / "Fruit becomes black when ripe. Used—as black dye" / "Wu Lik Kwo"

3503   *D. malabarica* Ceylon / D. B. Sumithraarachchi 477 / 1974 / "ripe fruits edible"

3504   *D. ferrea* U.S.: Hawaii / Melville, Melville, Degener & Degener 1145 / 1971 / "Fruit edible but with little flavor" / "Lana"

3505   *D. foliosa* Fiji Is. / A. C. Smith 6439 / 1947 / "seed edible" / " 'Ulalo' "

3506   *D. major* Fiji Is. / O. Degener 15388 / 1941 / "fruit edible" / " 'Mbama' "

07 *D. nitida* Philippine Is. / R. S. Williams 2682 / 1905 / "Leaves used for dyeing cloth a dark brown"

08 *Euclea divinorum* Kenya / Glover et al. 2615 / 1961 / "Goats browse leaves. Fruit edible. Roots used by Masai for soup" / "Uswet (Kips); Olkimyei (Masai)"

09 *Royena fischeri* Tanganyika / R. E. S. Tanner 1401 / 1953 / "Sukuma tribe: Roots pounded and mixed with butter as poultice to treat scabies; roots boiled with white mtama flour, taken with water for Bilharzia" / "Kisukuma: Fubatta"

10 *R. macrocalyx* Nyasaland / L. J. Brass 17915 / 1946 / "fruit orange, edible"

STYRACACEAE

11 *Styrax ovatum* Peru / Y. Mexia 8171 / 1936 / "(bat food)" / " 'Chichic-micuma' "

SYMPLOCACEAE

12 *Symplocos martinicensis* W.I.: Dominica / G. P. Cooper 143 / 1933 / "Birds eat fruits" / "Native name 'Cacarat' or 'Grine bleux' "

13 *S. theiformis* Colombia / J. M. Duque Jaramillo 2868 / 1946 / "Las hojas se pueden tomar como sustituto del té oriental" / " 'Té de Bogotá' "

14 *Symplocacea* indet. N. Rhodesia / W. R. Bainbridge 588 / 61 / "Infusion of roots (very bitter) used as medicine for V.D." / "Plateau Tonga name KATUNDA"

OLEACEAE

15 *Fraxinus xanthoxyloides* India / W. Koelz 1097 / 1930 / "Bark used as medicine for cows, goats and sheep. 2. usually heavily browsed" / "Dugmanyu (Lahuli)"

16 *Jasminum fluminense* Tanganyika / R. E. S. Tanner 2103 / 1955 / "Leaves used as vegetable . . ." / "Kibondei; Mwafu"
Tanganyika / R. E. S. Tanner 959 / 1952 / "Sukuma tribe: Roots used 'to increase someone's liking for you' " / "Kisukuma: Mbililikigyi"

17 *J. humile* var. *humile* India / W. Koelz 1889 / 1931 / "Make tea of leaves & bark" / "Raggul cha (Kumore)"

18 *J. simplicifolium* Fiji / A. C. Smith 1922 / 1934 / "Tea from leaves used for sore throats" / " 'Mothe ni vai' "
Fiji / A. C. Smith 42 / 1933 / "Leaf used as medicine for fever" / " 'Vono ni mbengga' "

3519   *Linociera nilotica* Tanganyika / R. E. S. Tanner 1569 / 1953 / "Sukuma tribe: Roots pounded, taken with water to treat enlarged spleen in children" / "Kisukuma: Ngongondi"

3520   *Schrebera oligantha* Tanganyika / R. E. S. Tanner 1365 / 1953 / "Sukuma tribe: Roots pounded, eaten raw to treat poisoning" / "Kisukuma: Mbudika; Njimya wa nghinda"

LOGANIACEAE

3521   *Geniostoma rupestre* Fiji / O. Degener 32116 / 1968 / ". . . excrement odor, liquid from macerated bark squeezed in water drunk for women's diseases" / " 'Boiboida' "

3522   *Spigelia pedunculata* Colombia / M. T. Dawe 796 / 1918 / "Root used as vermifuge" / " 'Quiteria' (sp.?)"

3523   *Spigelia* indet. Colombia / Scolnik & Palacio 1663 / 1949 / "Purgante y anti-helmintica" / "lombricera"

3524   *Strychnos diaboli* Brazil / Prance, Fidalgo et al. 21617 / 1974 / "Principal ingredient in Mayongong Indian arrow poison" / " 'Cumadua' (Mayongong)"

3525   *S. erichsonii* Surinam / A. M. W. Mennega 467 / 1954 / "used as fish poison"
Colombia / R. E. Schultes 3682 / 1942 / "Bark used in Kofán curare" / "ee-ru-che"

3526   *S. fendleri* Venezuela / L. Aristeguieta 5821 / 1965 / "La corteza hervida se usa contra el paludismo" / "Cruceto"
Venezuela / J. A. Steyermark 55559 / 1944 / "Used as source of quinine in this area supposedly and is type known to conocedores" / " 'quina' "
Venezuela / J. A. Steyermark 57680 / 1944 / "Used for treating malaria—said to be stimulating when having malaria and cures it according to natives" / " 'Cruceta real' "

3527   *S. glabra* Brazil / R. Fróes 21528 / 1945 / "Unico usado pelos Indios Baniuas para o veneno 'Irary' "

3528   *S. guianensis* Brit. Guiana / A. C. Smith 2836 / 1938 / "Said to be the most important component of Wai-wai Balauitu (arrow-poison); only outer bark used, preferably from roots"
Brazil / Prance, Fidalgo et al. 21618 / 1974 / "Ingredient of Mayongong Indian arrow poison" / " 'Quamitea' (Mayongong)"

3529   *S. jobertiana* Colombia / Schultes & Lopez 10091 / 1948 / "Kuripakas use this as principal ingredient of their arrow-poison"
Venezuela / J. A. Steyermark 57910 / 1944 / "Leaves mashed, placed in rum, and made into a perfume, with intent of attracting opposite sex"

Peru / R. Kayap 1230 / 1974 / "root used for dart venom" / "paigsi unweke"

530 **S. mitscherlichii** var. **mitscherlichii** Colombia / J. M. Idrobo 2628 / 1957 / "Bejuco leñoso, con corteza amarga. Con éste se prepara veneno para flechas. Se raspa, se separa la corteza se reune con otros varios bejucos y se cocina en largo proceso para preparar el veneno" / "Nombre Siona: 'Pux-ae-o' = 'Veneno caimán', 'Que-he-ae-o' = 'Bejuco carrasposo' "

531 **S. panamensis** Venezuela / J. de Bruijn 1424 / 1967 / "Seeds eaten by monkeys" / "Naranjuelo"

532 **S. peckii** Colombia / R. E. Schultes 3601 / 1942 / "Root used in curare preparation" / "se-he-pa"
Venezuela / Maguire, Wurdack & Maguire 41941 / 1957 / "Fruit brown, said to be edible" / " 'Cachete vieja' "

533 **S. rondeletioides** Brazil / Campbell, Nelson et al. P21207 / 1974 / "Bark & mashed leaves mixed with water & strewn in water by Paumaris Indians as a fish poison" / " 'Jadadakaikapihai' (Paumaris)"
Venezuela / J. A. Steyermark 60441 / 1944 / "Roots used for arrow poison; boil all day for juice to get thick" / " 'cumarawa-yek' "

534 **S. solimoesana** Brazil / Campbell, Nelson et al. P21254 / 1974 / "Ingredient of Jamamadi arrow poison. Scraped stem bark pounded into powder, boiled down with other constituents" / " 'Iha' (Jamamadi)"
Brazil / Prance, Maas, Atchley et al. 13929 / 1971 / "Bark used as principal ingredient in arrow and blow-gun dart poison. Mixed with Menispermaceae, Annonaceae and Sapindaceae" / " 'Irã' "

535 **S. tarapotensis** Peru / E. Ancuash 718 / 1974 / "comestible su fruta" / " 'tsacik' "
Peru / Y. Mexia 6180 / 1931 / "mature fruit brown, edible"

536 **S. henningsii** Tanganyika / R. E. S. Tanner 1588 / 1953 / "Sukuma tribe: Roots pounded and taken in food to treat ankylostomiasis" / "Kisukuma: Kaguha ngosha"

537 **S. innocua** Tanganyika / R. E. S. Tanner 1476 / 1953 / "Sukuma tribe: roots pounded and taken with food to treat ankylostomiasis . . ." / "Kisukuma: Mihundu"

538 **S. lucens** Tanganyika / R. E. S. Tanner 1557 / 1953 / "Sukuma tribe: Roots pounded, used orally in water to treat ankylostomiasis" / "Kisukuma: Ikome"

539 **S. madagascariensis** Tanganyika / R. E. S. Tanner 3710 / 1957 / "Zigua: For constipation, roots boiled and the water drunk" / "Zigua: Mtonga mfyala"
Tanganyika / R. E. S. Tanner 3927 / 1958 / "Bondei: For yaws, roots boiled and the water drunk" / "Bondei: Mkwakwa"

AFR. / L. J. Brass 16373 / 1946 / "fruit . . . edible" / "Matemi (Chinyanja)"

3540 *S. spinosa* Tanganyika / R. E. S. Tanner 1428 / 1953 / "Sukuma tribe: Roots cooked in porridge to treat coughs" / "Kisukuma: Mwage" Tanganyika / R. E. S. Tanner 3711 / 1957 / "Zigua: For asthma, roots boiled and the water drunk" / "Zigua: Mtonga ngombe"

3541 *S. cathayensis* China: Kwangtung / S. K. Lau 817 / 1932 / "fruit, edible"

BUDDLEJACEAE

3542 ***Buddleia marrubiifolia*** var. ***marrubiifolia*** Mexico / G. B. Hinton 16537 / 44 / "Used as a tea" / " 'Salvia' 'Azafrán' "

3543 ***B. brasiliensis*** Brazil / Y. Mexia 4710 / 1930 / "Leaves used in infusion to bathe swellings" / " 'Barbasso' "

GENTIANACEAE

3544 ***Chelonanthus acutangulus*** Peru / Killip & Smith 25402 / 1929 / "Used for cuts" / " 'Pichispan' (Chunchu); 'ojo manga' (Spanish)"

3545 ***Coutoubea spicata*** Brit. Guiana / D. H. Daris (sp.?) 777 / 1968 / "reputed poisonous to cattle"

3546 ***Exacum quinquenervium*** Tanganyika / R. E. S. Tanner 3664 / 1957 / "Swa: For gnawing pain in stomach, pounded up green, squeezed and the juice drunk" / "Swa: Mnaymati"

3547 ***Lisianthus alatus*** Brit. Guiana / S. G. Harrison 1345 / 1958 / ". . . fleshy infusion of leaves used by Arawaks for small pox" / "Wild Tobacco"

3548 ***Pleurogyne carinthiaca*** Himalaya / W. Koelz 1144 / 1930 / "mixed with other herbs used for fever and headache" / "Zaugti (Lahul)"

3549 ***Sabatia decandra*** U.S.: Fla. / P. P. Sheehan s.n. / 1919 / "Used as a specific for indigestion. The plant is brewed as a tea and the decoction then is drunk hot (?) or cold (?)" / "Marsh-pink"

3550 ***Sebaea hymenosepala*** S. Africa / R. D. A. Bayliss BS5287 / 1972 / "Fls bright yellow. Bitter taste. 'Reputed to be cure for snake bite by local natives' "

APOCYNACEAE

3551 ***Acokanthera longiflora*** N. Rhodesia / Brenan & Greenway 7982 / 1947 / "used as arrow poison" / "Musongwa (Chiwemba)"

3552 ***A. oppositifolia*** S. Africa / R. D. A. Bayliss BS-BRI 544 / 1973 / "*All parts* of this plant are highly toxic. Used by natives to poison their arrows"

553 *A. venenata* S. Rhodesia / W. C. Chase 1580 / '49 / "arrow poison"

554 *Aganonerion polyphorum* Indochina / Poilane 4540 / 1922 / "Elle donne un latex blanc tres collant les feuilles qui sont aigrelottes seraient comestibles"

555 *Alafia* indet. W. Africa / B. A. Krukoff 222 / 1950 / "Seeds used as arrow poison"

556 *Allamanda violacea* Tanganyika / R. E. S. Tanner 3411 / 1956 / "Zigua: For severe head colds, roots boiled and the water drunk" / "Zigua: Mtula-maama"

557 *Alstonia reineckeana* Fiji / O. Degener 15062 / 1942 / "Used similar to other species as chewing gum and as eye remedy" / " 'Mbuleki' in Gerua dialect"

558 *A. vitiensis* Fiji / O. Degener 15266 / 1941 / "For eye trouble scrape bark, and squash juice into eye. Used for chewing gum like the other species" / " 'Bule' in Serua"

559 *Alyxia* indet. Fiji / E. Ordonez 13615 / 1940 / "Fresh uncooked leaves eaten by Fijians"

560 *Amalocalyx yunnanensis* China: Yunnan / A. Henry 11756 / no year / "fruit edible"

561 *Ambelania acida* Brazil / W. A. Archer 7766 / 1942 / "frt. edible" / " 'pepino do mato' "

562 *A. grandiflora* Brazil / W. A. Archer 7625 / 1942 / "frt. edible" / " 'pepino do matto' "

563 *A. quadrangularis* Venezuela / Bernardi 920 / 1953 / "Frutos comestibles" / "arekuma: Azatani-yeh"
Brazil / R. Fróes 21149 / 1945 / "fruto comestivel"

564 *Carissa edulis* S. Africa / J. Gerstner 5581 / 1945 / "Fruits, when ripe, blackish, edible"

565 *C. opaca* India / W. Koelz 4319 / 1933 / "fruit blue black edible"
India / R. R. Stewart s.n. / 1934 / "Fruits eaten by children"

566 *C. spinarum* Ceylon / P. L. Comanor 593 / 67 / "fruit black oblate spheroidal, edible, tasty"

567 *Catharanthus roseus* Ceylon / P. L. Comanor 1013 / 68 / "flowers, red-violet; medicinal"

568 *Couma catingae* Venezuela / C. A. Blanco C. 1197 / 1971 / "Fruto globoso, comestible"

569 *C. macrocarpa* Brazil / Prance, Steward et al. 13592 / 1971 / "latex drinkable" / " 'Omnamshifi' (Uaicá)"

570 *C. rigida* Colombia / H. García y Barriga 13809 / 1951 / "frutos amarillos con látex blanco. (comestibles)" / " 'Baa' "

3571   *Ecdysanthera rosea* China: Kwangtung / W. T. Tsang 21471 / 1932 / "Leaves are edible" / "Suen T'ang Tsz"

3572   *Himatanthus attenuata* Brazil / Y. Mexia 4136 / 1929 / "Used as a purgative against fevers" / " 'Aguniada' "

3573   *H. bracteata* Brazil / Ratter, de Santos et al. R.1390 / 1968 / "milk edible but rather tasteless, supposed to be beneficial to women in pregnancy" / "Sucuúba"

3574   *H. obovata* Brazil / R. Fróes 2062 / 1933 / "Calices used for women troubles" / "Jan de Leite"

3575   *Hunteria zeylanica* Tanganyika / R. E. S. Tanner 3722 / 1957 / "Zigua: For vaginal hernia?, roots boiled and the water drunk" / "Zigua: Mvuti"

3576   *Lacmellea standleyi* Brit. Honduras / W. A. Schipp 1326 / no year / "known locally as Vaca or Milk tree on account of the latex being good to drink. Fruits yellow with the odor of mangoes" / "Vaca or Milk tree"

3577   *L. floribunda* Colombia / R. Romero-Castañeda 8408 / 1960 / "frutos amarillos, dulces y comestibles"

3578   *Lacmellea* indet. Brazil / M. Barbosa da Silva 123 / 1942 / "Frt. yellow, edible" / " 'pau de colher' "

3579   *Landolphia amaena* Tanganyika / R. E. S. Tanner 610 / 1952 / "Fruits eaten, stems used in basket weaving, sap used as drops to treat sore eyes" / " 'Kisukuma: Ikamila' "

3580   *L. petersiana* Tanganyika / R. E. S. Tanner 3439 / 1957 / "Fruits edible" / "Zigua: Mbungo"

3581   *Macoubea guianensis* Brazil / R. Fróes 22787 / 1947 / "edible fruit" / "Uarma gogo"
Brazil / G. Bondar 8 / 1946 / "It is often destroyed and felled for its savory fruit" / " 'Piquia' "

3582   *Macrosiphonia hypoleuca* Mexico / G. B. Hinton 7716 / 35 / "spice for atole" / "Flor de San Juan"

3583   *Melodinus suaveolens* China: Kwangsi / W. T. Tsang 24110 / 1934 / "fruit yellow, edible" / "Ma Lau Chang"

3584   *Ochrosia oppositifolia* Marshall Is. / F. R. Fosberg 26735 / 1946 / ". . . seed eaten" / " 'Kish par' "

3585   *Parahancornia amapa* Brazil / Prance & Pennington 2073 / 1965 / "Used locally as cure of tuberculosis" / " 'Amapá' "
Brit. Guiana / For. Dept. of Brit. Guiana F706 / 41 / "fr . . . edible, milky" / "Dukali"

3586   *Rauwolfia mombassana* Tanganyika / R. E. S. Tanner 3150 / 1956 / "Bondei: For breast abcess, leaves pounded up and applied as a paste" / "Swaheli: Mwembe mwito"

Tanganyika / R. E. S. Tanner 3955 / 1958 / "Zigua: For asthma and tuberculosis, roots boiled and the water drunk" / "Zigua: Mhelapogo"

87   *R. verticellata* Burma / Bisset 776b / 60 / "Leaves and/or whole root boiled with water and used as a bath for aches and pains; leaves (or sometimes roots) along with Andrographis paniculata (segagyi) leaves boiled and used as a bath for scabies"
Java / Government coll. s.n. / 1954 / "said to be a plant cultivated for medicine"

88   *Saba camosa* Tanganyika / R. E. S. Tanner 1992 / 1955 / "Fruits edible" / "Kibondei: Mabungo"
Tanganyika / R. E. S. Tanner 3098 / 1956 / "Fruits edible" / "Zigua: Mbungo"

89   *Schizozygia coffaeoides* Tanganyika / R. E. S. Tanner 3944 / 1958 / "Sambaa: For painful eyes, leaves boiled and the patient sits in the steam" / "Sambaa: Mbaika"

90   *Strophanthus eminii* Tanganyika / R. E. S. Tanner 590 / 1952 / "Many medicinal uses — abortifact, poison, etc." / "Kisukuma: Msungululu"

591   *S. intermedius* Angola / Brass & Woodward 20936 / 1950 / "A tea made from the roots is drunk by the natives as a cure for rheumatism"

592   *S. nicholsoni* Africa / R. B. Usher s.n. / 1950 / "arrow poison" / " 'Kanyanga' "

593   *S. caudatus* var. *marckii* Malaya / G. anak Umbai for A. H. Millard 1988 / 60 / "Stupefying" / "Akar Jangut Baung (Temuan)"

594   *Tabernaemontana sananho* Brazil / B. A. Krukoff 7840 / 1939 / "used extensively in Indian medicine"

595   *T. tetrastachya* Colombia / R. Romero Castañeda 4997 / 1954 / "Fruto y pulpa amarillos, comestibles"

596   *Tabernaemontana* indet. Venezuela / J. A. Steyermark 93052 / 1964 / "latex white, said to be poisonous" / " 'guachimacán' "
Ecuador / R. C. Gill 14 / 1938 / "emetic"

597   *Urechites lutea* Jamaica / D. J. Rogers J-9 / 1953 / "Said to be poisonous to cattle"
W.I.: Grand Cayman Is. / C. M. Maggs II.67 / 1938 / "Said by islanders to be very poisonous. Latex said to cause blisters"

598   *Wrightia pubescens* Hainan / C. I. Lei 209 / 1932 / "Leaves to feed pigs" / "Foo Yung"
Hainan / H. Fung 20314 / 1932 / "Lvs. boiled in water can cure fever, etc." / "Foo Yung"

3599 *Apocynacea* indet. Colombia / García y Barriga, Hashimoto et al. 18519 / 1965 / "Frutos verde amarillentos, comestibles, acidos" / " 'Lechemiel' "

3600 *Apocynacea* indet. Ecuador / C. C. Fuller 137 / 53 / "Raw juice of stems & leaves poured in dog's nose to make him better hunter. After treatment dogs appear intoxicated" / "Tzicta; Quichua"

3601 *Apocynacea* indet. Peru / J. Schunke V. 1062 / 1966 / "The root drinking in brandy for rheumatism" / "Ucho Sanango"

3602 *Apocynacea* indet. W. Africa / Krukoff & Letouzey 117 / 1949 / "seeds used as arrow poison"

3603 *Apocynacea* indet. China: Kwangtung / W. T. Tsang 21471 / 1932 / "Leaves are edible" / "Suen T'ang Tsz"

3604 *Apocynacea* indet. Philippine Is. / C. Miguel 1 / 1928 / "fruit is said to be edible" / "Tabu" (sp.?)

ASCLEPIADACEAE

3605 *Asclepias galioides* U.S.: Colo. / H. C. Cutler 2744 / 1939 / "This, according to several farmers, makes cattle sick, occasionally killing them"

3606 *A. rolfsii* U.S.: Fla. / P. P. Sheehan s.n. / 1919 / "Used as a remedy for snake-bite. The whole plant is boiled and the decoction is drunk" / "Butterfly-weed"

3607 *Asclepias* indet. U.S.: Utah / Herdhey 2440 / 41 / "Very poisonous to livestock"

3608 *A. curassavica* Mexico / G. B. Hinton 15398 / 39 / "foliage said to be deadly poisonous for goats"
Colombia / R. E. Schultes 3461 / 1942 / "Considered poisonous" / " 'vencenuco' "

3609 *Curroria volubilis* Tanganyika / R. E. S. Tanner 1274 / 1953 / "Sukuma tribe: Pounded roots put in drinking water to increase lactation in women, cattle" / "Kisukuma: Ibongote; Igilingwa"

3610 *Gymnea sylvestre* Tanganyika / R. E. S. Tanner 1481 / 1953 / "Sukuma tribe: leaves used to rub into small cuts for stitch in side; roots cooked in food for epilepsy" / "Kisukuma: Luhaga"

3611 *Gymnolaema tuberosa* Tanganyika / R. E. S. Tanner 608 / no year / "Stems used for arrow shafts, also medicinally to reduce body swellings" / "Kisukuma: Matala"

3612 *Margaretta rosea* Tanganyika / R. E. S. Tanner 3890 / 1957 / "Zigua: male aphrodisiac, leaves boiled and the water drunk" / "Zigua: gosi"

3613 *Marsdenia inelegans* Venezuela / L. Aristeguieta 3090 / 1958 / "considered a poisonous plant"

14   *Pentarrhinum insipidum* Tanganyika / R. E. S. Tanner 1408 / 1953 / "Sukuma tribe: Leaves used in poultice to treat boils" / "Kisukuma: Lundata"
Tanganyika / R. E. S. Tanner 1291 / 1953 / "Sukuma tribe: Roots soaked, used to wash sores; leaves used as poultice on sores" / "Kisukuma: Luduta lwa mshilugu"

15   *Sarcostemma cynanchoides* U.S. & CAN. / Dr. Gregg s.n. / 47 / "Said to be poisonous"

16   *Secamone parvifolia* Tanganyika / R. E. S. Tanner 1613 / 1953 / "Sukuma tribe: Roots pounded, mixed with honey, used orally to treat harsh cough, leaves pounded, used in wash water for scabies" / "Kisukuma: Melameja nkima"

17   *Secamone* indet. Tanganyika / R. E. S. Tanner 1277 / 1953 / "Sukuma tribe: Roots used in preparation put in porridge to treat anemia; also to increase lactation" / "Kisukuma: Melameja"

18   *Stathmostelma pedunculatum* Tanganyika / R. E. S. Tanner 3895 / 1957 / "Zigua: male aphrodisiac, roots pounded into flour, cooked, and eaten" / "Zigua: Dundilo"

CONVOLVULACEAE

19   *Astrochlaena volkensii* Tanganyika / R. E. S. Tanner 1505 / 1953 / "Sukuma tribe: Roots pounded and taken with food to treat ankylstomiasis" / "Kisukuma: Igalagala"

20   *Convolvulus hystrix* Libya / M. Drar 91/1062 / 1932 / "The roots are used in some districtes as a purge" / "Shubruq"

21   *Erycibe borneensis* Borneo / C. Wee-lek CWL.964 / 1966 / "flowers white, with strong lemon smell"

22   *E. pequensis* Bengal / V. S. Ras 5674 / 35 / ". . . with jasmine-scented fl.

23   *Ipomoea pandurata* U.S.: N.J. / B. Heritage 57 / 1891 / "Wild Potato, Man-of-the-earth"

24   *Ipomoea* indet. Dominican Republic / G. P. Cooper III   190 / 1933 / " 'Patat marron' meaning 'wild potato' "

25   *I. abutiloides* Ecuador / Játiva & Epling 109 / 1962 / " 'Goat-killer' "

26   *I. carnea* Venezuela / Agostini & Agostini 1008 / 1972 / "Se dice que produce 'arrellera' de los chivos, cuando seco es más tóxico" / " 'Aritibal' "

27   *I. crassicaulis* Brazil / R. Fróes 2028 / 1933 / "tea from leaves for stomach troubles" / "Algodaõ do Campo"

28   *I. indica* Mariana Is. / F. R. Fosberg 24943 / 1946 / "juice squeezed out and used to cure cracked feet—both human and cows" / " 'fófigo' "

Mariana Is. / F. R. Fosberg 25347 / 1946 / "leaves fed to pigs, young tips mashed and applied to bruises" / " 'Fofgo' "

Caroline Is. / F. R. Fosberg 32057 / 1950 / "said to be used medicinally against large swellings which are infected from scratching — swelling decreases, then increases and ruptures" / " 'Oleamad' "

3629 *I. littoralis* Caroline Is. / F. R. Fosberg 25775 / 1946 / "leaves eaten" / " 'toro' "

3630 *Jacquemontia curtissii* U.S.: Fla. / P. P. Sheehan s.n. / 1919 / "Esteemed as a leg-medicine. The whole plant is boiled. When tepid, the decoction is rubbed on the ailing member" / "White Jacquemontia"

3631 *J. paniculata* Tanganyika / R. E. S. Tanner 2940 / 1956 / "Leaves used as vegetable" / "Kizigua: Matako Mashano"

Tanganyika / R. E. S. Tanner 2059 / 1955 / "Leaves used as poultice to treat swellings; leaves soaked in water, liquid drunk to treat syphilis" / "Kizigua: Kilemba chajumbe; Kingindo: Mpurulu; Kirufigi: Mkundekunde"

3632 *Maripa panamensis* Colombia / J. M. Idrobo 1830 / 1955 / "Entre el pericarpio y la semilla hay abundante líquido espeso de color caramelo quemado y de sabor a caramelo, muy apreciado por los nativos para comerlo" / " 'Miel quemada' "

3633 *M. scandens* Peru / Robertson & Austin 242 / 1967 / "Apical end of fruits and seeds eaten by monkeys" / " 'Monkey syrup' "

3634 *Merremia reniformis* Burma / O. White 74 / 50 / "Burmese and Chinese eat it as greens — tastes like spinach — very good — I like it — had three or four messes more like chard in taste — could be an awful weed but excellent human fodder — perhaps does better where water stands for several weeks"

CUSCUTACEAE

3635 *Cuscuta hyalina* var. *nubiana* Tanganyika / R. E. S. Tanner 1270 / 1953 / "Sukuma tribe: Leaves pounded, mixed with butter as poultice to treat boils" / "Kisukuma: Kalitongo"

MENYANTHACEAE

3636 *Limnanthemum indicum* India / P. A. Waring 10 / 1954 / "for indigestion in children" / "Toptopi"

POLEMONIACEAE

3637 *Cantua quercifolia* Ecuador / I. L. Wiggins 10992 / 1944 / "Said to be used as medicine to combat malaria, the parts being used to make a tea"

Ecuador / F. Prieto AP-34 / 1944 / " 'Used by the natives of the region as a remedy for malaria' "

38 *Loeselia scariosa* Mexico / Stanford, Retherford et al. 168 / 1941 / "heavily grazed by goats"

39 *Polemonium pulcherrimum* var. *delicatum* U.S.: Ariz. / Clausen & Trapido 4651 / 1940 / "Odor fetid"

HYDROPHYLLACEAE

40 *Eriodictyon tomentosum* U.S.: Calif. / J. G. Lemmon s.n. / 1878 / "Medicinal"

41 *Phacelia* indet. Chile / W. Biese 690 / 1944 / "medicinal"

BORAGINACEAE

42 *Arnebia euchroma* Tibet / W. Koelz 2072 / 1931 / "Use dye from root for lama worship & (coloring butter)" / "Muktsi"

43 *Borago* indet. Argentina / J. E. Montes 14708 / "Es cultivada como medicinal . . . Uso: Flores de uso medicinal" / " 'Borraja azul' "

44 *Cordia panamensis* Costa Rica / A. Jimenez M. 4109 / 1966 / "Frutos blancos como perlas, con jugo pegajoso de sabor dulce amarroso"

45 *Cordia* indet. Guatemala / J. A. Steyermark 42186 / 1942 / "Bark contains water which exudes and quenches thirst" / " 'peshtillo' "

46 *C. laevigata* Jamaica / W. Harris 9360 / 1906 / "Wild Cherry"

47 *C. nitida* Puerto Rico / Britton, Britton & Brown 6009 / 1922 / "fruit red . . . eaten by wild doves"

48 *C. obliqua* Puerto Rico / E. L. Little, Jr. 16425 / 1954 / "pink, edible"
Puerto Rico / E. L. Little, Jr. 21685 / 1966 / "Fruits whitish, edible, but very mucilaginous"

49 *C. nodosa* Peru / F. Woytkowski 5818 / 1960 / "in the 'pockets' there live abundant, small ants; hairy parts of the plant burn painfully"
Bolivia / J. Steinbach 5400 / 1921 / "La fruta es blanca, del tamano de los testiculos de un gato (por eso el nombre) es dulce y fresco" / "Huevo de gato"

50 *Cordia* indet. Brazil / M. Barbosa da Silva 21 / 1942 / "For constipation: take drink made from leaf decoction mixed with egg yolk" / " 'carú-cáa' "

51 *Cordia* indet. Brazil / A. Silva 96 / 1944 / "Infusion of bark is taken for chest colds. One of the popular remedies" / " 'carú-cáa' "

52 *Cordia* indet. Brazil / A. Silva 203 / 1944 / "Frt. yellow, edible"

3653   *Cordia ovalis* Tanganyika / R. E. S. Tanner 4408 / 1959 / "Bark used to treat swollen spleen in children" / "Kizanaki: Musenua"

3654   *Cryptantha fulvocanescans* U.S.: Utah / A. H. Holmgren 3225 / 1944 / "The Navajo Indians use the root of this plant as a medicine. About 3 inches of the root is macerated and soaked in a cup of water. The medicine is used generally in curing many illnesses"

3655   *Ehretia coerulea* Tanganyika / R. E. S. Tanner 1268 / 1953 / "Sukuma tribe: Leaves pounded and soaked; odor attracts and intoxicates snakes, then easily killed" / "Kisukuma: Nembu"
Tanganyika / R. E. S. Tanner 1015 / 1952 / "Sukuma tribe: leaves chewed and spat into beer — prevents quarrelling while being drunk; to catch a snake hit with a stick from this tree and it won't hurt" / "Kisukuma: Nembu"

3656   *E. littoralis* Tanganyika / R. E. S. Tanner 3348 / 1956 / "Zigua: For stomach ache in children, roots pounded up and the water drunk" / "Zigua: Mshimbampuku"
Tanganyika / R. E. S. Tanner 3737 / 1957 / "Bondei: For stomach ache in children, roots boiled and the water drunk" / "Bondei: Mfuti"
Tanganyika / R. E. S. Tanner 3766 / 1957 / "Sambaa: For pain in the stomach, roots boiled and the water drunk" / "Sambaa: Mhasha"

3657   *E. microphylla* Hainan / H. Fung 20027 / 1932 / "fr. edible"

3658   *Heliotropium leavenworthii* U.S.: Fla. / P. P. Sheehan s.n. / 1919 / "Used as a remedy for yellow fever. The whole plant is boiled and the decoction is drunk" / "Yellow heliotrope"

3659   *H. indicum* Brazil / G. E. & L. T. Eiten 4808 / 1962 / "Use: all parts of plant boiled in water and given to children as a cold refreshing drink & purgative"

3660   *H. polyphyllum* Brazil / R. Fróes 21594 / 1946 / "usada como depurativo"

3661   *Heliotropium* indet. Brazil / A. Silva 115 / 1944 / "Used as wash for inflammation of eyes" / " 'fedegoso' "

3662   *H. ovalifolium* Tanganyika / R. E. S. Tanner 932 / 1952 / "Sukuma tribe: Used to treat fevers by drying plant, mixing with ghee, and using as stick poultice over painful areas" / "Kisukuma: Ikunggulu jike"

3663   *H. subulatum* Tanganyika / R. E. S. Tanner 656 / 1952 / "Leaves used to treat boils by pounding green and mixing with butter — used as poultice" / "Kisukuma: Igungurukima"
Tanganyika / R. E. S. Tanner 1236 / 1953 / "Sukuma tribe: Leaves and stems used to treat yaws, as poultice" / "Kisukuma: Igangula"

3664   *Plagiobothrys arizonicus* U.S.: Ariz. / R. & M. Spellenberg 3069 / 1973 / ". . . roots red-staining; stems dying paper purple on drying"

65 *Tournefortia hartwegiana* Mexico / H. S. Gentry 1157 / 1934 / "Decoction made of herbage for snake bites as a wash and as a potation" / "Confituria negra, Mex."

66 *T. johnstonii* Costa Rica / A. Jiménez M. 831 / 1963 / "Frutos blancos, basico alimento para muchas aves . . ." / " 'maicillo' "

67 *T. paniculata* Brazil / Y. Mexia 4221 / 1930 / "Used for mange or itch. Decoction of leaves for bathing, and tea of leaves" / " 'Cura Sarna' "

68 *T. psilostachya* Venezuela / Curran & Haman 605 / 1917 / "Medicinal — stomach trouble" / "Sanalostodo"

69 *T. argentea* Marshall Is. / F. R. Fosberg 26699 / 1946 / ". . . used medicinally . . ." / " 'Kirin' "
Mariana Is. / F. R. Fosberg 24953 / 1946 / ". . . wood used for diving goggles, leaves boiled and water used as remedy in base of poisoning from fish (tagofi, skipjack) which have eaten poisonous seaweed" / " 'buni' "

70 *Trichodesma zeylanicum* Tanganyika / R. E. S. Tanner 931 / 1952 / "Sukuma tribe: used to treat dry cough by burning plant and mixing ash with wild salt — eaten 3 times daily" / "Kisukuma: Ikunggulu"

AVICENNIACEAE

71 *Avicennia germinans* W.I.: Martinique / F. E. Egler 39–34 / 1939 / "palatable to stock" / " 'mangle blanc' "
Colombia / R. Romero Castañeda 5309 / 1955 / "Tiene una resina que irrita los ojos al aserrarlo" / " 'Comedero', 'Iguanero' "

72 *A. marina* Tanganyika / R. E. S. Tanner 3414 / 1957 / "Zigua: For fever, leaves boiled and the water drunk" / "Zigua: Mohu"

VERBENACEAE

73 *Acantholippia hastulata* Argentina / A. T. Hunziker 1334 / 1941 / "Muy empleado como planta medicinal; sus infusiones se recommendan especialmente contra el dolor de estómago"

74 *A. seriphioides* Argentina / J. Semper 118 / 1944 / "como té y condimento"
Argentina / A. T. Hunziker 4170 / 1944 / "Utilizado para aromatizos comidas" / " 'oregano' "

75 *Aegiphila hassleri* Paraguay / Balansa 2085a / 1874 / ". . . comestibles"

76 *A. integrifolia* Brazil / S. Lopez-Palacios 3654 / 1975 / "La hoja macerada da una especie de jabon que se usa contra erupciones" / " 'Arco iris' "

77 *A. peruviana* Peru / J. Schunke V. 936 / 1965 / "The leaves are use to cure the ulcers, they are also toxic" / " 'Ocuera Blanca' "

3678   *Aloysia gratissima* U.S.: N. Mex. / Blake s.n. / no year / "Fine perfume from lvs"

3679   *A. gratissima* var. *paraguariensis* Argentina / J. E. Montes 27627 / '58 / "medicina popular" / " 'evina rupa' "
Paraguay / W. A. Archer 4678 / 1936 / "leaves aromatic"

3680   *A. herrerae* Peru / P. C. Hutchison 4199 / 1964 / "Leaves very aromatic—used to make a tea" / " 'Cidron del Campo' "

3681   *A. polystachya* Argentina / A. Hunziker 4771 / 1944 / "Planta medicinal, se prepara un té muy agradable" / " 'té de burro' "
Argentina / A. Brizuela 389 / 1946 / "Planta de té de burro con flor"

3682   *A. pulchra* Argentina / J. E. Montes 12 / 1944 / "Medicinal" /
" 'Nino-rupa' "
Argentina / J. E. Montes 14841 / 1955 / "Hojas y flores medicinales" /
" 'Niña rupa', 'Poleo' "

3683   *A. sellowii* Argentina / J. E. Montes 1864 / '46 / "flores. Medi. popular" /
" 'Niña-rupá' "
Argentina / J. E. Eukontes 534 / 1945 / "Medicinal" / " 'Rina rupa' "

3684   *A. triphylla* Ecuador / J. A. Steyermark 53289 / 1943 / "Boil the leaves and drink with bicarbonate of soda to alleviate pains in the lower abdomen"
/ " 'cedrón' "
S. AM. / R. Espinosa E2812 / 1950 / "por considerarselo medicinal" /
"Cedrón"

3685   *A. virgata* var. *elliptica* Argentina / P. R. Legname 3052 / 1963 / "flores blancas (medicinal)" / "sacha piuscan"

3686   *Amasonia erecta* Peru / W. A. Archer 7606 / 1942 / "Used for stomach inflammations" / " 'mendoca' "

3687   *Burroughsia fastigiata* Mexico / H. S. Gentry 7768 / 1947 / "Decocted as a tea; reputed to have aphrodisiac properties" / " 'Damiana' "

3688   *Callicarpa erioclona* Philippine Is. / A. L. Zwickey 14 / 1938 / "berries edible; leaves mixed with cocunut oil in application to wounds" / " 'Alo-loy' "

3689   *Citharexylum subflavescens* Venezuela / J. A. Steyermark 56449 / 1944 /
". . . said to be poisonous to the skin and to cause swelling and itching like poison ivy" / " 'cuidadito' "

3690   *Clerodendrum aculeatum* St. Vincent I. / H. H. Smith & G. W. Smith 936
/ 1889 / " 'wild coffee' "
Brit. Virgin Is. / W. C. Fishlock 250 / 1919 / "medicinally in cases of gonorrhea" / " 'Privet wood' "
Brit. W.I. / W. Kings G. C. 148 / 1938 / "leaves boiled for cough"

3691   *C. anafense* Cuba / J. G. Jack 8724 / 1933 / "much eaten by birds"

692 *C. capitatum* N. Rhodesia / A. Angus 2796 / 1961 / "Stems hollowed out by ants. The hollow stems sought by natives for sucking of beer"

693 *C. discolor* var. *oppositifolium* Tanganyika / R. E. S. Tanner 1208 / 1953 / "Sukuma tribe: Roots used to treat stomach pains—pounded up, put in cold water, and drunk" / "Nghuwambu wa matongo"

694 *C. glabrum* Tanganyika / R. E. S. Tanner 3671 / 1957 / "Zigua: For reducing fever, leaves and twigs used as an infusion over which patient sits" / "Zigua: Kuwakilo"

695 *C. hildebrandtii* Tanganyika / R. E. S. Tanner 3412 / 1956 / "Sambaa: For heartburn, roots boiled and the water drunk" / "Sambaa: Kuwkilo" Tanganyika / R. E. S. Tanner 3767 / 1957 / "For pain in the stomach, root boiled and the water drunk" / "Zigua: Mkusakilo"

696 *C. rotundifolium* Tanganyika / R. E. S. Tanner 1485 / 1953 / "Sukuma tribe: Roots pounded, taken with water to treat diarrhoea" / "Kisukuma: Sekeseke"

697 *C. cyrtophyllum* China: Kwangtung / F. A. McClure 9489 / 1921 / "The Fus (Kwangtung aborigenes) eat this plant as a vegetable"

698 *C. lindleyi* China: Kwangtung / W. T. Tsang 21353 / 1921 / "leaf used as a tea; fruit is edible" / "Yuen Tau Fung"
Hainan / F. A. McClure 9207 / 1922 / "The roots when dried well for about 5 hours are then taken internally for weak muscles of the legs" / "Ch'au shi mit li"

699 *C. colebrokianum* India / F. Kingdon Ward 18855 / 1949 / "Invariably covered with small ants"

700 *C. inerme* Marshall Is. / F. R. Fosberg 26733 / 1946 / "used medicinally" / " 'ulij' "
Mariana Is. / F. R. Fosberg 25396 / 1965 / "Boiled and water drunk to remedy fevers" / " 'lodigao' "
Caroline Is. / F. R. Fosberg 47648 / 1965 / "Used as medicine and as fish poison" / " 'embrert' "

701 *Cornutia pyramidata* W.I.: St. Lucia / J. S. Beard 502 / 1945 / " 'Bois cassave' "

702 *Lantana glandulosissima* El Salvador / P. C. Standley 22492 / 1922 / "fr. black, eaten" / " 'Cinco negritos' "

703 *L. velutina* Mexico / G. B. Hinton 4429 / 1933 / "Fruit edible" / "Sarza mora"

704 *L. urtiaefolia* W.I.: Grand Cayman Is. / W. Kings GC 112 / 1938 / "Leaves used as a tea"

705 *L. glutinosa* Brazil / Williams & Assis 6430 / 1945 / "To make tea for grippe" / " 'Camara' "

3706   *L. trifolia* var. *rigidiuscula* Bolivia / R. F. Steinbach 768 / 1967 / "fruto comestible, dulce"

3707   *L. camara* var. *aculeata* Tanganyika / R. E. S. Tanner 3949 / 1958 / "Sambaa: for pain in the stomach, leaves pounded up and applied as a paste" / "Sambaa: Mfuti"

3708   *L. mearnsii* var. *latibracteolata* Tanganyika / R. E. S. Tanner 3610 / 1957 / "Bondei: For straining cough, leaves chewed green" / "Bondei: Mfufisa Mgosi"
Tanganyika / R. E. S. Tanner 1945 / 1955 / "Leaves chewed for sore throat" / "Kibondei: Fufurisa"

3709   *L. rugosa* var. *tomentosa* Tanganyika / R. E. S. Tanner 1028 / 1952 / "Sukuma tribe: leaves cooked and used as poultice to treat sores on lips; leaves pounded and mixed with salt, put in sore eyes in cattle" / "Kisukuma: Masangu gawa ndimi"

3710   *L. viburnoides* Tanganyika / R. E. S. Tanner 2802 / 1956 / "For cough: leaves pounded up and sap squeezed out and drunk" / "Kibondei: Mfufuisa"
Tanganyika / R. E. S. Tanner 3576 / 1957 / "Sambaa: For persistent stomach ache, leaves pounded up and rubbed into body" / "Sambaa: Mvuti"
Tanganyika / R. E. S. Tanner 3319 / 1956 / "Bondei: For cough, leaves pounded up, soaked and the water drunk" / "Bondei: Mfufuisa"

3711   *Lippia alba* Guatemala / J. A. Steyermark 50859 / 1942 / "Reputed to be effective in treating coughs"
Puerto Rico / Neves & Stimson 1610 / 1965 / "Natives believe this plant to be medicinal, it is used by them in baths" / " 'Poleo' "

3712   *L. libertensis* Costa Rica / Burger & Ramirez 4082 / 1966 / "leaves with strong sweet odor when crushed"

3713   *L. palmeri* Mexico / Carter & Moran 5381 / 1967 / " 'oregano' "

3714   *L. micromera* var. *helleri* Dominican Republic / Lavastre 1857 / 1964 / " 'Orégano' "

3715   *L. alba* var. *globiflora* Peru / M. B. da Silva 2 / 1942 / "Tea of leaves taken to counteract effects of purgative" / " 'sideira' "
Argentina / I. Morel 690 / 1945 / "Plant medicinal"
Brazil / B. Rambo S. J. 38012 / 1948 / "Species in R. G. S. generatim culta quia medicinalis, sed vera indigena quamquam rara"
Bolivia / R. F. Steinbach 413 / 1966 / "tomadas en infusion por los nativos para purificar la sangre y como bebida estimulante"

3716   *L. integrifolia* Argentina / Schreiter 1202 / 1920 / "medicinal"

3717   *L. origanoides* Venezuela / J. A. Steyermark 58601 / 1944 / "leaves boiled in water with guarapo (sugar drink), and is used for colds" / " 'chara-ceúr' "

3718   *L. turbinata* Argentina / T. Meyer 8337 / 1945 / "Medicinal"
Argentina / T. Meyer 3453 / 1941 / "planta medicinal"

3719   *L. oatesii* S. Rhodesia / J. C. Hopkins 10231 / 1943 / "keeps away snakes"

3720   *L. macromera* Hawaii / M. C. Neal s.n. / 1946 / "used to flavor meat loaf, stuffing, and gravy"

3721   *Phyla scaberrima* Dominican Republic / R. A. & E. S. Howard 9677 / 1946 / "ls. sweet smelling and eaten like mint"
Colombia / J. A. Ewan 16278 / 1944 / "pls. used medicinally"

3722   *Premna chrysoclada* Tanganyika / R. E. S. Tanner 2862 / 1956 / "Zigua: For stomache. Roots boiled and water drunk. Used for arrow shafts" / "Zigua: Nyandu Mwasanyoya"
Tanganyika / R. E. S. Tanner 3107 / 1956 / "Bondei: Stomach pains in children, roots boiled and water drunk" / "Bondei: Mhasanyoya" "Zigua: Mtulavuha"
Tanganyika / R. E. S. Tanner 3686 / 1957 / "Bondei: For improving quality of mothers milk, leaves pounded and put on breasts as paste" / "Bondei: Muhasanyoya"
Tanganyika / R. E. S. Tanner 2067 / 1955 / "Leaves pounded, juice dripped into sore eyes" / "Kisukuma: Muhasanyoya"
Tanganyika / R. E. S. Tanner 3917 / 1958 / "Bondei: for blood in the urine, roots boiled and the water drunk" / Zigua: Mkomakiema"

3723   *P. latifolia* var. *mollisima* Ceylon / P. L. Comanor 612 / 1967 / "fruit black, edible"

3724   *P. obtusifolia* var. *gaudichaudii* Caroline Is. / P. Anderson 2103 / 1950 / "Fresh leaves used for tea" / " 'Ior' "

3725   *P. odorata* Philippine Is. / Ahern 24 / 1904 / "somewhat used by the natives in the practice of medicine"

3726   *P. taitensis* var. *rimatarensis* Fiji Is. / A. C. Smith 26 / 1933 / "Leaf furnishes medicine for stomach ills" / " 'Nguronguru' "

3727   *Priva curtisiae* Tanganyika / R. E. S. Tanner 1269 / 1953 / "Sukuma tribe: leaves pounded to flour, used as poultice on sores" / "Kisukuma: Ikubya"

3728   *Recordia boliviana* Bolivia / W. M. A. Brooke 5958 / 1949 / "blossom with faint scent attracting quantities of butterflies and moths"

3729   *Rhaphithamnus venustus* Juan Fernandez Is. / J. P. Chapin 1083 / 1935 / "Visited by hummingbirds"

3730   *Stachytarpheta dichotoma* Argentina / J. E. Montes 14737 / 1955 / "*Uso:* Medicina populare"

3731   *Verbena recta* Mexico / G. B. Hinton 4607 / 1933 / "medicinal" / " 'Verbena' "

3732   *V. bonariensis* Argentina / E. Schmidt 37 / 1947 / "Uso medicinal populare"
Chile / G. F. S. Elliot 77 / 1903 / "eaten by cattle"

3733   *V. intermedia* Argentina / J. E. Montes 14866 / 1956 / "Uso: Medicinal" / " 'Verbena' "

3734   *V. littoralis* Brazil / R. Reitz C516 / 1943 / "Medicinal"
Colombia / F. R. Fosberg 20564 / 1943 / "used as a remedy for typhoid"

3735   *V. montevidensis* Brazil / R. Reitz C76 / 1944 / "Medicinal"
Argentina / J. E. Montes 765 / 1945 / "Medicinal"
Argentina / J. E. Montes 27659 / 1958 / "Uso: medicinal popular"

3736   *Vitex gaumerii* Honduras / Davidse & Pohl 2133 / 1970 / "flws. blue fragrant and being visited by numerous species of insects"

3737   *V. pyramidata* Mexico / G. B. Hinton 6956 / 1934 / "Fruit edible" / "Querenda"

3738   *V. cymosa* Bolivia / M. Cardenas 3527 / 1944 / "fruit edible; bark used as expectorant and anti-dysenteric" / " 'Taroma' "

3739   *V. litoralis* S. AM. / R. Espinosa 58 / 46 / "propiendades medicinales"

3740   *V. polygama* var. *bakeri* Brazil / R. M. Harley 10841 / 1968 / "Fruit soft, . . . sweet, edible flesh"

3741   *V. polygama* var. *glaziovii* Brazil / Irwin et al. 27322 / 1970 / "Fruit black-violet, edible"

3742   *V. amboniensis* Tanganyika / R. E. S. Tanner 3315 / 1956 / "Bondei: For sharp pains in stomach, roots boiled and the water drunk" / "Bondei: Mnegege"

3743   *V. keniensis* Tanganyika / R. E. S. Tanner 1119 / 1952 / "Uses: fruits edible. Used to add flavour to tobacco snuff" / "Kisukuma: Mpuru"

3744   *V. mombassae* Tanganyika / R. E. S. Tanner 1245 / 1953 / "Sukuma tribe: Pounded roots used for snake-bite to cause vomiting; pounded roots mixed with butter and put in cuts on temples to close fontanelle on newly-born infants" / "Kisukuma: Sungwi"

3745   *V. payos* Tanganyika / R. E. S. Tanner 3957 / 1957 / "Zigua: For pain in the stomach, roots boiled and the water drunk: also bark boiled and the water drunk" / "Zigua: Mgobe"
Tanganyika / R. E. S. Tanner 3877A / 1957 / "Uses. Bondei—for pain in lower stomach, roots boiled and the water drunk" / "Bondei: Mkoko"
Port. E. Africa / A. A. Pimenta 17230 / 1947 / "fruit edible" / " 'Mucoro' "

3746   *V. strikeri* Tanganyika / R. E. S. Tanner 630 / 1952 / "Fruits edible. Leaves used to treat swollen gums—cooked and water used to rinse mouth" / "Kisukuma: Mhamu"

Tanganyika / R. E. S. Tanner 1305 / 1953 / "Sukuma tribe: leaves pounded, juice taken orally, used to treat snake bite; juice used directly for cobra poison in eye" / "Kisukuma: Mpulu ngosha"

747 *V. negundo* var. *intermedia* China: Kwangtung / S. K. Lau 20133 / 1932 / "Used for medicine" / "Mai Tap Kong"

748 *V. leucoxylon* Ceylon / Nowiche, Fosberg & Jayasuriya 364 / 1973 / "fruits used as fish poison"

749 *V. siamica* Malaysia / V. Balgooy 2306 / 1975 / "Flower blue, visited by bees"

750 *V. trifolia* var. *bicolor* Samoa Is. / A. J. Eames 36 / 1921 / "Leaves used for fever medicine" / " 'gamolega' "

LAMIACEAE

3751 *Acrotome belckii* S.W. Africa / R. Seydel 3965 / 1964 / "Gelegentlich etwas befressen von Kuddu und Kleinvieh"

3752 *Agastache nepetoides* U.S.: Ill. / R. G. Mills s.n. / 40 / "Catnip"

3753 *A. rugosa* China: Kwangsi / R. C. Ching 6423 / 1920 / "Cultivated for its medical property"

3754 *Anisomeles indica* Philippine Is. / M. Ramos 228 / 1930 / "good for stomach trouble"

3755 *Basilicum polystachyon* Tanganyika / R. E. S. Tanner 1077 / 1952 / "Burnt in milking pots to give pleasant smell to milk" / "Kisukuma: Ilumba ya hasi"

3756 *Bystropogon mollis* Venezuela / H. Pittier 13213 / 1929 / " 'Oreganote' "

3757 *Cedronella canariensis* Canary Is. / Torrey Herb. s.n. / ca. 1840 / "Balm of Gilead"

3758 *Coleus amboinicus* Mexico / G. B. Hinton 3950 / 33 / "Orégano cimarron" Jamaica / W. Harris 9595 / 1907 / "whole plant with a strong odour of mint"
W.I.: Guadeloupe / P. Duss 2936 / 1892 / "Thym de l'Inde"

3759 *C. aquaticus* Tanganyika / R. E. S. Tanner 3296 / 1956 / "flowers . . . unpleasantly aromatic" / "Masai: Endani Narok"

3760 *C. esculentus* N. Rhodesia / W. R. Bainbridge 554 / 61 / "Roots edible, planted as a crop in parts of this Chieftaincy, (small patches sometimes seen near villages)" / "Plateau Tonga name LUSENSHA OR LUMENHYA"

3761 *Coleus* indet. Belgian Congo / J. P. Chapin 186 / 1927 / "plant has mint-like odor"

3762   *C. macranthus* Philippine Is. / A. D. E. Elmer 8529 / 1907 / "the green leaves are pounded into plasters for wounds" / " 'Buñbuñglit' "

3763   *Cunila polyantha* Honduras / C. & W. von Hagen 1221 / 1937 / "Tea used for fever" / " 'Poleo' "

3764   *C. pycnantha* Mexico / Y. Mexia 9037 / 1937 / "Used as a tea"

3765   *Dracocephalum moldavica* China: Shansi / C. F. Li 10629 / 1929 / "The young shoots are edible" / "Chinese name 'Little flowers' "

3766   *Dysophylla heterophyllum* Kashmir / W. Koelz 2272 / 1931 / "Delightful odor to plant"

3767   *D. moldavica* W. Himalayas / W. Koelz 9626 / 1936 / "used as condiment" / " 'Kololo' "

3768   *Elsholtzia densa* Himalay: Lahul / W. Koelz 1320 / 1930 / "Seeds ground with pepper" / "Makadeni"

3769   *E. pusilla* Kashmir / W. Koelz 2182 / 1931 / "Mint odor"

3770   *E. strobilifera* Punjab / W. Koelz 3081 / 1931 / "plants very fragrant, used to keep moths out of wool"

3771   *Eriope macrostachya* var. *hypoleuca* Brasil / Irwin, Harley & Onishi 28965 / 1971 / ". . . with pungent odor of eucalyptus oil"

3772   *Hedeoma drummondii* var. *drummondii* U.S.: Kans. / L. C. Hulbert 3287 / 1958 / "Strong mint odor"

3773   *H. drummondii* var. *reverchonii* U.S.: Tex. / M. C. Metz 83 / 1931 / "Lemon scented"

3774   *H. nanum* var. *nanum* Mexico / C. L. Lundell 5046 / 1934 / " 'Tornillo loco' "

3775   *Hoslundia opposita* Tanganyika / R. E. S. Tanner 3347 / 1956 / "Sambaa: For tuberculosis, roots boiled and the water drunk" / "Sambaa: Mshwee"
Tanganyika / R. E. S. Tanner 2772 / 1956 / "Fruits edible. Leaves squeezed into water, beaten up and drunk for stomach ache" / "Swaheli: Mrambaa Jongo"
Tanganyika / R. E. S. Tanner 569 / 1952 / "Fruits eaten; leaves used as poultice on gums, roots carved into balls" / "Kisukuma: Bugokorome"
Tanganyika / R. E. S. Tanner 2856 / 1956 / "Fruits eaten" / "Zigua: Molwe"
Tanganyika / R. E. S. Tanner 2801 / 1956 / "For stomach ache in children: Roots cooked and drunk when cool" / "Swaheli: Mtambaa Jongoo"
Tanganyika / R. E. S. Tanner 3757 / 1957 / "Bandei: For epilepsy in children, roots boiled and the water drunk" / "Bondei: Mswe"
Tanganyika / R. E. S. Tanner 3608 / 1957 / "Bondei: For difficult urination, roots boiled and the water drunk" / "Bondei: Msweo"

776   *Hyptis albida* Mexico / H. S. Gentry 1272 / 1935 / "Flowers put in ear to relieve ear-ache"

777   *H. escobilla* Dominican Republic / Marcano & Jimenez 5462 / 1962 / "ROMERILLO"

778   *H. arborea* Brit. Guiana / G. H. Tate 240 / 1928 / "Leaves with strong lemon-like odor" /

779   *H. atrorubens* Venezuela / J. A. Steyermark 75271 / 1953 / "herb boiled in water reputed effective for hot bath to take soreness out of muscles and to relieve congestion after one has been wet or has sore muscles"
Venezuela / A. L. Bernardi 1800 / 1954 / "Hierba buena"
Brazil / A. Silva 170 / 1944 / "commonly used as a tea to cure colds" /
" 'hortelã pimenta' "

3780   *H. brevipes* Peru / J. Schunke V. 1381 / 1966 / / "flores . . . fragantes" /
"Mala hierba"

3781   *H. excelsa* Venezuela / H. Pittier 12065 / 1925 / "infusion used as gargle to ease pain in 'corrimiento' " / " 'Corrimienta' "

3782   *H. fruticosa* Brasil / M. Lima 97 / 1968 / "(medicinal)"

3783   *Hyptis* indet. Peru / J. Schunke-Vigo 1076 / 1966 / "Leaves fragrant. En infusión dan de tomar a los ganados vacunos para la fiebre aftosa" / "Aya Albahaca"

3784   *H. pectinata* Tanganyika / R. E. S. Tanner 957 / 1952 / "Sukuma tribe: Leaves pounded in water, this drunk to treat diarrhoea; leaves used as poultice on boils" / "Kisukuma: Budumyo"
Mariana Is. / P. Nelson 18 / 1918 / "Miumutung lake 'Lake' means 'male' Mumutung means 'smell' or 'odor' "

3785   *H. suaveolens* Tanganyika / R. E. S. Tanner 1950 / 1955 / "Roots used to treat stomach-ache; boiled and water drunk" / "Swaheli: Kipumpasi Msitu"
E. Africa / O. B. Bakari AH8688 / 1940 / "A witch craft medicine"
Philippine Is. / A. L. Zwickey 154 / 1938 / "lvs. applied to wounds" /
" 'Mamarasagi' (Lan.); 'Ban-lo' (Bis.)"

3786   *Iboza riparia* S.W. Africa / R. Seydel 3465 / 1963 / ". . . unangenehmer Geruch"
Tanganyika / R. E. S. Tanner 1602 / 1953 / "Sukuma tribe: leaves used to treat cuts, roots for diarrhoea" / "Kisukuma: Misusu"

3787   *Leonitis cardiaca* U.S.: N.J. / J. Oehler s.n. / 1875 / "Herzenspannkraut"

3788   *L. nepetaefolia* U.S. & CAN. / College of Pharmacy Herb s.n. / no year / "Das kraut wird zu Bädern gegen Rheumatismus gebraucht"
W.I.: Virgin Is. / W. C. Fishlock 122 / 1918 / "Used as a remedy for fish poisoning"

Tanganyika / R. E. S. Tanner 1589 / 1953 / "Sukuma tribe: Flowers eaten for food; leaves pounded, used as poultice for treatment fresh scratches" / "Kisukuma: Zungzu"

3789 **L. sibiricus** W.I.: Martinique / P. Duss 1976 / 1880 / "employé dans le medicine populaire contre les refroidesement, les coliques et cours de ventre" / "Chandelier"
Puerto Rico / E. C. Montalvo 3859 / 1966 / "When animals have open wounds with worms, the natives of this area grind up this plant and put it on the open wound to kill the worms" / " 'cebadilla' "

3790 **L. dysophylla** S.W. Africa / R. Seydel 3958 / 64 / "Die Blueten Koepfe gelegentlich abgefressen"

3791 **Lepechinia speciosa** Brasil / Lindeman & de Haas 4158 / 1967 / "Leaves used for tea"

3792 **Leucas martinicensis** Dominican Republic / E. J. Valeur 253 / 1931 / " 'MANZANA' "
Tanganyika / R. E. S. Tanner 946 / 1952 / "Sukuma tribe: Leaves used as poultice in cuts" / "Kisukuma: Zonzo"

3793 **Lycopus virginicus** var. **pauciflorus** U.S.: Vt. / J. S. Harper 26 / 1932 / " 'Sugar bush' "

3794 **Marsypianthus hyptoides** Brazil / W. A. Archer 8386 / 1943 / "Juice expressed from leaves and taken internally for snake bite, or any poisonous bite" / " 'paracari' "

3795 **Mentha cardiaca** U.S.: Fla. / C. H. Johnson s.n. / 1945 / "Medicinal Plant Garden, Univ. of Florida, Gainesville" / " 'spearmint' "
U.S.: Penn. / J. K. Small s.n. / no year / "strong Spearmint odor"

3796 **M. viridis** U.S.: N.J. / A. Niederer s.n. / 1884 / "Spearmint"

3797 **Mentha** indet. U.S.: D.C. / E. S. Burgess s.n. / 1893 / "Portuguese spearmint — Hortela"

3798 **M. aquatica** Ecuador / W. H. Camp E-2480 / 1945 / "used as remedy for horse-worms in both horses and humans" / " 'Yerba buena' "

3799 **Mentha** indet. Brazil / M. Barbosa da Silva 33 / 1942 / "Leaves aromatic. Used in decoction as a wash to cure pains" / " 'cordão de frade' "

3800 **M. saturejoides** Australia: Queensland / L. S. Smith 3011 / 1947 / "Leaves pale green, emitting a strong peppermint-like odour when crushed" / "Pennyroyal"

3801 **Mesona chinensis** Formosa / A. Henry 82 / no year / "Used as cooling drink in summer"

3802 **Minthostachys mollis** Argentina / L. A. Dawson 53 / 1939 / " 'menta peperina' "

Colombia / L. Uribe U. 4587 / 1963 / "Reputado como medicinal en afecciones bronquiales" / " 'orégano' "

Ecuador / W. H. Camp E-2312 / 1945 / "Used to sweep house when full of fleas' (but they must come right back on dogs, pigs, etc). Also, 'dried' (toasted) lvs. pulverized and rubbed on parts affected with rheumatism" / " 'Guarmi poles' "

Peru / Sagástegui & Fabris 3050 / 71 / "Usada en alimentación" / " 'chancua' "

Ecuador / W. H. Camp E-4515 / 1945 / "Plants with heavy and somewhat unpleasant odor"

3803   *M. verticillata* Argentina / G. Hieronymus 14 / 1876 / "tl pié del Pan de azucar mas arriba de la Reduccion" / "Piperita ó Piperina"

3804   *Molucella laevis* U.S.: Utah / A. & R. A. Nelson 2738 / 1938 / " 'Shell balm' "

3805   *Monarda citriodora* var. *citriodora* U.S.: Tex. / V. L. Cory 53773 / 1943 / " 'purple horsemint' "
Mexico / Dr. Gregg s.n. / 47 / "Yerba buena"

3806   *M. clinopodia* U.S.: N.J. / J. Kezer s.n. / 36 / "Balm"
U.S. & CAN. / Columbia Univ. Herb. s.n. / no year / "Wild Bergamot"

3807   *M. clinopodioides* U.S./ Tex. / J. W. Lake 21 / 1890 / "Horsemint"

3808   *M. pectinata* U.S.: Neb. / H. Hapeman s.n. / 1935 / "Contain a valuable oil"

3809   *Monardella lanceolata* U.S.: Nev. / P. Train 3141 / 1939 / "Mint odor"

3810   *Mosla lanceolata* China: Kwangtung / T. K. Peng 625 / 1920 / "Soaked in wine & the wine used for sprains" / "Tuk Hang Lai"

3811   *Nepeta betonicaefolia* Kashmir / W. Koelz 2495 / 1931 / "Strong pennyroyal scent"

3812   *N. salviaefolia* Kashmir / W. Koelz 2495 / 1931 / "Plant catnip scented"

3813   *Ocimum basilicum* W.I.: Guadeloupe / P. Duss 215 / 1892 / "Cultivé dans les jardins de toutes les Antilles comme pl. aromatique, stimulante et antispasmodique" / "Basilie"
Colombia / J. M. Duque Jaramillo 2918 / 1946 / "Carminativa y estimulante" / " 'Albahaca' "

3814   *O. gratissimum* W.I.: Guadeloupe / P. Duss 4219 / 1905 / "Herbe annuelle, aromatique, employeé comme reméde contre les coliques, mauv de ventre" / "Framboisin"

3815   *O. micranthum* W.I.: Martinique / F. E. Egler 39-274 / 1939 / "Apparently not palatable to stock"
W.I.: Martinique / P. Duss 399 / 1878 / ". . . aromatique, medicinal" / "Herbe carreé"

W.I.: Guadeloupe / P. Duss 2155 / 1892 / "Rechesché contre les chumes, fievres, etc."

Venezuela / F. Bianco 108 / 76 / "MEDICINAL" / "ALBAHACA BLAN-CA"

Colombia / J. M. Duque Jaramillo 4091-A / 1946 / "Aromática y medicinal" / " 'Albahaca' "

3816   *Ocimum* indet. Dominican Republic / E. J. Valeur 224 / 1929 / " 'Alvahaca de burro' "

3817   *Ocimum* indet. Brazil / A. Silva 360 / 1946 / "plantas medicinães do mercado. Communmente utilizado nas gripas, toses, etc." / " 'Isturaque' "

3818   *O. americanum* Tanganyika / R. E. S. Tanner 2951 / 1956 / "Zigua: If snake in house plant burnt on house fire to remove it" / "Zigua: Luvumba Kuku"

Nyasaland / L. J. Brass 16343 / 1946 / "smelling rather like Eucalyptus oil" / "Chanzi (Chinyanja)"

Tanganyika / R. E. S. Tanner 1959 / 1955 / "Plants burned to produce smoke to treat sore eyes" / "Swaheli: Kifumbasi"

Tanganyika / R. E. S. Tanner 653 / 1952 / "Alleged mosquito repellent" / "Kisukuma: Irumba"

Tanganyika / R. E. S. Tanner 2090 / 1955 / ". . . strongly aromatic" / "Kizigua: Ufubbapuku"

Tanganyika / R. E. S. Tanner 3084 / 1956 / "Bondei: Leaves put into hair oil. For bad eyes, leaves pounded and the juice squeezed into them" / "Swaheli: Rihami"

Ceylon / S. Ripley 377 / 1971 / "Voucher: Primate Ecology Studies. leaves fed on by *P.e.t.* C troop"

3819   *O. hadiense* Tanganyika / R. E. S. Tanner 3197 / 1956 / "Zigua: For removal of evil spirits; burnt on piece of broken pot" / "Zigua: Uvumba upuku"

3820   *O. hanningtonii* Tanganyika / R. E. S. Tanner 2886 / 1956 / "Zigua: For sore throat leaves chewed" / "Zigua: Muhahazi"

3821   *O. kilimandscharicum* Tanganyika / R. E. S. Tanner 3906 / 1957 / "For stomach ache (Bondei) in children, roots boiled and the water drunk; for colds, leaves chewed" / "Bondei—Msagati Kivuma"

3822   *O. ranum* S.W. Africa / R. Seydel 377 / 1954 / "Aromatischer Geruch, beliebtes Teegetränk"

3823   *O. suave* Tanganyika / R. E. S. Tanner 3144 / 1956 / "Sambaa: For sharp pains in the stomach, leaves pounded and drunk in water while green" / "Sambaa: Mxumbasi"

Tanganyika / R. E. S. Tanner 696 / 1952 / "Sukuma tribe: used to treat swollen gums" / "Kisumukuma: Malumba"

Tanganyika / R. E. S. Tanner 2955 / 1956 / "Zigua: Leaves rubbed into a paste and put on sores" / "Zigua: Mfumbasa"

824 *Origanum cordifolium* Cyprus / E. W. Kennedy 1633 / 1948 / "Used locally for healing cuts & scratches"

825 *O. creticum* Attica / Heldreich s.n. / 1848 / "Alle Theile der Pflanze sind als Herba v. Summitates Origani cretia gebräuchlich"

826 *Orthosiphon parvifolius* Tanganyika / R. E. S. Tanner 1351 / 1953 / "Sukuma tribe: Roots used to treat syphilis—boiled, mixed with white mtama flower, water, then taken orally" / "Kisukuma: Ilumba lya mlutogoro"

827 *O. spiralis* Fiji Is. / A. C. Smith 7233 / 1953 / ". . . cultivated in village" / " 'kumi ni pusi' "

828 *Perilomia ocymoides* Ecuador / Rose, Pachano & Rose 22960 / 1918 / " 'Cardiaca' "

829 *Perovskia abrotanoides* Kashmir / W. Koelz 2539a / 1931 / "odor rather unpleasant"

830 *Phlomis tuberosa* U.S. & CAN. / Columbia Univ. Herb. s.n. / no year / "Jerusalem Sage"
Austria / Semry 6 / 1919 / "Brandkraut"

831 *Platystoma africana* Liberia / S. B. Freeman 3 / 1950 / "Used for stomach medicine"

832 *Pogostemon patchouly* W.I.: Guadeloupe / P. Duss 2169 / 1892 / ". . . cultivé comme plante sudorifique et stomachique / "Thym de l'Inde"
W.I.: Guadeloupe / P. Duss 1222 / 1882 / ". . . cultivé dans le jardins comme plante sudorifique et pour tuer les insectes" / "Patchouly"

833 *Poliomintha glabrescens* Mexico / Hinton et al. 16508 / 44 / "Tea, and a good one" / " 'Mentha' "
Mexico / Hinton et al. 16523 / 44 / "condiment sold in markets for 3.00 pesos kilo" / " 'Oregano' "

834 *Pycnanthemum aristata* U.S.: N.J. / J. Oehler s.n. / 1877 / "Mountain Mint"
U.S. & CAN. / Columbia Univ. Herb. s.n. / no year / "Basil-leaved Mountain Mint"

835 *P. beadlei* U.S.: N.C. / A. M. Huger s.n. / 1900 / "Horsemint"

836 *P. flexuosa* U.S.: N.Y. / W. M. s.n. / 1879 / "Mountain Mint. Basil"
U.S. & CAN. / Columbia Univ. Herb. s.n. / no year / "Narrow-leaved Virginian Thyme"

837 *P. incanum* U.S. & CAN. / College of Pharmacy Herb. s.n. / no year / "In Nordamerika gegen Schlangenbiss gebraucht"

3838   *P. lanceolatum* U.S.: N.Y. / E. H. Day s.n. / 78 / " 'Mountain Mint' 'Wild Basil' "

3839   *Pycnothymus rigidus* U.S.: Fla. / P. P. Sheehan s.n. / 1919 / "Medicinal Plants of the Seminole Indians. Used as a tea; also esteemed as a febrifuge. Plant boiled in water, and the infusion is usually drunk hot"

3840   *Raphiodon echinus* Brasil / T. Guedes 367 / 1957 / "Plant muito comestivel pelo gado" / " 'Betonia' "

3841   *Salvia ballotaeflora* U.S.: Tex. / Schott s.n. / 1852 / "Highly recommended as a tonic aromatic tea, for exterior & inner use" / "Majorano Mexic."

3842   *S. coccinea* U.S.: Tex. / Schott 70 / no year / "Coma (Mexic.)"
Mexico / Dr. Perrine s.n. / 1834 / "Flowers rubbed on the cheeks instead of Rouge" / " 'chacsits' "

3843   *S. clinopodioides* Mexico / Hinton et al. 13350 / 38 / "Root is brewed for toothache" / "Camote de Cuche"

3844   *S. helianthemifolia* Mexico / A. J. Sharp 441088 / 1944 / "penetrating odor when crushed"

3845   *S. hispanica* Mexico / College of Pharmacy s.n. / no year / "Die Samen dieser aus Mexico nach Südeuropa eingeschleppten Art, Semen Chia, geben Schleim, der wie Quitten Körner—und ähnlichen Schleime benützt wird"
Mexico / E. Palmer 659 / 1886 / " 'Chia' "

3846   *S. micrantha* Brit. Honduras / P. H. Gentle 376 / 1931–32 / "(medicinal)" / " 'catnip', 'Nassau bitters' "
Dominican Republic / A. H. Liogier 11509 / 1968 / "Near a house, said to be used as medicine"
Bahama Is. / E. G. Britton 6373 / 1907 / " 'Blue catnip' "

3847   *S. polystachya* Mexico / Hinton et al. 13241 / 38 / "Leaves have a strong pleasant smell"

3848   *S. riparia* Mexico / W. B. Gourlay 61 / 1938 / "very sticky plant, foetid odor"

3849   *S. palea* Puerto Rico / Hinton et al. 14971 / 39 / "Sap has a strong fishy smell"

3850   *S. serotina* Bahama Is. / W. Cerbin C 145 / 1973 / "Plants dried or fresh are brewed into a tea; used for indigestion, a shampoo, and pleasure" / " 'Catnip' "

3851   *S. grewiaefolia* Brazil / Harley, Souza & Ferreira 10994 / 1968 / ". . . with foetid smell"

3852   *S. hirtella* Ecuador / W. H. Camp E-2491 / 1945 / " 'Quindi sungaña' (Quindi—name of humming bird; sungaña—to sip)"

853   *S. macrostachya* Ecuador / W. H. Camp E-5285 / 1945 / "Raised by natives in region of Taday" / "This is 'Duc-Duc' "

854   *S. occidentalis* Peru / J. Schunke V. 3505 / 1969 / " 'Sacha hierbabuena' "

855   *S. palaefolia* Colombia / J. M. Duque Jaramillo 3187 / 1946 / "Popularmente recomendado como tónico cardiaco y contra la arterioesclerosis" / " 'Mastranto' "

856   *S. petiolaris* Colombia / R. A. Toro 175 / 1927 / " 'Yerba de sapo' "

857   *S. scutellarioides* Ecuador / W. H. Camp E-1928 / 1945 / "Infusion of lvs. and fls. taken for kidney trouble" / " 'Manga-paqui' (azul)"

858   *Salvia* indet. Argentina / J. E. Montes 744 / 45 / "Uso: Med. pop." / " 'Abasa montes' "

859   *S. glutinosa* India / W. Koelz 1374 / 1930 / "plant emits strong odor rather disagreeable"

860   *S. lanata* Punjab / W. Koelz 1938 / 1931 / "Flowers eaten by Chaman pheasants"

861   *S. plebeia* Punjab / W. Koelz 2003 / 1931 / "plant with mint odor"

862   *S. triloba* TROP. ASIA / College of Pharmacy Herb. s.n. / no year / "Gallen werden zum Gärben und Färben verwendet" / "Dreilappige Salbei"

863   *Satureja arkansana* U.S.: Tenn. / E. H. Hemann 11178 / 1939 / "Odor of pennyroyal"

864   *S. douglasii* U.S.: Calif. / L. T. Chamberlain s.n. / 1899 / " 'Yerba Buena' "
       U.S.: Calif. / B. Trask s.n. / 1900 / "used as a tea by Mexicans"

865   *S. glabella* U.S. & CAN. / Herb. of A. Wood s.n. / no year / "Niagara Thyme"

866   *S. nepeta* U.S.: N.Y. / S. Hartstall s.n. / '87 / "Basil Thyme"
       Bermuda Is. / W. M. Rankin s.n. / 1897 / " 'Catnip' "
       England / P. J. Greenway 133 / 23 / "Calamint"

867   *S. rigida* U.S.: Fla. / E. West 1128 / 1953 / "This is a good honey plant"
       U.S.: Fla. / Cooley, Monachino & Eaton 5651 / 1958 / "Bruised plant has scent of pennyroyal"

868   *S. vulgaris* U.S. & CAN. / Herb. Columbia Univ. s.n. / no year / "Wild Basil"
       U.S.: N.J. / J. Oehler s.n. / 1876 / "Basil   Gemeine Wirbeldoste"

869   *S. alpestris* Dominican Republic / A. H. Liogier 21913 / 1968 / "OREGANILLO"

870   *S. brownei* Jamaica / G. P. Proctor 9964 / 1955 / "Aromatic herb. Said to be abortifacient" / "Pennyroyal"
       Colombia / H. St. John 20765 / 1944 / "Spearmint odor" / " 'polea' "

Colombia / J. M. Duque Jaramillo 2926 / 1946 / "Contimenticio carminativo y tónico polmonar. Cultivado y silvestre" / " 'Poléo' "

3871 *S. vimenea* W.I.: Guadeloupe / P. Duss 4218 / 1905 / "médicinal" / "Gombeyre"
W.I.: Guadeloupe / H. Stehlé 10 / 1934 / ". . . forte odeur de menthol"

3872 *S. darwinii* Patagonia / O. A. Peterson s.n. / 1896 / " 'Pampas Tea' "

3873 *S. mathewsii* Peru / Hutchison & Bennett 4529 / 1964 / "Leaves pungent, minty. Used as a tea for colds or for bellyaches" / " 'Oregano de Monte' "

3874 *S. nubigona* Ecuador / E. Heinrichs 670 / 1934 / "starker Geruch de Blätter" / "Blüku blapkila"
Ecuador / W. H. Camp E-2090 / 1945 / "used as stomach stimulant" / " 'Tipo' "
Colombia / M. T. Darve 750 / 1918 / " 'Mejorana' "

3875 *S. taxifolia* Peru / J. J. Wurdack 1656 / 1962 / " 'Romero' "

3876 *Satureja* indet. Ecuador / J. A. Steyermark 52979 / 1943 / "Plant boiled in water for use as tea in treating chills" / " 'pleo' "

3877 *S. juliana* Italy / College of Pharmacy Herb. s.n. / no year / "Im südlichen Europa häufig als Heilmittel verwendet"

3878 *Scutellaria* indet. Colombia / Garcia, Hashimoto & Ishikawa 18606 / 1965 / "Se usa para el flujo de las mujeres"

3879 *Scutellaria* indet. EUR.: Morea; Bassae / Rev. D. D. Robertson s.n. / 1829 / "Ruins of the Temple of Apollo"

3880 *Sphacele conferta* Ecuador / W. H. Camp E-4369 / 1945 / "The people of the region esteem the plant as a cure for rheumatism, using the lvs. both as tea and as host compresses applied locally"

3881 *Stachys hyssopifolia* U.S.: Mass. / Oakes s.n. / no year / "Hyssop-leaved Woundwort"

3882 *S. parviflora* Rawalpindi / R. R. Stewart 7059 / 1922 / "evil odor" / " 'bui' "

3883 *Stachys* indet. Afghanistan / E. Bacon 79 / 1939 / "Used as fodder" / "butrang Ru"

3884 *S. arvensis* Australia: Queensland / L. S. Smith 3075 / 1947 / "Stagger Weed"

3885 *Teucrium japonicum* China: Kwangtung / Lingnan (To & Ts'ang) 12690 / 1924 / "Ch'an to ts'o (Bad smelling knife herb)"

3886 *T. viscidum* China: Kwangtung / F. A. McClure 23 / 1921 / "whole plant steeped in water to make wash for feet that have itch from wading in dirty water" / "Chue Mo Soh"

387   *T. integrifolium* Australia: Queensland / S. L. Everist 7308 / 1963 / "Severely grazed and suspected of causing deformation of leg bones in young rams"

388   *Thymus* indet. U.S.: D.C. / E. S. Burgess s.n. / 1893 / "Portuguese Thyme"

389   *T. capitatus* Dalmatia / College of Pharmacy Herb. s.n. / no year / "Ein Halbstrauch, dessen Blätter mit der Blüthen als Herba Thymi cretici gebräuchlich waren, und in dessen Väterlande auch wohl noch in Anwendung kommen" / "Kopfiger Thymian. Cretischer Thymian"

390   *Tinnea physalis* Tanganyika / R. E. S. Tanner 1596 / 1953 / "Sukuma tribe: Leaves soaked, water used to treat red eyes; roots pounded, water drunk for cough"

391   *T. vestita* N. Rhodesia / W. R. Bainbridge 739 / 1963 / "Roots rubbed into cuts to relieve neuralgia" / "Toka name KACHOLOLA"

392   *Trichostema brachiatus* U.S.: N.Y. / Columbia Univ. Herb. s.n. / no year / "False Pennyroyal"

393   *T. laxum* U.S.: Calif. / Constance & Baker 3016 / 1942 / " 'Turpentine-scented' "

394   *T. parishii* U.S.: Calif. / F. W. Johnson 1367 / 10 / " 'Romero' "

395   *T. rubisepalum* U.S.: Calif. / R. F. Hoover 2716 / 1937 / "Turpentine Weed"

396   *Lamiacea* indet. Mexico / O. & T. Degener 26781 / 1959 / "Uxmal . . . clearing about Mayan temple ruins"

397   *Lamiacea* indet. Ecuador / G. W. Prescott 160 / 53 / "medicinal plant" / " 'SUNFO' "

398   *Lamiacea* indet. Brazil / G. & L. T. Eiten 9807 / 1969 / "Previously planted by Base Camp personnel. Makes excellent tea" / " 'mangericão' "

399   *Lamiacea* indet. Colombia / Killip & Smith 16713 / 1927 / "(used for indigestion)" / " 'Mejorana' "

900   *Lamiacea* indet. Argentina / J. E. Montes 27576 / 58 / "Uso: medicina popular" / " 'Verbena' (Variedad)"

901   *Lamiacea* indet. Argentina / J. E. Montes 14717 / 1955 / "Raiz medicinal" / " 'Guaycurú' "

902   *Lamiacea* indet. Ecuador / J. A. Steyermark 53084 / 1943 / "Used in making teas to cure chills" / " 'tipo' "

903   *Lamiacea* indet. Argentina / J. E. Montes 14657 / 1955 / "Uso: hojas y raices medicina popular" / " 'Albahaca silvestre' 'A. del Campo' "

904   *Lamiacea* indet. Argentina / J. E. Montes 650 / 1945 / "Uso: Semillas: Medicina popular" / " 'Albaca del campo' "

3905   *Lamiacea* indet. Turkey / Lambert & Thorp 560 / 1968 / "Used by natives to make tea"

SOLANACEAE

3906   *Acnistus arborescens* Peru / Killip & Smith 27501 / 1929 / "sap very poisonous" / " 'Catahui' "

3907   *Atropa arborescens* W.I.: Martinique / College of Pharmacy Herb. s.n. / no year / "Die Pflanze enthält narkotisch giftige Stoffe"

3908   *Bouchetia arniatera* Mexico / A. Dugès s.n. / 1904 / "said to poison sheep" / " 'Maradilla' "

3909   *Browallia americana* Colombia / J. M. Duque Jaramillo 3686 / 1946 / " 'Pensamiento de pobre' "

3910   *Brunfelsia grandiflora* Peru / T. Plowman 2594 / 1969 / "Plant inhabited by stinging ants" / " 'chirú sanango' "
Peru / Killip & Smith 26859 / 1929 / "Roots used to brew a cooling tea"

3911   *B. grandiflora* var. *schultesii* Peru / T. Plowman 2495 / 1969 / "bark of root scraped with cold water and taken for rheumatism" / " 'chiric sanango' "

3912   *B. mire* Venezuela / O. E. White 535 / 1921 / "Possesses remarkable physiological properties" / " 'Mire' "

3913   *B. pauciflora* Venezuela / T. Plowman 1918 / 1968 / "Used for 'banos' by healers"

3914   *B. uniflora* Brazil / H. C. Cutler 12422 / no year / "Part of a collection of 42 remedies from one herb stall. Used as a tea or a cold infusion for venereal diseases, especially syphilis. Twice this amount for 1 cent U.S. money"
Argentina / T. M. Pedersen 10210 / 1972 / "Horses browse freely on the foliage of this shrub, said to be harmful to them"

3915   *Capsicum baccatum* U.S.: Tex. / A. Schott s.n. / no year / "To hipi tenoo Dohibiden Indians. Mexicano"
U.S.: Tex. / V. L. Cory 51407 / 1945 / " 'chillipiquin' "
U.S.: Tex. / H. R. Reed 432 / 1947 / " 'pepper' "
Honduras / C. & W. von Hagen 1269 / 1937 / " 'chilepete' "
Mexico / E. Matuda 17590 / 1950 / "Cultivated for cooking" / "Siete Cardo"
Dominican Republic / E. J. Valeur 275 / 1929 / "red berries very hot to taste" / " 'Ajiscito montecino' "

3916   *C. annuum* Mexico / E. Matuda 17642 / 1948 / "Cultivated" / "Chile chocolate"

3917   *C. annuum* var. *conoides* El Salvador / S. Calderon 1197 / 1922 / " 'chile chocolate' "

3918   *C. macrophyllum* Colombia / F. R. Fosberg 19463 / 1942 / " 'aji indio' "

919   *C. microcarpum* Paraguay / A. L. Woolston 1632 / 53 / "Fruits red, piquant" / "n.v.: ki-ii-ca'agui"
Brazil / Y. Mexia 4493 / 1930 / "very hot fruit used" / " 'Pimenta' "

920   *Capsicum* indet. Brazil / G. & L. T. Eiten 10140 / 1970 / " 'pimenta malagueta' "

921   *Capsicum* indet. Argentina / J. E. Montes 14727 / 1955 / "Frutos comestibles, muy picantes" / " 'Cunubari' "

922   *Cestrum aurantiacum* Guatemala / J. A. Steyermark 51721 / 1942 / "The crushed leaves, applied with soap in cold water, are said to remove stains from clothes" / " 'pay-ti' "

923   *C. laurifolium* El Salvador / P. C. Standley 21933 / 1922 / " 'Palo hediondo, Huele de noche' "

924   *C. laxum* Mexico W. H. Camp 2241 / 1936 / "plant ill-smelling"

925   *C. racemosum* Panama / A. Gentry 6444 / 1972 / "Leaves eaten by leaf cutter ants"

926   *C. thyrsoideum* Mexico / Hinton et al. 13411 / 38 / "Herba del Zopecate"

927   *C. daphnoides* Dominican Republic / J. de Js. Jimenez 5187 / 1967 / "MATA GALLINA"

928   *C. megalophyllum* W.I.: Dominica / G. P. Cooper 154 / 1933 / " 'Cafe marron' or 'wild coffee' "

929   *C. parqui* Uruguay / G. Herter 187a / 1925 / "Duraznillo negro"
Argentina / Mathews S124996 / 1943 / "Planta muy tóxica para el ganado" / " 'Hediondillo' "

930   *C. parvifolium* Venezuela / Gehriger 40 / 1930 / "Uvita"
Colombia / T. Plowman 3194 / 1972 / "Foul-smelling"

931   *C. reflexum* Peru / J. Schunke V. 2217 / 1967 / " 'Huasca hierba santa' "

932   *C. sendtnerianum* Peru / J. Schunke V. 3700 / 1970 / " 'Hierba Santa' "

933   *C. stipulatum* Peru / F. Woytkowski 7806 / 1962 / "very toxic plant for animals & humans"

934   *Cestrum* indet. Peru / T. B. Croat 20821 / 1972 / "used for coughs and headaches" / " 'santa maria' "

935   *Chamaesaracha* indet. Mexico / Dr. Gregg s.n. / 47 / "wash for wounds" / "Yerba del cáncer (one species)"

936   *Cyphomandra endopogon* Colombia / A. Fernández-Pérez 6863 / 1964 / "Reacción para alcaloides + debil"
Peru / J. Schunke V. 3323 / 1969 / "Las hojas . . . con olor nauseabundo" / " 'Chup sacha masha' "

937   *C. obliqua* Peru / J. Schunke V. 1883 / 1967 / "Las hojas tiene olor nauseabundo" / " 'chupo sacha' "

3938   *C. splendens* Peru / Y. Mexia 8235 / 1936 / "leaves have disagreeable odor"
/ " 'Tomato del campo' "

3939   *Cyphomandra* indet. Brazil / Prance, Maas et al. 16402 / 1971 / "the leaves
heated in water used to bathe babies to keep healthy & when in fever" /
" 'Tsetsepere' (Deni)"

3940   *Cyphomandra* indet. Colombia / T. Plowman 2332 / 1969 / "Whole plant
gives strong narcotic odor"

3941   *Cyphomandra* indet. Bolivia / R. F. Steinbach 436 / 1966 / "fruto comestible . . . ácido dulson muy jugoso" / "tomate de monte"

3942   *Datura inoxia* U.S.: Ohio / G. M. Hocking s.n. / 1944 / "Cultivated — used
in medicine"

3943   *D. inoxia* var. *quinquecuspida* U.S.: Ariz. / W. N. Clute 110 / 1920 /
"Used by Indians who may be responsible for its occurrence"

3944   *D. candida* El Salvador / P. C. Standley 19332 / 1921 / "fls . . . placed on
pillow to cure insomnia" / " 'Floripundio, Florifundio, Floricundia' "
Mexico / F. J. Lipp s.n. / 1978 / "Mije use: 13 leaves taken in infusion
during divinatory rites . . . fed to turkeys to insure rapid growth" / "Mije
'po:p pu:ix' Sp. 'flor de compañia' "

3945   *D. stramonium* Mexico / B. MacBryde 768 / 1967 / "used for
inflammation" / " 'Tolvache' "
Haiti / E. C. Leonard 8012 / 1925 / "Used for whitening in washing
clothes" / " 'Bergine moroine' "

3946   *D. affinis* Colombia / J. M. Duque Jaramillo 3127 / 1946 / "Semillas ven-
enosas. Cultivado y silvestre" / " 'Borrachero' 'Cacao sabanero' "

3947   *D. aurea* Colombia / F. R. Fosberg 20406 / 1943 / " 'culebra borrachero' "
Colombia / García y Barriga, Hashimoto & Ishikawa 18601 / 1965 / "Medicinal" / " 'Amarón' "

3948   *D. insignis* Colombia / G. Klug 1889 / 1930 / " 'Borrachera' "

3949   *D. tatula* EUR. / College of Pharmacy Herb. s.n. / no year / ". . . wird im
südlichen Europa und in Amerika, wohin sie aus Asien verschleppt zu sein
scheint, wie D. stramonium gebraucht"

3950   *Hyoscyamus albus* EUR.: Aegina Is. / Heldreich s.n. / 1848 / "Hat
dieselbes Heilkräfte wie H. niger, ist jedoch milder"

3951   *H. aureus* Lebanon / H. C. Stutz 2926 / 1967 / "Byblos . . . common in
old Phoenician ruins"

3952   *Iochroma* indet. Colombia / García y Barriga, Hashimoto & Ishikawa
18602 / 1965 / "La corteza es medicinal para heridas" / " 'Guatillo'; 'Teta-
janse' en Kansa"

3953   *Lycianthes heteroclita* Guatemala / J. A. Steyermark 49399 / 1942 /
" 'chilete"

54   *L. cyathocalyx* Peru / J. Schunke V. 4576 / 1970 / " 'Aji sillo' "

55   *Lycium berlandieri* U.S.: Tex. / C. H. Muller 8437 / 1945 / " 'tomatillo' "

56   *L. pallidum* U.S.: Ariz. / W. N. Clute 4 / 1920 / "The Indians use the berries after proper preparation"

57   *L. carolinianum* Mexico / Y. Mexia 1011 / 1926 / "Edible" / " 'Camasime' 'Pinolo blanco' "

58   *L. umbellatum* Mexico / H. S. Gentry 7131 / 1945 / "Fruit edible" / "Chamiso blanco"

59   *L. bosciifolium* S.W. Africa / R. Seydel 3984 / 64 / "von Voegeln gefressen"

60   *L. chinense* China: Kwangsi / W. T. Tsang 24564 / 1934 / "fruit edible" / "Kau Kei Tsai"
Hainan / F. A. McClure 7994 / 1921 / "drug plant" / "Kau ki ts'oi"

61   *Lycopersicum pimpinellifolium* Peru / Y. Mexia 8328 / 1936 / "Fruit is used and sold by the natives" / " 'Tomate' "

62   *Nectouxia formosa* Mexico / Balls & Gourlay B4434 / 1938 / "Strong 'snuff' scent"

63   *Nicandra physalodes* U.S. & CAN. / College of Pharmacy Herb. s.n. / no year / "Apple of Peru"
Bolivia / H. H. Rusby 367 / 1921 / "Called 'Belladona' "
EUR. / Oberneder 4640 / 1915 / "judenkirschenartige Giftbeere"

64   *Nicotiana bigelovii* U.S.: Calif. / S. A. Barrett s.n. / 1907 / ". . . plants used for smoking by the Miwok Indians"

65   *N. multivalvis* U.S.: Calif. / Herb. of University of Calif. s.n. / no year / "from seed . . . originally obtained in 1825 by David Douglas (from Indian cultivation along the Columbia River in Oregon)"

66   *N. repanda* Mexico / Dr. Gregg s.n. / 47 / "Wild tobacco"

67   *N. trigonophylla* Mexico / Aguirre & Reko 110 / 1946 / " 'tobacco loco' "

68   *N. acuminata* Chile / P. Aravena 33315 / 1942 / " 'Tobacco del Diablo' "

69   *N. pusilla* Bolivia / O. E. White 546 / 1921 / " 'Tobaco del monte' "

70   *N. sylvestris* Argentina / J. West 8376 / 1937 / " 'Tobaco del Campo', 'Sacha-tobaco', 'Tabaquilla' "

71   *N. tabacum* Peru / J. West 8082 / 1936 / "Used medicinally rather than as a narcotic" / " 'Tobaco' "
Brazil / Prance et al. 13928 / 1971 / "Cultivated by Indians as ingredient of narcotic snuff. Leaves dried rapidly, pulverized & mixed with ashes of Theobroma bark (13933 or 13939) . . . Vicinity of Jamamadi Indian village"

72   *N. thyrsiflora* Peru / Stork & Horton 10018 / 41 / "Leaves are used for smoking" / " 'Tabac sylvestre' "

3973   *N. tomentosa* Peru / Stork & Horton 9879 / 1938 / "A peon says you get a headache if you smoke the leaves"

3974   *Nicotiana* indet. Brazil / W. A. Archer 8159 / 1943 / "Lv. cured for smoking" / " 'fumo' "

3975   *Nicotiana* indet. S.W. Africa / R. Seydel 605 / 55 / ". . . das saftige Merk von Pavianen gefressen"

3976   *N. goodspeedii* Austrialia: Queensland / S. L. Everist 3141 / 1947 / "With 3140 responsible for death of 19 travelling cattle"

3977   *Nicotiana* indet. Brit. New Guinea: Papua / L. J. Brass 5459 / 1933 / "The tobacco most generally grown by natives of the AUGA River FUYUGE tribe. Said to have been introduced from the southwest by early missionaries: probably about 20 years ago" / "KUKU (Introduced name)"

3978   *Physalis angulata* U.S.: N.J. / College of Pharmacy Herb. s.n. / 1882 / "Ground Cherry"

3979   *P. elliottii* U.S.: Fla. / P. P. Sheehan s.n. / 1919 / "Medicinal Plants of the Seminole Indians. Used for coughs and colds. The whole plant is boiled, and the infusion is applied to the affected parts, generally for two or three days" / " 'Ground cherry' "

3980   *P. philadelphica* Mexico / Hinton et al. 5823 / 34 / "edible but not cultivated" / "Tomate"
Mexico / E. Palmer 11452 / 1896 / "Sold in Mexican markets, for sauces, etc."
El Salvador / P. C. Standley 20244 / 1922 / "Fr. of this & other spp. boiled or fried & eaten" / " 'Huevitos, Millomate' "

3981   *P. peruviana* Haiti / C. V. Morton 824 / 1941 / "Fruits yellow when mature, edible"
Colombia / J. H. Duque J. 2896 / 1946 / "Frutos . . . son comestibles"
Peru / Fosberg, Ferreyra & Cerrate 28218 / 1947 / "Fruits eaten" / " 'capuli de la costa' 'tomatilla' "

3982   *P. hygrophila* Peru / Y. Mexia 6299 / 1931 / "mature fruit yellow, edible" / " 'Muyaca' "

3983   *Physalis* indet. Peru / Killip & Smith 24715 / 1929 / "fruit yellow, edible"

3984   *Physalis* indet. Brazil / A. Silva 13 / 1944 / "Frt. edible" / " 'camapú' "

3985   *Physalis* indet. Brazil / M. B. da Silva 3 / 1942 / "Infusion of root taken for bladder inflamations" / " 'camapú' "

3986   *Physalis* indet. Bolivia / R. F. Steinbach 29 / 1966 / "fruto comestible, anarenjado de gusto dulce"

3987   *Physalis* indet. Argentina / J. E. Montes 27730 / 58 / "Uso: medicina popular" / " 'Camambú' "

88 *Solanum elaeagnifolium* U.S.: Tex. / Schott s.n. / 1852 / "With the Mex-
ican in use against Rheumatism" / "Trompillo Mexic."
U.S.: Ariz. / B. S. Klinger s.n. / 34 / "Eat berry, make cheese from its
powder" / "Indian: Haxlza"
U.S.: N.M. / M. C. Stevenson 2 / 1902 / "Used by the Zuni Indians"
Mexico / Dr. Gregg s.n. / 47 / "Sudorific, sternutatory" / "Trompillo"

89 *S. atitlanum* Honduras / C. & W. von Hagen 1042 / 1937 / " 'Quema
nariz' "

90 *S. hernandezii* El Salvador / P. C. Standley 19661 / 1922 / "lvs. ill-scented
when crushed" / " 'Huistomate' "

91 *S. houstonii* Brit. Honduras / P. H. Gentle 60 / 1931–32 / "medicinal; used
for sores" / " 'sosumbra' "

92 *S. nigrum* Mexico / G. B. Hinton 3855 / 33 / "Concoction for Erisipelas" /
"Yerba Mora"
S.W. Africa / R. Seydel 3783 / 63 / "Neuerdings Abkochung als
Asthmamittel gebraucht"

93 *S. torvum* Guatemala / J. A. Steyermark 49207 / 1942 / "Fruit edible for
man, considered poisonous to dogs, reputed to be used in treating and curing mumps (papéras)" / " 'lavoplato' "
Colombia / F. R. Fosberg 20293 / 1943 / "immature fruit said to be used
as a soap"

94 *S. verbascifolium* Mexico / Dr. Gregg s.n. / 47 / "Poultice for ulcer, bilis,
&c." / "Yerba de San Pedro"

95 *S. verrucosum* Mexico / Balls & Gourlay B5010 / 1938 / ". . . the tubers are
eaten, but the general hoeing out from among the crops and long the edges
of the fields would suggest that the plant was not considered very
favourably" / " 'Papa Morda' "

96 *S. yucatanum* Honduras / J. B. Edwards P-750 / 1934 / "Fruit edible"

97 *Solanum* indet. Mexico / P. S. Ortiz 49 / no year / "Es buen plato de ensalada; pero sancochada y con sal es como el campesino la apetece. Sus hojas, con aguardiente, en cataplasmas o cinturas deshacen las hinchazones" /
"Yerba mora"

998 *S. antillarum* Dominican Republic / Gastony et al. 175 / 1967 / "juice
smells skunk-like"

999 *S. bahamense* Bahama Is. / O. Degener 18823 / 1946 / "fruit — squashed,
put in cloth and rubbed over inside of mouth of babies suffering from
thrush" / " 'Cankerberry' "
Bahama Is. / W. Cerbin C112 / 1973 / "Used to 'clean trash out of mouth'
(mouthwash)" / " 'Wild Berry' "
Bahama Is. / R. A. & E. S. Howard 9986 / 1948 / "Used medicinally"

4000 *S. ficifolium* Puerto Rico / W. R. Stimson 2985 / 1966 / "used as a medicinal for rheumatism, also used to clothing in wash" / " 'beregona cimarona' "
W. I.: Dominica / W. H. Hodge 805 / 1937 / " 'wild eggplant' "

4001 *S. nodiflorum* W.I.: Guadeloupe / P. Duss 2604 / 1893 / "L'emploie en infusion contre les fievres" / "Agouman"
Peru / J. Schunke V. 1448 / 1967 / "hojas medicinales para cólicos. Mala hierba" / " 'Hierba mora' "

4002 *S. robustum* Bermuda / L. M. Andrews 337 / 1957 / "Cockroach Poison"

4003 *Solanum* indet. Cuba / E. H. Day s.n. / 81 / "Will bear the 'eggplant' if grafted"

4004 *S. acuminatum* Brazil / Harley et al. 10914 / 1968 / "The fruit is edible"

4005 *S. albidum* Brazil / H. C. Cutler 12427 / no year / "Take tea made with hot or cold water for diseases of liver, kidneys, spleen. Twice this amount for ½ cent U.S. Part of a collection of 42 remedies from one herb stall" / " 'Jurubeba branca' "

4006 *S. asperum* Brazil / Y. Mexia 4786 / 1930 / "Pubescence on stems irritating to skin"

4007 *S. chacoense* Argentina / A. T. Hunziker 1310 / 1941 / "cult . . . con objeto de utilizarla como material genético resistente a enfermedades" / "pepe de yuto (yuto = perdiz in Quicha)"

4008 *S. cinnamomeum* Brazil / Y. Mexia 4131 / 1929 / "Used medicinally as a tonic" / " 'Mercurunho' "

4009 *S. coconilla* Peru / J. Schunke V. 1474 / 1967 / "Los frutos son comestibles. Cultivado" / " 'Coconilla' "

4010 *S. curtipes* Argentina / J. E. Montes 778 / 45 / "Uso: med. pop." / "Tintórea"

4011 *S. gardneri* Venezuela / H. Pittier 11887 / 1926 / "Children gather and eat the fruits" / " 'Santa Maria' "

4012 *S. grandiflorum* Brazil / W. A. Archer 7707 / 1942 / "Frt. grated and used as poultice for inflammations" / " 'jurubeba' "

4013 *S. grandiflorum* var. *macrocarpum* Bolivia / H. H. Rusby 814 / 1921 / "Fruit edible"

4014 *S. granuloso-leprosum* Argentina / J. E. Montes 14816 / 1955 / " 'Fumo bravo' 'Fumito' "

4015 *S. hirtum* Colombia / H. H. Smith 1147 / 1898–99 / "The fine hairs of the stem and fruit cause considerable irritation to the skin, coming off when touched"

016   **S. hispidum** Peru / Hutchison & Tovar 4146 / 1964 / "The fruit contains saponin and is used by natives as a substitute for soap" / " 'Nahui-tojyachi' ('blinds-your-eyes')"

017   **S. hypomalacophyllum** Venezuela / Gehriger 1 / 1930 / "N.v. Borrachera"

018   **S. mammosum** Venezuela / Nee & Mori 4179 / 1971 / "said to be poisonous" / " 'una de gato"
Venezuela / F. Bianco 77 / 76 / "Muy VENENOSO" / "Manzanillo, Manzana del Diablo"

019   **S. micranthum** Colombia / E. L. Core 1529 / 1944 / "leaves said to be used to combat malaria"

020   **S. minutobaccatum** Bolivia / J. Steinbach 5362 / 1921 / "fruta negra comible" / "Guaparmillo"

021   **S. molle** Venezuela / G. Agostini 1513 / 1973 / "Pruebas de Dragendorff, Meyer e Iodo-platinico: positivo"

022   **S. multiinterruptum** var. **longipilosum** Peru / Correl & Smith P950 / 1960 / "tubers . . . eaten by natives" / " 'Coro' "

023   **S. nollanum** Colombia / J. M. Duque Jaramillo 4633 / 1947 / "Frutos . . . utiles para hacer bebidas refrescantes" / " 'Lulo'; de Castillo y para sorbetes"

024   **S. pseudolulo** Peru / R. Kayap 813 / 1973 / "Comestible la fruta" / "betság"

025   **S. riparium** Argentina / E. Villa 39 / 45 / "Tabaquillo"

026   **S. rugosum** Brazil / A. Silva 60 / 1944 / "Frt. poisonous, used by fishermen to kill and catch fish" / " 'cajuçára' "
Brazil / W. A. Archer 7919 / 1942 / "Juice from leaves mixed with Clibadium as fish poison" / " 'cajú-sal' "

027   **S. sessiliflorum** Brasil / R. de Lemos Fróes 21274 / 1945 / "Planta, servindo para doces e como legume, usado pelos indios e civilizados da região Vaupés" / " 'Cubiu' "

028   **S. sisymbriifolium** Argentina / G. J. Schwarz 768 / 45 / "la raiz de esta planta se usa como remedio para los riñones" / "Yuti á"
Argentina / R. Alvarez 187 / no year / "medicinal para reuma" / " 'Espina colorado' "
Argentina / A. Krapovickas 674 / 1944 / " 'tomate del campo' 'tu tia' "

029   **S. stramonifolium** Colombia / R. R. Castañeda 3777 / 1952 / "Frutos rojas, comestibles" / " 'Etopaa' (Tokano)"
Brazil / W. A. Archer 773 / 1942 / "frt. red and edible" / " 'Joá' "

030   **S. subinerme** Brazil / M. Barbosa da Silva 134 / 1942 / "Root infusion taken three times daily for inflamations of kidney or liver" / " 'juúna preta' "

031   **S. trachycyphum** Colombia / E. L. Little 7415 / 1944 / " 'yerba de muerto del negro', 'pepito' "

4032   **S. tripartitum** Bolivia / M. Bang 2062 / 1894 / "used as medicine for cleaning the blood" / "Cuti cuti"

4033   *Solanum* indet. Tanganyika / R. E. S. Tanner 2078 / 1955 / "Roots soaked in water used to bathe child with fever" / "Kibondei: Mtuwa"

4034   *Solanum* indet. Tanganyika / R. E. S. Tanner 1265 / 1953 / "Sukuma tribe: roots chewed to treat stomach pains, roots and fruits ground, mixed with water, drunk to treat gonorrhoea with sores / "Kisukuma: Stulla'

4035   **S. macranthum** Ceylon / Maxwell & Fernando 988 / 1972 / "Fruits eaten"

4036   *Solanum* indet. Ceylon / R. H. Maxwell 992 / 1972 / "stems thorny, covered with red ants. Fruits eaten" / "Titta tibbutu (S.)"

4037   *Solanum* indet. Ceylon / R. H. Maxwell 995 / 1972 / "Eaten; sold in market"

4038   *Solanum* indet. Ceylon / R. H. Maxwell 993 / 1972 / "Eaten. Found in the market"

4039   **Witheringia solanacea** Panama / Kirkbride & Duke 1362 / 1968 / "leaves used in a tea to kill pain" / " 'Mata dolor' "
        Peru / Hutchison, Wright & Straw 5995 / 1964 / "Leaves used as a tea to put one to sleep" / " 'Mullaca Rojo' "

4040   **Witheringia** indet. Peru / J. Schunke V. 2105 / 1967 / "Las hojas utilizan en infusion para lavar heridas infecciosas" / " 'Tabaco mullaca' "

4041   **Solanacea** indet. Bolivia / O. E. White 535 / 1921 / "Remarkable physiological!" / " 'Mire' "

4042   **Solanacea** indet. Colombia / G. Klug 1810 / 1930 / "medicinal" / " 'Zanango' "

4043   **Solanacea** indet. Bolivia / J. Steinbach 5200 / 1920 / "La raiz contiene una apreciado tintura que se emplea en alimentos" / "Palillo"

SCROPHULARIACEAE

4044   **Alonsoa meridionalis** Ecuador / W. H. Camp E-2650 / 45 / "Decoction in hot water used to bathe vermin-infested babies" / " 'Giz-giz' or 'Guizeh-guizeh' "
        Ecuador / Grubb et al. 1004 / 1960 / "Made into paste with bad cheese and applied to cuts" / "Napan"

4045   **Angelonia biflora** Brit. Guiana / J. S. de la Cruz 1077 / 1921 / "Used for fever"

4046   **Artanema longifolia** var. **amplexicaule** Tanganyika / R. E. S. Tanner 2773 / 1956 / "Leaves used as vegetables" / "Zigua: Mzmbarau"

47 *Buchnera elongata* var. *elongata* U.S.: Fla. / P. P. Sheehan s.n. / 1919 / "Medicinal Plants of the Seminole Indians. Used as an emetic, as a head-wash in case of dizziness. The plant is used to make an infusion in warm water" / "Blue-hearts"

48 *Calceolaria* indet. Argentina / Schulz-Varela 5080 / 1944 / "planta con fuerte olor desagradable"

49 *Capraria biflora* Cuba / College of Pharmacy Herb. s.n. / 1840 / "considered as emmenagogue" / "Cuban name: escala biosa"
S.AM. / Rusby & Squires 37 / 1896 / " 'Venezuelan Tea' "

50 *Castilleja arvensis* Peru / Mr. & Mrs. F. E. Hinkley 58 / 1920 / "Used as a medicine" / "Violeta"

51 *Craterostigma hirsutum* Tanganyika / R. E. S. Tanner 1295 / 1953 / "Sukuma tribe: Roots pounded, used in gargle to treat dry cough" / "Kisukuma: Malaba ga muluguru"

52 *Escobedia grandiflora* Colombia / Killip & Smith 15435 / 1926 / "Roots used for yellow coloring matter" / " 'Asofrin' "

53 *E. scabrifolia* Peru / Hutchison & Wright 3899 / 1964 / "Used as a dye in cooking, and according to Ferreyra, medicinal"

54 *Euphrasia borneensis* Brit. N. Borneo / D. L. Topping 311 / 1915 / "Part of these were gathered by Sumpat after the sacrifice was offered. Sacrificial altar of Kina above Paka cave"

055 *Gratiola peruviana* Ecuador / W. H. Camp E-2581 / 45 / "used for intestinal fevers" / " 'Yaco-muyo' "

056 *Halleria lucida* S. Africa / W. G. Barnard 222 / 34 / "Fruits edible" / "Mokwayne (Bakone Sesuto)"

057 *Hebenstretia dentata* Tanganyika / R. E. S. Tanner 3801 / 1957 / "Masai—for syphyllis in women, roots boiled and the water drunk" / "Masai—Osigerai"

058 *Lindernia diffusa* Brit. Guiana / D. M. Davis 199 / 1967 / "Used for sickness"
Brazil / W. A. Archer 7695 / 1942 / "Poisonous to chickens" / " 'douradinha' "

059 *Mecardonia procumbens* Ecuador / Rose & Rose 23822 / 1918 / "medicinal" / " 'yerba del Taylor' "

060 *Melasma hispidum* Mexico / G. B. Hinton 2202 / 32 / "Root sold as spice for meats" / "Azafrán"

061 *Mimulus glabratus* Ecuador / W. H. Camp E-2697 / 45 / "Infusion taken for irritations of kidneys and urinary tract" / " 'Yaco-muyo' (Water-herb)"
Ecuador / W. H. Camp E-1935 / 1945 / "Infusion of flowers taken for chest colds or bronchitis" / " 'Violeta' "

4062 **M. glabratus** var. *fremontii* Mexico / Dr. Gregg s.n. / 1847 / "good emolient poultice" / "Yerba amarilla"

4063 **Rhamphicarpa recurvata** Tanganyika / R. E. S. Tanner 585 / 1952 / "Flowers used on baby's gums in teething" / "Kisukuma: Mboula ya mbde"

4064 **R. tenisecta** Tanganyika / R. E. S. Tanner 3809 / 1957 / "Masai—fruits eaten" / "Masai—Olagoley"

4065 **Scoparia dulcis** Mexico / Carter & Chisaki 1250 / 1959 / "used for bruises" / " 'hierba de pajarito' "
Ecuador / W. H. Camp E-3852 / 1945 / "Said to be used in healing wounds"
Brit. Guiana / D. H. Davis 196 / 1967 / "herb dried for tea 'to purify blood' "
Argentina / J. E. Montes 14806 / 1955 / "Uso: medicinal" / " 'Miel de tierra' "
Peru / J. Schunke-Vigo 1052 / 1966 / "Medicinal: utilizan como purgante, por colicos"

4066 **Scoparia** indet. Brazil / M. B. da Silva 12 / 1942 / "For retention of urine; make tea of leaves and seed, take 3 times daily" / " 'vassourinha' "

4067 **Scoparia** indet. Brazil / M. B. da Silva 17 / 1942 / "Stems boiled and tea taken for fevers. Aromatic leaves used for refreshing baths"

4068 **Stemodia maritima** Bahama Is. / W. Cerbin C119 / 1972 / "Smoked as tobacco when dried. Boiled, the oils are used to bathe in; brewed it eases the discomfuture of overeating" / " 'Poor Man's Strength' "

4069 **S. suffruticosa** Colombia / J. M. Idrobo 2641 / 1957 / "Corteza del tronco y raices, muy amarga. Se usa como purgante muy drástico" / "Nombre Kamsa: 'Hoxa-mda-na-ixa'; 'Purgux' "

4070 **Striga asiatica** S. Rhodesia / Norlindh & Weimarck 5102 / 1931 / " 'Witch seed' "

4071 **S. pubiflora** Tanganyika / R. E. S. Tanner 2850 / 1956 / "Swaheli: For persistent cough, leaves pounded up green, mixed with water and used as gargle" / "Swaheli: Kidivu Ya Mbuzi"

4072 **Scrophulariacea** indet. Venezuela / J. A. Steyermark 75227 / 1953 / "flowers white, when boiled in water, they serve as a remedy for diarrhea" / " 'fregosa' "

4073 **Scrophulariacea** indet. Argentina / J. Semper 273 / 1944 / "se usa como ensalada"

LENTIBULARIACEAE

4074 **Utricularia humboldtii** Venezuela / J. A. Steyermark 58619 / 1944 / "Leaves boiled in water for use in stomachache" / " 'ere-da' "

OROBANCHACEAE

75   *Conopholis americana* Canada: Nova Scotia / Fernald & Long 24475 / 1921 / "Fresh bruised plant with odor of cider"
U.S.: N.J. / A. Niederer s.n. / 1884 / "Squaw-root"

76   *Epifagus virginiana* U.S.: Mass. / F. C. MacKeever MV 480 / 1961 / "Cancer-root"

77   *Orobanche uniflora* U.S.: N.Y. / no collector s.n. / 76 / " 'One-flowered Cancer-root"

GESNERIACEAE

78   *Besleria* indet. Peru / D. McCarroll 8 / 1939 / "used as a purgative" / " 'acala' "

79   *Chirita heterotricha* Hainan / S. K. Lau 1872 / 1933 / "The leaves are edible" / Kap Yin Ip"

80   *Codonanthe calcarata* Venezuela / Aristeguieta & Lizot 7424 / 1970 / "La raiz la utilizan para curar heridas. Indios Guaicas (Yanomanö)"

81   *Columnea picta* Ecuador / E. Heinrichs 739 / 1934 / "Tee gegen Blutkrankheit" / "Puntalanza od. Yaguarpanga"

82   *Drymonia turrialve* Panama / Kirkbride & Duke 684 / 1968 / "Plant used medicinally by Guarmi"

83   *Isanthera discolor* Philippine Is. / A. D. E. Elmer 10436 / 1908 / "its leaves are boiled and applied to wounds" / " 'Handalomog' "

84   *Kohleria bogotense* W.I.: Martinique / P. Duss 1937 / no year / "il pousse a poison"

85   *Kohleria* indet. Peru / D. McCarroll 18 / 1939 / "used for wounds" / " 'foutilla' "

86   *Nautilocalyx whitei* Bolivia / R. F. Steinbach 511 / 1966 / "infusion de las flores como analgésico contra el dolor de cabeza y neuralgias"

BIGNONIACEAE

087   *Adenocalymna impressum* Peru / O. Woytkowski 5343 / 1957 / "People say it is toxic"

088   *Crescentia alata* Mexico / G. B. Hinton 755 / 1932 / "The seeds are sold in the market and are used boiled for colds and pneumonia"

089   *Dolichandrone spathacea* Caroline Is. / F. R. Fosberg 32104 / 1950 / "leaf and fruit used as substitute for Piper betle in chewing betel nut" / " 'Narin' "

4090   *Kigelia pinnata* Nyasaland / L. J. Brass 17444 / 1940 / "visited by bats"

4091   *Markhamia obtusifolia* Tanganyika / R. E. S. Tanner 1250 / 1958 / "Sukuma tribe: Roots used to make fermented drink to treat advanced syphilis with sores" / "Kisukuma: Nghubu"

4092   *Pleonotoma jasminifolia* Brazil / Prance et al. 15596 / 1971 / "Lvs. squeezed and liquid applied to burn in Makú" / " 'Yououbiden' "

4093   *Stereospermum kunthianum* Tanganyika / R. E. S. Tanner 1251 / 1952 / "Sukuma tribe: Pounded roots soaked in water, this drunk to treat syphilis, constipation" / "Kisukuma: Nhelela"
Tanganyika / R. E. S. Tanner 1565 / 1953 / "Sukuma tribe: Roots used as poultice to treat syphilitic sores" / "Kisukuma: Muntelela"

PEDALIACEAE

4094   *Sesamum angolense* Tanganyika / R. E. S. Tanner 1561 / 1953 / "Sukuma tribe: for sores: roots pounded up and put on sores after they bleed" / "Kisukuma: Ikonda lya hisanga"

4095   *S. angustifolium* Tanganyika / R. E. S. Tanner 1012 / 1953 / "Sukuma tribe: roots and stems pounded, mixed in water, this drunk to treat diarrhoea; stems once used as soap substitute" / "Kisukuma: Ikonda"
Tanganyika / R. E. S. Tanner 564 / 1952 / "Leaves used as vegetable; roots ground up and soaked and then drunk to treat diarrhoea" / "Kisukuma: Ikonda"
Tanganyika / R. E. S. Tanner 2572 / 1956 / "Kibondei: Mgukiwa"

4096   *S. orientale* Liberia / G. P. Cooper 133 / 1928 / "Fruit eaten"
Philippine Is. / R. S. Williams 3050 / 1905 / "fruit used as medicine"

ACANTHACEAE

4097   *Asystasia gangetica* Tanganyika / R. E. S. Tanner 1375 / no date / "Sukuma tribe: Roots pounded, used orally to treat swollen stomach (bengo) in child" / "Kisukuma: I shingisha"

4098   *Barleria grandicalyx* Tanganyika / R. E. S. Tanner 1097 / 1952 / "Sukuma tribe: seeds mixed with grain seeds before cultivation to encourage profuse growth. Analogy from clumps into which it grows" / "Kisukuma: Kakuli"

4099   *B. priontis* Tanganyika / R. E. S. Tanner 3012 / 1956 / "Ngindo: for cough, leaves cooked and mixed with maize flour and eaten. Bondei: for syphilis, leaves pounded up and mixed with water and blown into penis" / "Bondei: Msunguguyu", "Ngindo: Lihomanga"

Tanganyika / R. E. S. Tanner 2000 / 1955 / "In disease that blisters hands and feet, thorns used to prick blisters, leaves rubbed on them" / "Kizigua: Luningulu"

00 *Barleria* indet. Tanganyika / R. E. S. Tanner 1495 / 1953 / "Sukuma tribe: Roots pounded, used in food to treat poisoning" / "Kisukuma: Lweja bugota"

01 *Blechum brownei* Marian Is. / E. Y. Hosaka 3007 / 1946 / "used for medicine" / " 'Lasaga' "

02 *Blepharis maderaspatensis* Tanganyika / R. E. S. Tanner 1382 / 1953 / "Sukuma tribe: Used to treat swollen legs — plant burned, pounded, mixed with ground nut oil, used as poultice" / "Kisukuma: Ilamata; ilamatilwa"

03 *Blepharis* indet. Tanganyika / R. E. S. Tanner 1247 / 1953 / "Sukuma tribe: Dried plant burned, smoke used to treat smallpox, infected legs" / "Kisukuma: Yinza"

04 *Dyschoriste hirsutissima* Mexico / H. S. Gentry 7151 / 1945 / "Browsed by goats"

05 *Dyschoriste* indet. Tanganyika / R. E. S. Tanner 1380 / 1953 / "Sukuma tribe: Roots pounded, soaked, taken orally to treat stomach pain"

06 *Elytraria imbricata* Mexico / C. & E. Seler 1783 / 1896 / " 'un pié' "

07 *Hygrophila phlomoides* Indo-Chine / D. Bois 2211 / 1903 / "aromatique, employée comme condiment et en medicine, por les indigenes"

08 *Hypoëstes forskalii* S.W. Africa / R. Seydel 4122 / 65 / "gelegentlich etwas befressen"

09 *H. verticillaris* Tanganyika / R. E. S. Tanner 200 / 1955 / "Whole plant soaked in water, used to bathe children with fever" / "Kizigua: Mnyota"
Tanganyika / R. E. S. Tanner 2017 / 1955 / "Used as a vegetable" / "Kizigua: Mnyago"
Tanganyika / R. E. S. Tanner 2935 / 1956 / "Zigua: For swellings, leaves pounded up and put on as a paste" / "Kizigua: kiviza"

10 *Jacobinia mohintli* Mexico / G. B. Hinton 3714 / 1933 / "Concoction of leaves & flowers used as a tonic" / "Muicle"
Mexico / Dr. Perrine s.n. / no year / "Leaves yield a purple color — a substitute for indigo" / "Titi or Añil de Cucury"

11 *J. sericea* Peru / Hutchison & Wright 4376 / 1964 / "Dr. E. Cerrate at USN reports that at Chiquian . . . it is used for menstrual regulation. An infusion is used as a cure for venereal disease" / " 'Arzobispo' or 'Sin' "

112 *Justicia carthaginensis* El Salvador / P. C. Standley 19333 / 1921 / "Remedy for fits & spasms in children" / " 'Hierba del susto' "

113 *J. pectoralis* W.I.: Martinique / P. Duss 2007 / 1881 / ". . . cultivé pour ses vertus medicinales" / "Herbe aux charpentiers"

Peru / Killip & Smith 22914 / 1929 / "Used for washing"

Brazil / J. A. Steyermark 104040 / 1970 / "Herb used in dried powdered state as one of three ingredients in preparation of 'yopo' by local Guaica Indians"

Brazil / Prance et al. 11174 / 1971 / "Used by Indians for hallucinogenic & snuff" / " 'Paxararok' (Uaicá-Macajai)"

Brazil / Prance et al. 10531 / 1971 / "Whole plant used as an additive to narcotic snuff. It is added to *Virola* bark ('Mashfaraá')" / " 'Mash farhenak (Uaicá)"

Colombia / Schultes & Cabrera 15584 / 1952 / "Cultivated. For binding on arms for dances. Infusion for conjunctivitis" / "Makuna = see-teé-hoo tahoo ('fragrant leaf')"

Brazil / R. Souza s.n. / no year / "A planta toda, torrada e pulverizada, é usada no preparo do Rapé de Indios pelos indios Uaica. Voucher for materials under chemical studies by Prof. Norman A. LeBel"

4114   *J. pharmacodes* Colombia / O. Haught 3999 / 1944 / "Probably poisonous—not eaten by stock, dries with peach-kernel odor"

4115   *J. betonica* Tanganyika / R. E. S. Tanner 1282 / 1953 / "Sukuma tribe: Entire plant burned, ground up with butter to treat scaling of flesh" / "Kisukuma: Kagugu wa mlutogaro"

4116   *J. genistifolia* S.W. Africa / R. Seydel 330 / 54 / "Wertvoller Futterbusch"

4117   *J. glabra* Tanganyika / R. E. S. Tanner 1096 / 1952 / "Sukuma: for attacks of boils: green leaves pounded up mixed with butter, put on hot iron and then put on boils as a plaster" / "Kisukuma: Digwanoni"

4118   *J. sansibarensis* Tanganyika / R. E. S. Tanner 2762 / 1956 / "Zigua: for headache—roots pounded up and put on forehead as a paste" / "Zigua: Mkoma uilema"

4119   *Justicia* indet. Tanganyika / R. E. S. Tanner 1062 / 1952 / "Used as vegetable" / "Kisukuma: kalandi"

4120   *Justicia* indet. Tanganyika / R. E. S. Tanner 633 / 1952 / "Leaves used as a vegetable" / "Kisukuma: Makuwiata"

4121   *Louteridium donnell-smithii* Guatemala / E. Contreras 5510 / 1966 / " 'Tabaco Silvestre' "

4122   *Megalochlamys marlothii* S.W. Africa / R. Seydel 2962 / 61 / "Abkochung und Heilmittel gegen Ekzeme—Futterbusch"

4123   *Monechma* indet. Tanganyika / R. E. S. Tanner 2024 / 1955 / "Used as vegetable" / "Kizigua: Mkobo"

4124   *Petalidium setosum* S.W. Africa / R. Seydel 2945 / 66 / ". . . stark befressen von Wild und Vieh"

4125   *Phaylopsis* indet. Tanganyika / R. E. S. Tanner 1554 / 1953 / "Sukuma tribe: Roots used to treat sore lips" / "Kisukuma: Masangu ga badimi"

26   *Pseuderanthemum* indet. Peru / Killip & Smith 29826 / 1929 / "Narcotic herb" / " 'Dormidero' "

27   *P. hildebrandtii* Tanganyika / R. E. S. Tanner 3382 / 1957 / "Zigua: For pains in the joints, leaves burnt and the ash rubbed into cuts near the pain"

28   *P. atropurpureum* Gilbert Is. / E. T. Moul 8038 / 1951 / "Used for head leis" / " 'teroti' "

29   *Ruellia heteromorpha* U.S.: Fla. / P. P. Sheehan s.n. / 1919 / "Medicinal Plants of the Seminole Indians. Used for stomach disorders. The whole plant is boiled and the decoction is drunk hot"

30   *R. inundata* El Salvador / P. C. Standley 20665 / 1922 / "strong goat odor" / " 'Chancho de monte' "

31   *R. tuberosa* W.I.: Martinique / P. Duss 2013 / 1883 / "la racine est purgative et peut remplacer levintable Ipeca" / "Ipeca batard, herbe caraibe, patate macaque"
W.I.: Cayman Is. / W. Kings GC 103 / 31 / "Tuberous roots used for heart trouble" / "Heart Bush"

32   *R. haenkeanum* Peru / D. McCarroll 17 / 1939 / "Lung remedy" / " 'Floripondio'?"

33   *Ruellia* indet. Tanganyika / R. E. S. Tanner 1298 / 1953 / "Sukuma tribe: Roots and leaves pounded, mixed with ghee, made into hot poultice for boils" / "Kisukuma: Kubita wa mulugurú"

34   *Stenandrium floridanum* U.S.: Fla. / P. P. Sheehan s.n. / 1919 / "Medicinal Plants of the Seminole Indians. Used to cure snake bite. Plant is boiled, and the decoction drunk while hot" / "Rattlesnake-flower. Called Sin-ty-etsee by the Big Cypress Seminoles"

135   *Trichanthera gigantea* Colombia / W. A. Archer 523 / 1930 / "Used as remedy for cattle" / " 'Rompebarriga', 'Quebrabarriga' "
Venezuela / H. Pittier 12056 / no year / " 'Naranjillo' "

136   *Acanthacea* indet. Ecuador / C. C. Fuller 108 / 53 / "Indians claim small piece of crushed leaf tied on fish hook will attract fish. Cultivated herb around house" / "Yacu aicha pihui: Quichua"

MYOPORACEAE

137   *Bontia daphnoides* W.I.: Guadeloupe / H. Stehlé 138 / no year / "Peu refrandu"
W.I.: Martinique / F. E. Engler 3910 / 1939 / "Fruit reported to be used as antidote to poison of *Hippomane mancinella*"

138   *Myoporum boninense* Mariana Is. / F. R. Fosberg 24841 / 1946 / "flowers white with honey-like odor"

4139   *M. rapense* Rapa / F. R. Fosberg 11525 / 1934 / "strong sweet odor"

4140   *M. stokesii* Austral Is. / St. John & Fosberg 15964 / 1934 / "heavy fragrance"

PLANTAGINACEAE

4141   *Plantago major* Puerto Rico / W. R. Stimson 1472 / 1965 / ". . . said to be used medicinally for fever" / " 'Yaneten' "
Cuba / R. A. Howard 5317 / 1941 / "Used medicinally for fever"
Colombia / J. M. Idrobo 4651 / 1961 / " 'Llantén' "

4142   *P. goutiana* Colombia / J. M. Duque Jaramillo 2674 / 1964 / "Medicinal"

4143   *P. major* var. *paludosa* Peru / G. Tessman 3181 / no year / "Used in medicine"

4144   *Plantago* indet. Argentina / A. Scala s.n. / 1928 / "Cura-malal grande hasta"

4145   *P. princeps* Hawaii / D. P. Rodgers s.n. / 1946 / " 'Ale' "

RUBIACEAE

4146   *Adina rubella* China: Kwangtung / Lingnan, (To & Ts'ang) 12834 / 1924 / "Sai shek lau, (Small ponegranate)"

4147   *Agathisanthemum bojeri* Tanganyika / R. E. S. Tanner 3879 / 1957 / "Zigua: for pain in left side of stomach, roots soaked in water which is drunk" / "Zigua: Mnama"
Tanganyika / R. E. S. Tanner 1941 / 1955 / "Roots chewed to treat stomach-ache . . . aromatic" / "Sibondei: Nyawaiya kumwendi"

4148   *Aïdia cochinchinensis* China: Kwangtung / Lingnan (To & Ts'ang) 12056 / 1924 / "Yeung shi fu (Sheep excrement bitter)"
Caroline Is. / R. Kanehira 850 / 1950 / "Bark used for medicine"

4149   *Alibertia edulis* Honduras / C. & W. von Hagen 1312 / 1938 / "Fruit eaten" / "Sul-sul tree"
Brit. Honduras / P. H. Gentle 3565 / 1941 / " 'wild guava': 'sulsul chico' "
W.I.: Martinique / P. Duss 484 / 1880 / "cultivé pour ses fruits vous plusiers endroit"
Brazil / J. A. Ratter et al. 194 / 1967 / "Marmelada do Cerrado"
Brazil / D. Philcox et al. 3266 / 1967 / "Used when fruit ripe for jam-making" / "Marmelada"

4150   *A. curviflora* Brazil / B. A. Krukoff's 4th Exped. to Braz. Amaz. 4753 / 1933 / "Fruits are edible" / "Apurur"

4151   *A. elliptica* Brazil / J. A. Ratter et al. 2592 / 1972 / " 'MARMELADINHA' "

Brazil / M. Claussen 676 / 1838 / "Vulgo: 'Marmelada' "

52   *A. myrciifolia* S. AM. / R. Spruce 1092 / 1850 / "sw.-scented"

53   *A. sessilis* Brazil / G. Eiten 1642 / 1959 / "Frt . . . with flesh tasting of quince"
Paraguay / E. Hassler 627 / 1907 / "Segun datos es una fruta exquisita"
Brazil / G. C. Argent in Richards 6644 / 1968 / " 'MARMELADINHA' "

54   *A. verrucosa* Brazil / Harley & DeCastro 11299 / 1968 / "Pulp sweet pale brown. Fruit edible" / "Marmelada verdadeira"

55   *Alibertia* indet. Brazil / F. C. Hoehne 28307 / 1931 / " 'veludo' 'Marmelada Macho' "

56   *Alibertia* indet. Brazil / Killip & Smith 30032 / 1929 / "fruit edible"

57   *Alibertia* indet. Peru / J. Schunke V. 5364 / 1972 / "La yema terminal secreta poia resina pegajosa al frotar con la mano"

58   *Alseis yucatanensis* Mexico / J. Chavelas P. s.n. / 1965 / " 'Tabaquillo' "

59   *A. labitioides* Venezuela / C. Blanco 503 / 1965 / "Flores fragrantes" / "Carutillo"

60   *Amaioua corymbosa* Brit. Honduras / P. H. Gentle 2648 / 1938 / "bastard coffee"
Brit. Honduras / P. H. Gentle 3654 / 1941 / "wild coffee"
W.I.: Trinidad / D. Philcox et al 7374 / 1973 / "Fls. white, very fragrant"
Cuba / Leon & Fortun 8651 / 1919 / "Café cimarron"
Colombia / R. Jaramillo-Uejia 7109 / 1965 / "N.V. 'VARA DE SARNA': 'PALO SARNA' "

161   *A. guianensis* Surinam / C. E. B. Bremekamp LBB9358 / 1962 / "Nom. vern. 'manmarmadosae' "
Brazil / D. Philcox et al. 3576 / 1967 / "v.n. marmelada-de-mato"
Brazil / J. A. Ratter et al. 869 / 1968 / "Marmelada Braba"
Guyane Francaise / R. A. A. Oldeman B870 / 1967 / ". . . visitée par des fourmis"

162   *Arcytophyllum setosum* Ecuador / Cazalet & Pennington 5412 / 1961 / "sweet scented"

163   *Bathysa altiscandens* Brazil / A. Ducke 1126 / 1942 / "floribus alba odoratis"

164   *B. cuspidata* Brazil / Y. Mexia 4240 / 1930 / " 'Cabo da Colher' "

165   *Bikkia palauensis* Caroline Is.: Pelew Is. / R. Kanehira 222 / 1929 / "fragrant"

166   *Borojoa* indet. Costa Rica / P. H. Raven 21613 / 1967 / "apple-like"

167   *Borreria ocimoides* Dominican Republic / A. H. Liogier 10917 / 1968 / " 'JUANA LA BLANCA' "

Brazil / W. A. Archer 8416 / 1943 / "leaf infusion taken for kidney diseases"

4168　*B. corymbosa* Peru / S. G. E. Saunders 968 / 1964 / "Fls . . . strong disagreeable odor"

4169　*B. macrocephala* Venezuela / C. K. & B. Maguire 41664 / 1957 / "Voucher for Eli Lilly drug sample"

4170　*B. verticillata* Brazil / W. A. Archer 7677 / 1942 / "Plant has odor of pepper"

4171　*Borreria* indet. Tanganyika / R. E. S. Tanner 659 / 1952 / "aromatic"

4172　*Borreria* indet. Tanganyika / R. E. S. Tanner 579 / 1952 / "bitterly aromatic"

4173　*Bouvardia bouvardioides* Mexico / H. S. Gentry 5683 / 1940 / ". . . with repellant odor from broken stem"

4174　*Calycophyllum candidissimum* Panama / T. B. Croat 13007 / 1971 / "flowers with sweet aroma"

4175　*Canthium burttii* Tanganyika / R. E. S. Tanner 1366 / 1953 / "Fruits edible. Roots cooked, eaten in porridge to treat ankylostomiasis; bark burned, made into poultice for sores . . . aromatic" / "Kisukuma: Nkamu"

4176　*C. zanzibaricum* Tanganyika / R. E. S. Tanner 1552 / 1953 / "Roots pounded, chewed with water to treat ankylostomiasis . . . aromatic" / "Kisukuma: Ihoja"

4177　*Casasia clusiifolia* U.S.: Fla. / H. N. Moldenke 5654 / 1930 / " 'Seven year apple' "

4178　*Chimarrhis hookeri* Peru / R. Lao Magin 51 / 1963 / "cortera para soga" / " 'Papelillo' "

4179　*Chiococca phaenostemon* Panama / G. & P. White 38 / 1937 / "Odor resembling that of orange blossoms"

4180　*Chomelia protracta* Brit. Honduras / P. H. Gentle 2131 / 1937 / " 'Wild Coffee' "
Brit. Honduras / R. Howard et al. 541 / 1951 / "very aromatic"
Brit. Honduras / P. H. Gentle 3759 / 1941 / "pine ridge coffee"

4181　*C. spinosa* Costa Rica / H. E. Stork 4018 / 1932 / "Flowers very fragrant, fruit black, edible, used in remedy of fever" / " 'Malcahuite' "

4182　*Cinchona* indet. Colombia / H. García y Barriga 17293 / 1960 / "Flores . . . muy fragrantes" / " 'Quino' "

4183　*Cosmibuena grandiflora* Colombia / F. R. Fosberg 21407 / 1943 / "very fragrant with a heavy cloying sweet odor"

4184　*Coussarea paniculata* Venezuela / J. A. Steyermark 86639 / 1960 / "flowers very fragrant" / " 'Naranjillo' "

4185   *Craterispermum laurinum* Liberia / P. M. Daniel 81 / 1950 / "used for medicine"

4186   *Cremaspora triflora* Tanganyika / R. E. S. Tanner 3703 / 1957 / "strongly aromatic"

4187   *Crossopteryx febrifuga* Tanganyika / R. E. S. Tanner 3925 / 1958 / "Bondei: for syphilis and tuberculosis, roots boiled and water drunk . . . strongly aromatic" / "Bondei: 'Mkanyakya' "
Tanganyika / R. E. S. Tanner 3009 / 1956 / "Ngindo: Bark boiled and drunk as an abortifact, or to bring on menstruation"

4188   *Crusea longiflora* Mexico / Anderson & Lankowski 3918 / 1966 / "faintly fragrant, reminiscent of carnations"

4189   *Didymaea mexicana* Mexico / Balls & Gourlay B4479 / 1938 / "Sweet scented as 'woodruff' when drying"

4190   *Dioicodendron dioicum* Ecuador / J. A. Steyermark 54450 / 1943 / "Flowers very fragrant"

4191   *Duroia bolivarensis* Venezuela / B. Maguire et al. 53872 / 1962 / "fruit edible (?)"
Bolivia / J. A. Steyermark 75393 / 1953 / "fruit edible"

4192   *D. eriopila* Surinam / Lanjouw & Lindeman 827 / 1948 / " 'Marmeldoosje' "
Surinam / Wood Herb. 324 / 1945 / "fruits edible" / "KAMARAMARA"
Brit. Guiana / J. S. de la Cruz 1042 / 1921 / "fruit edible" / " 'Comaramara' "

4193   *D. hirsuta* Peru / Killip & Smith 29817 / 1929 / "ant-inhabited"

4194   *Elaeagia karstenii* Colombia / H. S. Kernan 139 / 1944 / " 'Quina' "

4195   *E. utilis* Colombia / F. R. Fosberg 21078 / 1943 / " 'barniz de Pasto' used to make a lacquer for wood"

4196   *Erithalis fruticosa* Bahamas / R. A. & E. S. Howard 10001 / 1948 / "pigeon berry"

4197   *Exostema caribaeum* Mexico / S. S. White 5072 / 1943 / "Infusion of bark used locally in place of quinine for malaria" / " 'Quina, quina blanca' "
Mexico / Hinton et al. 693 / 1934 / "medicinal" / " 'Copalche' "

4198   *Fadogia grandiflora* Tanganyika / R. E. S. Tanner 1369 / 1953 / "Sukuma: Roots pounded, used in porridge to treat cough; leaves pounded, used as poultice for sores" / "Kisukuma: Suppa' "

4199   *Faramea occidentalis* var. *occidentalis* Guatemala / J. A. Steyermark 37176 / 1940 / "Flowers very fragrant"

4200   *F. killipii* Venezuela / J. A. Steyermark 61133 / 1945 / "Birds eat fruit" / " 'cenicero' "

4201   *F. lourteigiana* Fr. Guiana / C. Sastre 1387 / 1972 / "racine tonique"

4202   *F. mayensis* Ecuador / Steere & Camp 8267 / 1944 / ". . . very sweetly fragrant"

4203   *Ferdinandusa speciosa* Brazil / J. G. Kuhlman 422 / 1930 / ". . . casca amarga" / " 'poqui' "

4204   *Galium bolandri* U.S.: Calif. / W. A. Dayton 428 / 1913 / "grazed by sheep" / "Bedstraw"

4205   *G. mollugo* U.S.: N.J. / H. N. Moldenke 1826 / 1931 / " 'Wild Madder' "

4206   *G. mexicanum* Mexico / G. B. Hinton 1509 / 1932 / "put under the bed to catch flees . . . very sticky" / "Yerba de la pulga"
Mexico / G. B. Hinton 13147 / 1938 / "Flor de Pulga"

4207   *Gardenia augusta* China: Kwangsi / W. T. Tsang 24686 / 1934 / "fruit yellow, edible" / "Wong Kwo Tze Shue"

4208   *G. tubifera* Malaysia / v. Balgooy 2603 / 1975 / "very fragrant"

4209   *G. ovularis* Australia: Queensland / S. F. Kajewski 1345 / "very strong sweet overpowering perfume"

4210   *G. jasminoides* Samoa / A. J. Eames 78 / 1921 / "fl . . . very fragrant, used in medicine and in scenting oil" / " 'pua' "

4211   *G. manii* Hawaiian Is. / H. St. John 11090 / 1931 / "very fragrant"

4212   *Gardenia* indet. New Guinea / D. Frodin 26419 / 1966 / "core supplies dye for clothes and artifacts. Dye very strong" / " 'Tzopi' "

4213   *Genipa americana* var. *americana* Panama / R. S. Williams 658 / 1908 / "used by indians to stain the face" / " 'Jagua' "
Colombia / W. A. Archer 2111 / 1931 / "Indians anoint various parts of body with grated fruit, soon skin turns deep blue-black. Color serves as decoration, also protection against sunburn . . . good indelible ink for manuscripts" / " 'Jagua' "

4214   *G. americana* var. *caruto* Brit. Honduras / P. H. Gentle 3367 / 1940 / " 'wild mammy apple' 'Ha wa' "

4215   *G. spruceana* Brit. Guiana / Maguire & Fanshawe 23369 / 1944 / ". . . edible"

4216   *Gonzalagunia rudis* Panama / M. Nee 7607 / 1973 / ". . . with faint wintergreen flavor"

4217   *G. cornifolia* Peru / C. M. Belshaw 3378 / 1937 / "with honeysuckle scent"

4218   *G. ovatifolia* Colombia / J. A. Duke 11306 / 1967 / "berries white (called 'mentolia') said to be edible"

4219   *Guettarda scabra* W.I.: Antigua / J. S. Beard 285 / 1944 / " 'Candlewood' 'Pigeon Foot' "

4220   *G. viburnioides* Surinam / R. S. Cowan 38998 / 1954 / "sweet-scented"

4221   ***G. speciosa*** Marshall Is. / F. R. Fosberg 33740 / 1951 / "very fragrant"
Marshall Is. / F. R. Fosberg 33667 / 1951 / "very fragrant" / " 'wut' "
Austral Is. / F. R. Fosberg 12120 / 1934 / "very sweet odor"
Caroline Is. / Fosberg & Evans 47093 / 1965 / "used medicinally and wood
for canoe paddles" / " 'wut' "

4222   ***Hamelia nodosa*** Honduras / C. & W. von Hagen 1332 / 1938 / "leaves
dried and used as tobacco" / " 'Pikwa kakma' "
Panama / Woodson, Allen & Seibert 1613 / 1938 / "leaves cooked and used
for mange on dogs"

4223   ***H. patens*** Peru / J. Schunke V. 1472 / 1967 / "Las hojas desinfectante en
infusion" / " 'Arco sacho' "
Peru / J. Schunke V. 2528 / 1968 / "Las hojas utilizan en infusion para
lavar heridas infectadas" / " 'Chirapasacha' "

4224   ***Hamelia*** indet. / Peru / J. Schunke V. 1107 / 1966 / "Leaves medicinal;
utilizan en infusion para lavar ulceras" / " 'Archo sacha' "

4225   ***Hamelia*** indet. Bolivia / R. F. Steinbach 294 / 1966 / "Usado como col-
orante el fruto maduro"

4226   ***Hamiltonia suaveolens*** India / W. Koelz 10299 / 1936 / "very fragrant"

4227   ***Hedyotis auricularia*** var. ***melanesica*** Fiji / A. C. Smith 7054 / 1953 /
"leaves used in preparing a medicine for eye-troubles" / " 'Kauvoro na
langi' "

4228   ***H. biflora*** Caroline Is. / D. Anderson 904 / 1949 / "Pounded and applied
as poultice for rash"

4229   ***H. rapensis*** Rapa / F. R. Fosberg 11631 / 1934 / "sweet odor"

4230   ***Heinsia pulchella*** AFR. / T. E. Edwardson 114 / 1937 / "Bark peeled,
dried, ground used by women for rubbing on skin"

4231   ***Hydnophytum formicarium*** Sarawak / J. W. Purseglove P.4892 / 1956 /
"sphaerical swollen chambered base . . . inhabited by ants"

4232   ***H. grandiflorum*** Fiji / A. C. Smith 7605 / 1953 / "tuber large, with ant-in-
habited canals"
Fiji / A. C. Smith 7144 / 1953 / ". . . with ant-inhabited canals" /
"Ndatokaikai"

4233   ***H. tenuiflorum*** Fiji Is. / A. C. Smith 7425 / 1953 / "tuber large, with ant-
inhabited canals"

4234   ***Hydnophytum*** indet. Sarawak / J. & M. S. Clemens 20394 / 1929 / "sym-
biotic with ants"

4235   ***Hymenodictyon*** indet. Tanganyika / R. E. S. Tanner 1197 / 1953 /
"Sukuma: Used to treat headaches or other pains — preparation rubbed into
small cuts in temples" / "Kisukuma: Juguji"

4236   ***Isertia hypoleuca*** Panama / N. Bristan 1056(6) / 1967 / "very aromatic"

4237   *Ixora ferrea* W.I.: St. Lucia / J. S. Beard 522 / 1945 / " 'Cafe grand bois' "

4238   *I. carolinensis* var. *typica* Caroline Is. / R. Kanehira 636 / 1929 / "Wood very hard and strong, used for suspender of canoe; young leaves eaten by wild deers"

4239   *I. casei* Yap Is. / F. R. Fosberg 25539 / 1946 / "flowers 'gacheu' used for garlands put around head when dancing; wood for scoop nets for fish"

4240   *I. philippinensis* Philippine Is. / J. C. Ondrada 31352 / 1931 / "fruit is edible"

4241   *Ixora* indet. Philippine Is. / A. L. Zwickey 189 / 1935 / " 'Bialui a kakua' or possibly 'Bialoikakua', the name meaning pseudo-coffee"

4242   *Joosia dichotoma* Peru / J. Schunke V. 1100 / 1966 / "Bark antipaludico, drinking in brandy"

4243   *Lachnosiphonium obovatum* Tanganyika / R. E. S. Tanner 3028 / 1956 / "Zigua: For leprosy in early stage, leaves and roots boiled and sufferer bathed in the water" / "Zigua: Mhangala ndasha"

4244   *Ladenbergia amazonensis* Brazil / R. de Lemos Fróes 20871 / 1945 / " 'Quinarana' "
Brazil / R. de Lemos Fróes 20946 / 1945 / " 'Quina' "

4245   *L. lambertiana* Brit. Guiana / C. L. & S. S. Tillett 45276 / 1960 / "strong sweet fragrance"

4246   *L. macrocarpa* Ecuador / Fosberg & Prieto 22730 / 1945 / "very fragrant"

4247   *L. moritziana* Venezuela / J. A. Steyermark 91525 / 1963 / "flowers with scent of orange (Citris) blossoms"

4248   *L. undata* Venezuela / J. A. Steyermark 56107 / 1944 / "flowers with similar odor to that of *Cinchona, Narcissis poeticus,* and *Polianthes tuberosa*"

4249   *Ladenbergia* indet. Peru / J. Schunke V. 1090 / 1966 / "The bark utilizan las mujeres para lavado vaginal"

4250   *Lamprothamnus zanguebaricus* Tanganyika / R. E. S. Tanner 3987 / 1958 / "Bondei: for stomache ache, roots boiled and drunk" / "Bondei: Mtemakanya"
Tanganyika / R. E. S. Tanner 3773 / 1957 / "Zigua: For syphillis, external application as well as drinking water in which roots boiled" / "Zigua: Mf yonzi"
Tanganyika / R. E. S. Tanner 2294 / 1955 / "Used in protective medicines against witchcraft . . . strongly aromatic" / "Kibondei: Kiviza"

4251   *Leptactinia benguelensis* Rhodesia / W. R. Bainbridge 415 / 1959 / "Fruit edible . . . with a very strong sweet scent"
Tanganyika / R. E. S. Tanner 1564 / 1955 / "Sikuma tribe: Roots pounded, used in water as drink to treat Bilharzia; leaves pounded, used as poultice for sores . . . aromatic" / "Kisukuma: Ngobole ngoshai"

4252 *Luculia pinceana* India / F. Kingdon Ward 18068 / 1948 / "very fragrant"

4253 *Machaonia martinicensis* Colombia / J. A. Duke 9718 / 1967 / ". . . aromatic, very attractive to ants and bees"

4254 *Maguireothamnus speciosus* Venezuela / J. A. Steyermark 74881 / 1953 / "very fragrant"

4255 *Malanea macrophylla* Brazil / R. Fróes 11737 / 1939 / " 'Fruta de periquito' "

4256 *Mitracarpum verticillatum* Tanganyika / R. E. S. Tanner 1235 / 1953 / "Sikuma tribe: Roots eaten, cooked for food; also made into fine flour and used directly to treat gonorrhoea" / "Kisukuma: Itama lya bagikulu"
Tanganyika / R. E. S. Tanner 655 / 1952 / "Used to cure erysipelas" / "Kisukuma: Itama lya gikuru"

4257 *Morinda citrifolia* New Guinea / G. Leach 34202 / 1972 / "very strong 'cheese' smell. Fruit edible" / " 'PANGI' "
Fiji / O. Degener 14974 / 1941 / "Ripe fruit edible; juice from root mixed with earth and applied to hair as black dye" / " 'Kura' "
New Hebrides / S. F. Kajewski 260 / 1928 / "sap used for red dye" / " 'Noah-i-rat' "
Caroline Is. / R. J. Alvis 88 / 1957 / "leaves and bark reported of medicinal value, yellow dye derived from bark; edible fruit" / " 'Maalueg' "
Fiji / A. C. Smith 6607 / 1947 / "fruit edible" / " 'Kura' "
Marshall Is. / F. R. Fosberg 24335 / 1946 / "with a powerful unpleasant odor when decomposing"
Marshall Is. / F. R. Fosberg 26687 / 1946 / "leaves used medicinally . . . roots for blue dye" / " 'nin' "
Caroline Is. / M. Evans 443 / 1965 / "used for medicine" / " 'LEL' "
Guam / F. R. Fosberg 46257 / 1965 / "young budding heads chewed with betel" / " 'lado' "
Guam / M. Evans 729 / 1965 / "Fruits and sometimes leaves boiled and drunk to relieve body pains" / " 'LADA' "

4258 *M. lanata* Philippine Is. / R. S. Williams 1558 / 1904 / "cranberry fragrance"

4259 *Mussaenda oreadum* Brit. New Guinea / L. J. Brass 5040 / 1933 / " 'honeysuckle scented' "

4260 *M. raiatensis* Fiji / A. C. Smith 1030 / 1935 / "leaves and bark used as medicine" / "Mbovo"
Fiji / A. C. Smith 9344 / 1953 / "infusion of leaves considered an 'all purpose' medicine" / "Mbombo"

4261 *Myrmeconauclea strigosa* Borneo / J. and M. S. Clemens 21756 / 1929 / "inhabited by ants"

4262   *Neonauclea* indet. Borneo / J. & M. S. Clemens 21432 / 1929 / "branchlets hollow and occupied by ants"

4263   *Oxyceros esculenta* Indo-China / R. W. Squires 906 / 1932 / ". . . with strong pleasant odor"

4264   *Paederia foetens* N. Rhodesia / A. Angus 2911 / 1961 / " 'Stinks!' "

4265   *P. scandens* Hainan / C. I. Lei 657 / 1933 / "ill-smelling"

4266   *Palicourea crocea* Venezuela / Wurdack & Adderley 42681 / 1959 / "Voucher for Eli Lilly Drug Sample"

4267   *P. nitidella* Brazil / Wurdack & Adderley 42680 / 1959 / "Voucher for Eli Lilly Drug Sample"

4268   *P. platypodina* Brazil / Prance & Silva 58238 / 1964 / "Very poisonous. Child in village died yesterday from eating one leaf"

4269   *P. rigida* Brazil / Prance & da Silva 58226 / 1964 / "leaves used locally as a cure for diarrhoea and for liver ailments"
Venezuela / J. A. Steyermark 7534 / 1953 / "boiled plant used for diarrhea"

4270   *Palicourea* indet. Argentina / T. M. Pedersen 4443 / 1957 / "(Strawberry-ice')"

4271   *Palicourea* indet. Colombia / Schultes & Cabrera 16286 / 1952 / "Makuna = *wa-kö-weý* ('tapir's paint'): Yukuna = see-yĕ-tĕ-lo-ro ('color of fire')"

4272   *Pavetta harbouii* Africa / L. E. Codd 8662 / 1954 / "Proved poisonous to stock"

4273   *P. terminataria* var. *glabra* Tanganyika / R. E. S. Tanner 1276 / 1953 / "Sukuma: Pounded leaves and roots used to bathe smallpox patients; also used to tan skins . . . aromatic" / "Kisukuma: Koganyaga"

4274   *Pentagonia donnell-smithii* Costa Rica / R. W. Lent 868 / 1967 / "Vile smelling red liquid"

4275   *Pentas bussei* Tanganyika / R. E. S. Tanner 3205 / 1956 / "Sambaa: For diarrhoea with blood, leaves pounded up and mixed with water and drunk" / "Sambaa: Mnyampome"

4276   *P. purpurea* Tanganyika / R. E. S. Tanner 3889 / 1957 / "Zigua: for gonorrhoea, leaves boiled and the water drunk" / "Zigua: Mwangare"

4277   *Perama spathacea* Venezuela / C. W. Wood 265 / 1975 / "very fragrant"

4278   *Polysphaeria dischistocalyx* Tanganyika / R. E. S. Tanner 3942 / 1958 / "Sambaa: for swellings, roots pounded up and used as paste" / "Sambaa: mshungunlungu"

4279   *P. parvifolia* Tanganyika / R. E. S. Tanner 3453 / 1957 / "Zigua: As a male aphrodisiac, roots boiled and the water drunk" / "Zigua: Mkalakala"
Tanganyika / R. E. S. Tanner 3898 / 1957 / "Zigua: for pain in left side of stomache, roots boiled and the water drunk" / "Zigua: Mchakachaka"

Tanganyika / R. E. S. Tanner 3008 / 1956 / "Ngindo: leaves pounded up and put on boils" / "Ngindo: Mambala"

4280 *Posoqueria panamensis* Venezuela / Steyermark & Rabe 96535 / 1966 / " 'café montañera' "

4281 *Posoqueria* indet. Fr. Guiana / Fr. Guiana For. Service 7167 / 1956 / "pulpe jaunâtre interieure contenant des graines comestibles"

4282 *Psychotria capitata* Panama / T. B. Croat 14959 / 1971 / "visited by small greenish black hummingbirds"
Panama / T. B. Croat 14932 / 1971 / "being pollinated by black and white butterfly"

4283 *P. chiapensis* Brit. Honduras / P. H. Gentile 3711 / 1941 / " 'wild lime' "

4284 *P. microdon* Panama / Croat & Porter 15523 / 1973 / "very sweetly aromatic"

4285 *P. muscosa* W.I.: Martinique / P. Duss 134 / 1878–79 / "Bois des Deun chohn"

4286 *P. nervosa* Bahama Is. / S. R. Hill 2248 / 1974 / " 'wild coffee' "

4287 *P. pleeana* W.I.: St. Lucia / J. S. Beard 513 / 1945 / " 'Bois mal estomac' "

4288 *P. amita* Venezuela / J. A. Steyermark 55298 / 1944 / "Used for dyeing articles, also boiled for drawing out Chimo, a tobacco juice; when mixed with the tobacco it helps draw out the juice"

4289 *P. aubletiana* Venezuela / J. A. Steyermark 55256 / 1944 / "Used to give a color to Chimo, a tobacco juice made here" / " 'cafecito' "

4290 *P. bahiensis* Surinam / Lanjouw & Lindeman 1241 / 1948 / "boffroe cassava"

4291 *P. barbiflora* Brazil / L. B. Smith 15089 / 1965 / "kills cattle"

4292 *P. carthaginensis* Venezuela / J. A. Steyermark et al. 02016 / 1968 / " 'martiño' "

4293 *P. erecta* Brazil / R. de Lemos Fróes 20835 / 1945 / " 'Cafe brabo' "

4294 *P. hoffmannseggiana* Venezuela / Wurdack & Adderley 42665 / 1959 / "Voucher for Eli Lilly Drug Sample"

4295 *P. ipecacuana* Venezuela / J. de Bruijn 1091 / 1966 / "Roots used by chemists against amoebas and vomits"
Brazil / H. C. Cutler 12403 / 1944 / "For teething and as a depurative or tonic, in tea. Also can be placed on pacifier or teething ring when prepared in form of syrup for coughing and colds"

4296 *P. poeppigiana* Venezuela / Wurdack & Adderley 42922 / 1959 / "Voucher for Eli Lilly Drug Sample"
Peru / G. Klug 2227 / 1931 / "remedy for burns" / "Huitoto: Usiya-puiño"

4297 *P. prunifolia* Brazil / J. A. Ratter et al. 1140 / 1968 / "Poisonous to cattle"

4298   *P. pseudoaxillaris* Colombia / F. R. Fosberg 21206 / 1943 / "(said to be called 'capitano' . . . and to be used for fever)"

4299   *P. tepuiensis* Venezuela / J. A. Steyermark, G. C. K., & E. Dunsterville 92909 / 1964 / "Tapir (danta) eat leaves of this plant" / " 'sere-curyek' "

4300   *P. viridis* Brazil / G. T. Prance et al. 7302 / 1968 / "leaves used as one ingredient of local hallucinogen"
Brazil / G. T. Prance et al. 7302 / 1968 / "sterile voucher for leaves of local halluceogenic brew, used mixed with bark of a liana"

4301   *Psychotria* indet. Peru / J. Schunke V. 1104 / 1966 / "Bark bitter"

4302   *P. neurodictyon* Nigeria / P. W. Richards 3300 / 1935 / "Used as a medicine for sores on the feet by natives" / " 'Alaifo' "

4303   *Psychotria* indet. Tanganyika / R. E. S. Tanner 3182 / 1956 / "Zigua: for fever in children, leaves soaked in water and used for bathing the child" / "Zigua: Mynota"

4304   *Psychotria* indet. Tanganyika / R. E. S. Tanner 1480 / 1953 / "Sukuma tribe: Roots pounded, used as poultice to treat sores; roots boiled, mixed with white mtama flour, taken with water for syphilis" / "Kisukuma: Ngawanugwa"

4305   *P. olivacea* Melanesia / J. H. L. Waterhouse 64 / 1933 / "Vine, used in fertility cult"

4306   *Randia aculeata* El Salvador / Allen & Armour 6851 / 1958 / "intensely fragrant"

4307   *R. armata* Brit. Honduras / P. H. Gentle 2674 / 1939 / " 'wild lime' "

4308   *R. echinocarpa* Mexico / G. B. Hinton 4271 / 1933 / "fruit edible"

4309   *R. formosa* Panama / N. Bristan 1025 / 1967 / "edible"
Guyana / Omawale & Persaud 51 / 1970 / "Taste like molasses" / "Bird shit"

4310   *R. coriacea* Port. Congo / J. Gossweiler 7714 / 1919 / "tree . . . frequented by the tree ant; fragrant" / "Quinomba"

4311   *R. spinosa* Hainan / C. I. Lei 199 / 1932 / "fruit . . . edible"

4312   *Relbunium microphyllum* Mexico / L. R. Stanford et al. 163 / 1941 / "heavily grazed by goats"

4313   *R. longifolia* Venezuela / Wurdack & Adderley 42658 / 1959 / "Voucher for Eli Lilly Drug Sample"

4314   *Retiniphyllum pilosum* Venezuela / Maguire, Wurdack & Keith 41742 / 1957 / "Voucher for Eli Lilly Drug Sample"

4315   *R. truncatum* Venezuela / Maguire, Wurdack & Keith 41743 / 1957 / "Voucher for Eli Lilly Drug Sample"

316 *Rondeletia stereocarpa* W.I.: St. Lucia / J. S. Beard 185 / 1943 / " 'Bois d'amand' "

317 *Rothmannia engleriana* Tanganyika / R. E. S. Tanner 1573 / 1954 / "Sukuma tribe: Roots pounded, taken in water for stomach pains" / "Kisukuma: Kabelele"

318 *Rothmannia* indet. Sarawak / J. W. Purseglove P5466 / 1956 / "Sap from fruits applied to sore eyes by Dyaks"

319 *Rudgea citrifolia* W.I.: St. Lucia / J. S. Beard 514 / 1945 / " 'Café marroni' "

320 *R. hostmanniana* Venezuela / Aristeguieta, Liogier & Guevara 7257 / 1969 / " 'Casco de mula' "

321 *R. jacobinensis* Brazil / Y. Mexia 5561 / 1931 / "yellow fruit eaten by birds" / " 'Anjý-gerý' "

322 *R. racemosa* Peru / J. Schunke V. 5140 / 1971 / " 'Sacha Café' "

323 *Sabicea colombiana* Venezuela / Steyermark & Raba 96052 / 1966 / "fruit eaten by birds" / " 'azucarita' "

324 *Sabicea* indet. Colombia / Schultes & Cabrera 15586 / 1952 / "sweet fruits edible"

325 *Sarcocephalus nervosus* Nigeria / P. W. Richards 3407 / 1935 / "Nodes inhabited by ants" / "Opepe Ira"

326 *Sickingia maxonii* Panama / R. Foster 2345 / 1971 / "fls . . . heavily visited by hummingbirds"

327 *Simira erythroxylon* var. *meridensis* Venezuela / Steyermark et al. 103595 / 1970 / "wood very hard . . . used in making wooden spoons" / " 'cucharo' "

328 *S. rubescens* Brazil / B. A. Krukoff 1544 / 1931 / "fls . . . with strong and pleasant odor of vanilla" / "Angustura"

329 *Sipanea pratensis* Venezuela / Wurdack & Adderley 42916 / 1959 / "Voucher for Eli Lilly Drug Sample"

330 *Spermacoce natalensis* Swaziland / R. M. Hornsby 2828 / 1948 / "Roots used by natives as stomach medicine"

331 *S. pilosa* Liberia / S. B. Freeman 1 / 1950 / "Used for treating sores" / "Button grass"

332 *Squamellaria wilsonii* Fiji Is. / A. C. Smith 8191 / 1953 / ". . . with ant-inhabited canals"

333 *Tarenna* indet. Tanganyika / R. E. S. Tanner 2688 / 1956 / "Zigua: For gonorrhoea; root boiled and drunk" / "Kizigua: Mdavendave"

334 *Tocoyena pittieri* Panama / Kirkbride & Duke 1386 / 1968 / "seeds embedded in a brownish-black pulp with somewhat the odor and taste of molasses"

4335   *Tricalysia bagshawei* Tanganyika / R. E. S. Tanner 1300 / 1953 / "Sukuma tribe: Roots used to treat sores, stomach pain" / "Kisukuma: Kaguha"

4336   *T. jasminiflora* S. Africa / L. E. Codd 8688 / 1954 / "Fruits black providing the blue pigment used in tatooing tribal markings on women's faces"

4337   *T. nyassae* Tanganyika / R. E. S. Tanner 1558 / 1953 / "Sukuma tribe: Roots pounded, used as a drink to treat diarrhoea" / "Kisukuma: Ngeke" Tanganyika / R. E. S. Tanner 1570 / 1953 / "Sukuma tribe: Roots cooked in porridge to treat Bilharzia, used as poultice on sores" / "Kisukuma: Luhanya"

4338   *T. ovatifolia* var. *acutifolia* Tanganyika / R. E. S. Tanner 2687 / 1956 / "Zigua: For headache, leaves pounded up, and used as pultice" / "Kizigua: Lukiluki"

4339   *T. dalzelli* Ceylon / Tirvengadum, Cramer & Waas 481 / 1974 / " 'Attacked by big black ants' " Ceylon / Tirvengadum, Cramer & Waas 480 / 1974 / "whole tree frequented by ants"

4340   *Uncaria guianensis* Venezuela / Wurdack & Adderley 42656 / 1959 / "Voucher for Eli Lilly Drug Sample"

4341   *Vangueria tomentosa* Tanganyika / R. E. S. Tanner 1027 / 1952 / "Sukuma tribe: leaves cooked and used as poultice to treat sores on lips; roots used to make fermented drink to treat gonorrhoea" / "Kisukuma: Ngubaru"

4342   *Vangueria* indet. Tanganyika / R. E. S. Tanner 1619 / 1953 / "Sukuma tribe: Roots pounded, taken internally with water for diarrhoea; leaves boiled, water used as mouth rinse for swollen gums" / "Kisukuma: Ngubalu"

4343   *Warscewiczia cordata* Peru / Y. Mexia 6080 / 1931 / "Indians extract juice from bracts to dye skin" / " 'Shambosisa' "

4344   *Xeromphis nilotica* Tanganyika / R. E. S. Tanner 3762 / 1957 / "Bondei: Male aphrodisiac; fruits chewed" / "Bondei: Mdauwa" Tanganyika / R. E. S. Tanner 3876 / 1957 / "Zigua: for syphilis, roots boiled with maize flour and drunk as porridge" / "Zigua: Mdasha"

CAPRIFOLIACEAE

4345   *Diervilla lonicera* Canada: Ontario / M. N. Zinck 5633 / 1939 / "blueberry"

4346   *Lonicera conjugalis* U.S.: Calif. / Univ. of Calif. s.n. / 1893 / "Medicine"

4347   *L. confusa* Hainan / F. A. McClure 8937 / 1932 / "drug plant, water in which leaves and flowers have been steeped is good for itchy rash; tea made from the same and honey good for boils" / "Kam ngan fa"

348   *L. orientalis* W. Himalaya / W. Koelz 3116 / 1931 / "fruit blackish, edible"

349   *Sambucus orbiculatus* U.S.: Mich. / Torrey Coll. s.n. / no year / "Indian Currant"

350   *S. intermedia* W.I.: Trinidad / M. Fleming 60 / 1960 / "Flowers brewed as a tea for colds"

351   *S. mexicana* var. *bipinnata* Colombia / F. R. Fosberg 19104 / 1942 / "infusion of fresh fls. used for fever or pneumonia in infants and for liver trouble" / " 'Sauco' "
Colombia / F. R. Fosberg 19999 / 1943 / "Bark used as purgative" / " 'sauco' "
Ecuador / W. H. Camp E5026 / 1945 / "An infusion of flowers used for chest cholds" / " 'Sauco blanco' "

352   *S. peruvianus* Colombia / J. M. Duque Jaramillo 4589 / 1947 / "Cultivado como medicinal"

353   *Viburnum fordiae* China: Kwangtung / Fung Hom A628 / 1931 / "eatable red fruit" / "Oo Tsai Shue"

354   *V. luzonicum* China: Kwangtung / W. T. Tsang 21588 / 1932 / "Fr. red, edible" / "Fo Chai Tsz Shue"
China: Kwangtung / W. T. Tsang 21383 / 1932 / "fruit yellow, edible" / "Sai Ip Fo Chai Tsz Shue"
China: Kwangtung / W. T. Tsang 21084 / 1932 / "fruit red, edible" / "Fo Chai Tse"
China: Kwangtung / W. T. Tsang 20789 / 1932 / "fruit red, edible" / "Fo Tsai Sze"

355   *V. mullaha* China: Kwangsi / W. T. Tsang 24127 / 1934 / "fr. red, edible" / "Fo Chai Tze Shue"
China: Kwangsi / W. T. Tsang 24156 / 1934 / "fr. edible" / "Tai Yip Fo Chai Tze Shue"

356   *V. odoratissimum* China: Kwangtung / W. T. Tsang 21692 / 1932 / "Fr. reddish purple and black; edible" / "Tse Woh Shue"

357   *V. sempervirens* China: Kwangsi / W. T. Tsang 24263 / 1934 / "fr. red, edible" / "Cheung Ip Fo Chai Tsz Shue"
China: Kwangtung / W. T. Tsang 21630 / 1932 / "fruit red and black, edible" / "Shan Taai Fo Chai Tsz Shue"

358   *Viburnum* indet. China: Kiangsi / T. Tsiang 10037 / 1932 / "frt. bright red, edible when black"

359   *Viburnum* indet. TROP. ASIA: Khaiyang / F. Kingdon Ward 17509 / 1948 / "Fruits translucent, deep red, quite pleasant to eat"

ADOXACEAE

4360   *Adoxa moschatellina* Switzerland / Columbia Univ. Herb. s.n. / no year / "Deutsch: Bilsam Kraut, Walmuster. English: musk crowfoot. Franz.: Moschatellino, herbe musquée"

VALERIANACEAE

4361   *Patrinia villosa* China: Kwangtung / W. T. Tsang 21654 / 1932 / "fr. edible" / "Fu Chai T'so"
China: Kwangtung / W. T. Tsang 20814 / 1932 / "Fl. white, edible" / "Tin Tong Tsze-Tin"

4362   *Valeriana occidentalis* U.S.: Nev. / Nichols & Lund 96 / 1937 / "dry plant and root smell like snuff"

4363   *V. pauciflora* U.S.: Ind. / Schweinitz s.n. / 1831 / "Supposed by the inhabitants of Indiana to be the potato in the wild state"

4364   *V. microphylla* Ecuador / W. H. Camp 2120 / 1945 / "The root is used as a heart stimulant and in nervousness" / " 'Valeriana' "
Ecuador / W. H. Camp 1772 / 1945 / "Plants pungent-aromatic. A commonly used medicinal plant of the region"

4365   *V. spicata* Venezuela / L. Aristeguieta 2609 / 1956 / "Es planta medicinal" / "Lunaria"

4366   *Valeriana* indet. Ecuador / W. H. Camp 3997 / 1945 / "The root . . . for a 'heart remedy' or 'for nervousness' " / " 'Valeriana' "

4367   *Valeriana* indet. Colombia / F. R. Fosberg 22356 / 1944 / "odor unpleasant, like sweaty feet"

4368   *Valerianella locusta* U.S.: Penn. / H. N. Moldenke 2373 / 1925 / " 'Cornsalad' "

4369   *V. stenocarpa* U.S.: Tex. / V. L. Cory 51524 / 1946 / " 'Cornsalad' "

DIPSACACEAE

4370   *Cephalaria syriaca* Israel / I. Amdursky 187 / 1931 / "It has poisonous properties, and a small quantity of it is sufficient to spoil the taste of flour. Sometimes cultivated for its oil"

CUCURBITACEAE

4371   *Anguria vogliana* Venezuela / J. A. Steyermark 61005 / 1945 / "witch doctors crush leaves and use for their curing" / " 'bejuco de patilla de agua' "

4372   *Benincasa hispida* Hainan / F. A. McClure 20147 / 1932 / "Fruits used as a vegetable" / "Tung Kwa"

373   *Cayaponia attenuata* Mexico / Y. Mexia 975 / 1926 / "Berry black when ripe. Used for washing clothes" / "Mata raton"
El Salvador / P. C. Standley 21704 / 1922 / "Used for washing clothes" / " 'Retámara' "
El Salvador / P. C. Standley 22551 / 1922 / "Remedy for bits of tamagaz" / " 'Herba del tamagaz' "

374   *Ceratosanthes hilariana* Brazil / W. A. Archer 4035 / 1936 / "frt. small, edible"

375   *Coccinia grandis* Tanganyika / R. E. S. Tanner 2691 / 1956 / "Zigua: For delayed childbirth, leaves boiled and drunk" / "Kisukuma: lukewja"

376   *Corallocarpus* indet. Tanganyika / R. E. S. Tanner 1402 / 1953 / "Fruits edible. Sukuma tribe: leaves pounded and mixed with butter as poultice to treat boils" / "Kisukuma: Ibombolwa"

377   *Cyclanthera brachybotrys* Peru / Hutchinson & Wright 4308 / 1964 / "Fruit eaten in salad" / " 'Caigua' "

378   *Gymnopetalum penicaudii* Hainan / S. K. Lau 359 / 1932 / "fr. red. Edible" / "Po Kwa"

379   *Luffa acutangula* El Salvador / P. C. Standley 22424 / 1932 / "Fr. used like sponge" / " 'Paiste, Paxte' "
El Salvador / S. Calderon 1192 / 1922 / " 'Paste' "
W. Himalayas / W. Koelz 3052 / 1931 / "fruit used to induce vomiting"

380   *L. cylindrica* Mexico / G. B. Hinton et al. 5710 / 1934 / "sponge for washing"
Hainan / S. K. Lau 1088 / 1933 / "fruit edible" / "Shau Kai"

381   *Melothria cucumis* Argentina / W. A. Archer 4621 / 1936 / "frt . . . edible raw or as pickles" / " 'Popino' "

382   *M. punctata* Tanganyika / R. E. S. Tanner 1370 / 1953 / "Sukuma tribe: leaves used green, heated to produce steam to treat fever" / "Kisukuma: Kabindillji"

383   *Microsechium helleri* Mexico / G. B. Hinton 1572 / 32 / "The tuberous root is used and sold for washing, especially blankets" / "Sanacoche"

384   *Momordica charantia* Bahama Is. / S. R. Hill C125 / 1973 / "Brought from Nassau for the purpose of combating the flu"
Brazil / Y. Mexia 4543 / 1930 / "leaves used for stomach-ache" / " 'Sao Caetanu' "

385   *Peponium kilimandscharicum* Tanganyika / R. E. S. Tanner 1180 / 1953 / "Fruits edible when fully ripe"

386   *Cucurbitacea* indet. Bolivia / W. M. A. Brooke 5567 / 1949 / "used for making mattresses and from the fruit a dye is obtained"

CAMPANULACEAE

4387   *Centropogon cornutus* Brazil / Y. Mexia 5109 / 1930 / " 'Bico de papagago' "

4388   *Isotoma longiflora* Brit. Honduras / W. A. Schipp 173 / 1929 / " 'Poisonous' "
Peru / Y. Mexia 6385 / 1932 / "milky juice said to irritate skin and to produce blindness" / " 'Flor del sapo' "
Caroline Is. / C. A. Salsedo 273 / 1969 / "Milky sap present in stem and roots which is poisonous causing blindness"

4389   *Lightfootia tanneri* Tanganyika / R. E. S. Tanner 965 / 1952 / "Sukuma tribe: roots cooked, liquid and drunk for dry cough" / "Kisukuma: Kakehe gosha"

4390   *Lobelia decurrens* Peru / W. J. Eyerdam 10755 / 1958 / "exudes a nauseous odor which impedes breathing"

4391   *Siphocampylus columnae* Colombia / J. M. Duque Jaramillo 3463 / 1946 / "usado como antisifilítico" / " 'Gallito-cascabel' "

PENTAPHRAGMATACEAE

4392   *Pentaphragma spicatum* Hainan / F. A. McClure 9337 / 1921 / "leaves sometimes eaten by Lois as vegetable"

GOODENIACEAE

4393   *Scaevola frutescens* Fiji Is. / O. Degener 15109 / 1941 / "root used medicinally" / " 'Vevendu' (Serua)"

4394   *Velleia paradoxa* Australia: Queensland / S. L. Everist 2890 / 1947 / "Reputed to cause urine of sheep to become red" / " 'Pestnebed' "

STYLIDIACEAE

4395   *Stylidium plantanifolia* Bolivia / O. E. White 897 / 1921 / "bark used for writing on, 'writing lasts as long as ston', bark is very durable"

ASTERACEAE

4396   *Achillea fragrantissima* Egypt / M. Drar 1393 / 1933 / "A native medicinal plant" / "Quysum"

4397   *Achyrocline satureioides* Paraguay / W. A. Archer 4737 / 1936 / "Used as intestinal disinfectant, especially for appendicitis. Even prescribed by local doctors" / " 'Marcela' "

398 *Adenostemma lavenia* Peru / D. McCarroll 95 / 1940 / "Used for cure for eye troubles" / " 'Flor de Seda' "

399 *Ageratum echoides* Guatemala / J. A. Steyermark 5150 / 1942 / "Used in treating malaria; boil plant and drink infusion for the treatment of malaria; also take baths in the infusion for malaria" / " 'frijolito' "

400 *Ambrosia deltoidea* U.S.: Ariz. / F. J. Lipp s.n. / 1975 / "Used by Papagos to cure rheumatism & sores of the legs. Entire plant is placed in hot water & utilized in a sweat bath for affected legs. Body should be well covered" / " 'tatcakia' or 'Donkey's hay' "

401 *A. hispida* Bahama Is. / W. Cerbin 106 / 1973 / "The dried leaves are steeped as a tea, which is drunk for ailments of the respiratory system. Some say it acts as a stimulent to sluggish appetites" / " 'Baysurina, Bay-jahrina, or Bayjoreena' "

402 *A. arborescens* Ecuador / W. H. Camp 2489 / 1945 / "Used as a broom to get rid of fleas; also it seems to be the only thing used to sweep out ovens after heating and before the bread is put inside" / " 'Altamiso' "
Ecuador / W. H. Camp 5030 / 1945 / "Used as a local application in skin infections . . ." / " 'Altamisa' "

403 *Archibaccharis androgyna* Mexico / J. D. Jackson 1034 / 1968 / "odor of dill strong from freshly crushed leaves"

404 *Artemisia herba-alba* Egypt / M. Drar 1402 / 1930 / "Decoction employed by natives for colic" / "Sheeh"

405 *A. judaica* Egypt / M. Drar 1403 / 1932 / "Employed for colic" / "Judean Wormwood"

406 *A. vulgaris* Hawaii Is. / O. Degener 9694 / 1933 / "Said to be used as tea"

407 *Baccharis articulata* Brazil / W. A. Archer 4138 / no year / "Formerly used in brews for beer. Also used as rheumatism remedy in decoctions" / " 'Czr-quija' "

408 *B. melastomaefolia* Brazil / G. M. Barroso 67439 / 1949 / "Planta usada para secar heridas" / " 'Herva Santa' "

409 *B. polyantha* Ecuador / W. H. Camp 2301 / 1945 / " 'The roots ground and mixed with water to form a shampoo to get rid of dandruff' " / " 'Chillca negra' "

410 *B. prunifolia* Venezuela / J. A. Steyermark 55486 / 1944 / "Use leaves for headaches" / " 'frilejoncito' "

411 *B. venosa* Bolivia / F. R. Fosberg 28470 / 1947 / ". . . Said to be used as a remedy for fevers"

412 *Bidens andicola* Equador / W. H. Camp 2459 / 1945 / "—an infusion of the flowers used for nervousness and mild spasms" / "Ñachay"

413 *B. pilosa* Peru / G. Klug 115 / 1930 / "Cure for pains" / " 'Pacunga' "

4414 **B. squarrosa** Venezuela / J. A. Steyermark 61255 / 1945 / "Cook root for fevers" / " 'pega-pega' "

4415 **B. menziesii** Hawaiian Is. / A. Greenwell 21395 / 1957 / "Considered best tea by old time Kona people" / "kokoolau lauliilii"

4416 **Blepharispermum** indet. Tanganyika / R. E. S. Tanner 3672 / 1957 / "For gonorrhoea, leaves boiled and the water drunk; for stomach pain, leaves pounded up and rubbed into body" / "Zigua: Mlenga"

4417 **Calea divaricata** Venezuela / J. A. Steyermark 59075 / 1944 / "Boiled leaves placed in water and the resultant water used to bathe in for curing leprosy"
Venezuela / B. Maguire 33288 / 1952 / "Infusion used by Arecuna Indians as cold remedy"
Venezuela / B. Maguire 33586 / 1952 / "Said to be good for colds"

4418 **C. oliverii** Venezuela / J. A. Steyermark 58560 / 1944 / "Plant placed over fire makes smoke reputed to drive insects away (appears to be poisonous or obnoxious to insects)" / " 'huoro-dán tumupér' "

4419 **Cenia sericea** S. Africa / R. D. A. Bayliss 4699 / 1971 / "Boiled leaves a cure for fever"

4420 **Chrysanthemum partheniun** Equador / W. H. Camp 2504 / 1945 / " — The leaves cooked and the infusion used to bathe nervous children" / " 'Sta. Maria' "

4421 **Chrysothamnus graveolens** U.S.: Ariz. / W. W. Clute s.n. / 1920 / ". . . used in basketry by the Pueblo Indians. Flowers used to dye yellow" / " 'Big rabbit brush' "

4422 **Chuquiragua insignis** Equador / W. H. Camp 2095 / 1945 / "An infusion of the flowers used in the treatment of fevers" / " 'chuquiragua' "

4423 **Chuquiragua** indet. Peru / A. Weberbauer 7593 / 1926 / ". . . used for gonorrhoea" / " 'Alacasa' "

4424 **Conyza bonariensis** Ecuador / W. H. Camp 2494 / 1945 / " — the root used with 'culin' for a stomach which has 'played out' " / " 'Garruchuela' "

4425 **C. canadensis** Peru / F. R. Fosberg 28806 / 1947 / ". . . said to be boiled and used for diarrhoea" / " 'huamantaro' "

4426 **C. dioscoridis** Egypt / M. Drar 112/1328 / 1933 / "Used by the natives for treating colds" / "Barnúf"

4427 **C. floribunda** China: Kwangtung / S. Y. Lau 20129 / 1932 / "Used as medicine"

4428 **Diplostephium adenachaenium** Colombia / E. L. Core 1437 / 1944 / " — said to be medicinal" / "romero"

4429 **Dubautia** indet. Hawaii / L. M. Andrews 368 / 1968 / " — fragrant root was used to perfume Hawaiian tapa" / "kupaoa"

430   *Egletes viscosa* Brazil / Prance & da Silva 58546 / 1946 / "Used locally as a cure of stomach disorders"

431   *Elephantopus scaber* Tanganyika / R. E. S. Tanner 1555 / 1953 / "Sukuma: For boils—roots pounded up and put on swelling after it has been washed in hot water" / "Kisukuma: Kasalila wa hiseni"
China: Quantei / A. Chevalier 1117 / 1920 / ". . . les parties de la plante serait médicinal, elle serait employée"

432   *Emilia coccinea* Tanganyika / R. E. S. Tanner 3592 / 1957 / "Zigua: For swollen throat, leaves squeezed and the sap eaten with vegetable leaves" / "Zigua: Limi lya-mgombe"
Tanganyika / R. E. S. Tanner 2917 / 1956 / "Zigua: For syphilis, leaves boiled in water and drunk when cool" / "Zigua: Limidyangombe"

433   *Emilia* indet. Tanganyika / R. E. S. Tanner 964 / 1962 / "Sukuma tribe: leaves beaten up and mixed with ghee to treat sore ears" / "Kisukuma: Lulimi wa ngombe"

434   *Epaltes brasiliensis* Brazil / Fróes & Krukoff 1756 / 1932 / "Used for pain, as tea" / " 'pitoco' "

435   *E. alata* Tanganyika / R. E. S. Tanner 1395 / 1953 / "Sukuma tribe: Roots and leaves pounded into poultice to treat sores" / "Kisukuma: Ibongobongo"

436   *Erigeron patentisquama* Himalaya / W. Koelz 1203 / 1930 / "Flower mixed with four other herbs and used as poultice for bringing boils to head" / "Khanpa karpo"
Himalaya / W. Koelz 1139 / 1930 / "Flower mixed with fifteen other herbs used for stomachaches" / "Deva"

437   *Erigeron* indet. Fiji Is. / O. Degener 15045 / 1941 / "Said to be a Fijian medicine"

438   *Erlangea tomentosa* Tanganyika / R. E. S. Tanner 1241 / 1953 / "Sukuma tribe: Roots pounded, boiled, made into drink to relieve stomach pain" / "Kiskuma: 'Sulla' "

439   *Erythrocephalum zambesianum* Tanganyika / R. E. S. Tanner 3637 / 1957 / "For sores, leaves placed green on sores"

440   *Ethulia conyzoides* Tanganyika / R. E. S. Tanner 2101 / 1955 / "Juice of rubbed leaves used to treat sore eyes"
Tanganyika / R. E. S. Tanner 3482 / 1957 / "Bondei: for painful eyes, bloodshot, leaves pounded up and the sap squeezed in" / "Bondei: Devu la mbuzi"
Tanganyika / R. E. S. Tanner 3039 / 1956 / "Zigua: leaves pressed on fresh cuts" / "Zigua: Kungu julu"

4441 **Eupatorium rotundifolium** var. **saundersii** U.S. & CAN. / Columbia College of Pharmacy s.n. / no year / "Die bittere Pflanze ist in ihrem Vaterland ein geschätztes Arznei mittel"

4442 **E. rugosum** U.S.: N.Y. / H. N. Moldenke 7460 / 1932 / " 'White Snakeroot' "
U.S.: Minn. / E. A. Mearns 115 / 1889 / "Collected for the Army Medical Museum" / "White Snake-root"

4443 **E. serotinum** U.S.: Ill. / R. A. Evers 252 / 1940 / "White Snakeroot"

4444 **E. solidaginifolium** U.S.: Tex. / L. C. Hinckley 2323 / 1941 / "Ill-smelling composite"
U.S.: Ariz. / F. J. Lipp s.n. / 1975 / "Smoked by Papago shamans in curing seances" / " 'Pihol' "

4445 **E. collodes** Mexico / T. Macdougall 21979 / 1950 / "Medicinal para dolor de estomago" / " 'Al guá kipela (Amarga hoja)' "

4446 **E. daleoides** Honduras / C. & W. von Hagen 1083 / 1937 / "Sometimes use the gum for incense" / " 'Copalillo' "

4447 **E. cacaloides** Ecuador / W. H. Camp 2514 / 1945 / " 'The leaves are especially good to tie around the head if you have gotten a headache' " / " 'Guichilca' "

4448 **E. longipetiolata** Bolivia / Britton & Rusby 2381 / 1894 / "the leaves and bark very bitter, remedy for sand flies and lice on animals" / "Tui"

4449 **E. collinum** Honduras / C. & W. von Hagen 1043 / "Leaves put in warm water and drunk for rheumatism"

4450 **E. pichinchense** Ecuador / W. H. Camp 2478 / 1945 / "used in 'those annoying kinds of diarrhoea which are mostly wind, the kind which,' as the old medico del campo graphically explained, 'you wish you *had* taken your pants down' " / " 'Pedorrera' "

4451 **E. sternbergianum** Peru / Hutchinson & Wright 5195 / 1966 / "A stomatic" / " 'Huarmi' "

4452 **Galinsoga urticaefolia** Colombia / H. St. John 20687 / 1944 / "lvs. eaten in potato soup" / " 'guaria' "

4453 **Geigeria acaules** S.W. Africa / R. Seydel 801 / 56 / ". . . gefuerchtet als Erreger der 'vermeersiekte'. Wahrscheinlich sind kleinere Mengen zusammen mit anderem Futter aufgenommen, den Schafen unschaedlich"

4454 **G. africana** var. **africana** S.W. Africa / R. Seydel 334 / 1954 / ". . . als Erreger der 'Vermeersiekte' gefuerchtet. Nach suedwester Erfahrungen sind Kleinere Mengen, zusammen mit anderem Futter, wahrscheinlich unschaedlich; vielleicht sogar gut"

4455 **Gnaphalium gracile** Mexico / Y. Mexia 2655 / 1929 / "Used as a remedy in chest affections" / " 'Gordolobe' "

4456 *G. oxyphyllum* Guatemala / P. C. Standley 61173 / 1938 / "Used medicinally" / " 'Samalotodo' "

4457 *G. polycephalum* Mexico / G. B. Hinton 2569 / 1932 / "Decoction to wash wounds" / "Gordo Lobo"

4458 *Gnaphalium* indet. Mexico / Rose, Standley & Russell 13992 / 1910 / "Bought in market. Used medicinally"

4459 *G. spicatum* Colombia / Killip & Smith 19659 / 1930 / "Remedy for indigestion" / "Lechiguilla"

4460 *Gnaphalium* indet. Colombia / F. R. Fosberg 19578 / 1942 / "infusion very bitter, used for fever" / " 'vitavita' "

4461 *G. hawaiiense* Hawaii Is. / Degener & Bromaghim 18464 / 1929 / "Plant is very aromatic (somewhat disagreeably so). According to E. B. was used by Hawaiians to put in calabash with their feather cloaks, etc., as insect repellent"

4462 *Guizotia oleifera* Bengal / College of Pharmacy Herb. s.n. / "Cultivated for its oil"

4463 *Helenium integrifolium* Guatemala / J. A. Steyermark 48246 / 1942 / "Reputed to be poisonous to goats if they feed on it" / " 'machul' "

4464 *Helichrysum bracteatum* Colombia / J. M. Duque Jaramillo 29841 / 1946 / "Inmortales"

4465 *H. foetidum* Zululand / J. Gerstner 4828 / 1944 / "used by witch doctors to get into a trance" / "imPepo"

4466 *H. pedunculare* S. Africa / T. Cooper 415 / 1860 / "Used to cure wounds, and to cure after circumcision" / "Fingo name 'Ery'kue' "

4467 *H. stenopterum* Zululand / J. Gerstner 4841 / 1944 / "herb used by witchdoctors for inhaling to get their trance" / " 'imPepo' "

4468 *H. tomentosulum* var. *aromaticum* S.W. Africa / Merxmüller & Giess 2851 / 1963 / "Geruch nach Ziegenkäse"

4469 *H. zeyheri* S.W. Africa / R. Seydel 3138 / 1961 / ". . . aromatischer Geruch"

4470 *Heterotheca inuloides* var. *inuloides* Mexico / G. B. Hinton 847 / 1932 / "An alcoholic extract of the flower, is used for wounds" / " 'Arnica' "

4471 *Inula racemosa* India / W. Koelz 1323 / 1930 / "Root used as medicine and perfume" / "manroota"

4472 *Jurinea macrocephala* Kashmir / R. R. Stewart 19517 / 1940 / "roots used in incense" / " 'Dhup' "

4473 *Lactuca biennis* U.S. & CAN. / Fr. Fabius 489 / 1946 / "Blue Lettuce"

4474 *L. ludoviciana* U.S.: Tex. / V. L. Cory 53823 / 1947 / " 'wild lettuce' "

4475 *L. serriola* U.S.: Tex. / V. L. Cory 53822 / 1947 / " 'wild lettuce' "

4476   *Lagascea angustifolia* Mexico / McVaugh & Koelz 262 / 1959 / "very fragrant"

4477   *Laggera alata* Tanganyika / R. E. S. Tanner 3481 / 1957 / "Bondei: For sore eyes, leaves pounded and the paste put on" / "Bondei: Puisi"

4478   *L. pterodonata* Tanganyika / R. E. S. Tanner 973 / 1952 / "Reported lethal to tobacco plants if grown adjacent" / "Kisukuma: Nyabasagyi"

4479   *Layia heterotricha* U.S.: Calif. / Keck & Hiesey 2267 / 1933 / "Plants smell of banana ester"

4480   *Lessingia germanorum* U.S.: Calif. / Alexander & Kellogg 2783 / 1942 / "Strong odor"

4481   *Liabum verbascifolium* Ecuador / W. H. Camp 2739 / 1945 / "—an infusion used to bathe bruises, etc., on animals" / " 'Cotaj' or 'kotah' "

4482   *Liatris acidota* U.S.: Tex. / V. L. Cory 50726 / 1945 / " 'button snakeroot' "

4483   *L. punctata* U.S.: Tex. / V. L. Cory 51296 / 1945 / " 'button snakeroot' " U.S.: Tex. / V. L. Cory 50417 / 1945 / " 'button snakeroot' "

4484   *L. scariosa* var. *novae-angliae* U.S.: Conn. / W. Lucian 47 / 1939 / "Large Button-snakeroot"

4485   *Lychnophora ericoides* Brazil / Duarte 8279 & Mattos 563 / 1964 / "Empregada na medicina popular" / " 'Arnica' "

4486   *Madia citriodora* U.S.: Wash. / Cronquist & Jones 5869 / 1949 / "Plants strongly lemon-scented"

4487   *M. gracilis* U.S.: Nev. / W. A. Archer 6324 / 1952 / "Aromatic odor"

4488   *Melanthera brownei* Tanganyika / R. E. S. Tanner 1047 / 1952 / "Stems used to clean teeth"

4489   *Microglossa pyrifolia* Liberia / M. Daniel 85 / 1951 / "Used for worm medicine"
Liberia / P. V. Konnel 616 / 1953 / "Use: headache medicine"
Tanganyika / R. E. S. Tanner 1207 / 1953 / "Sukuma tribe: roots smoken for coughs, chest trouble; eaten as porridge to aid conception" / "Kisukuma: Budingwahimbi"

4490   *Mikania scandens* U.S. & CAN. / Herb. of A. Wood s.n. / no year / "Die Pflanze ist in ihrem Vaterlande offizinell"

4491   *M. houstoniana* Brit. Honduras / W. A. Schipp 28 / 1929 / "Sweetly scented flowers"

4492   *M. leiostachya* Panama / T. B. Croat 13245 / 1971 / "sweet aroma"

4493   *M. punctata* Brit. Honduras / W. A. Schipp 745 / 1931 / "flowers sweetly perfumed"

4494   *M. guaco* Ecuador / W. v. Hagen 44 / 1934 / "Used for snake bite" / "Huaco"

4495   *M. glomerata* Brazil / F. C. Hoehne 32101 / 1934 / " 'Guaco' "

4496   *M. officinalis* Brazil / Harley, Souza & Ferreira 10457 / 1968 / ". . . used as an infusion of the fresh leaves when one is feeling under the weather" / " 'Corroga do Gato' "

4497   *M. sessilifolia* Brazil / A. P. Duarte 2708 / 1950 / "Usada na medicina popular como sendo poderese estomacal" / " 'Quassia' "
Brazil / Maia & Kuhlmann 54395 / 1945 / "Arbustinho campestre, medicinal, estomago e figado" / " 'chá de porrete' "

4498   *M. vitifolia* Ecuador / W. v. Hagen 8 / 1934 / "Much used for snake bite. They dip leaves in hot water and wash entire body or affected part at intervals of five minutes until cured" / "Huaco"

4499   *M. chenopodifolia* Liberia / P. V. Konneh 604 / 1952 / "Pregnant women eat it"

4500   *Mutisia lanata* Bolivia / W. M. A. Brooke 5459 / 1949 / "Flower used in an infusion as a remedy for the heart"

4501   *Nicolletia edwardsii* Mexico / Ripley & Barneby 14192 / 1965 / "Strong-scented"

4502   *N. trifida* Mexico / Carter & Ferris 4106 / 1960 / "Aromatic"
Mexico / I. L. Wiggins 5372 / 1931 / "Herbage with disagreeable odor"

4503   *Odontospermum imbricatum* S.W. Africa / R. Seydel 1115 / 1957 / "gelegentlich gefressen"
S.W. Africa / R. Seydel 623 / 1955 / "unangenehmer Geruch"

4504   *Parastrephia lepidophylla* Bolivia / F. R. Fosberg 28697 / 1947 / ". . . aromatic, resinous. Harvested and baled and sent to La Paz for fuel" / " 'tola' "

4505   *Parthenium hysterophorus* Bahama Is. / W. Cerbin C132 / 1973 / "Powdered leaves put on open sore as medicine; M-nor fever medicine" / " 'Pound Cake Bush' "
Trinidad / Herb. Collegii Pharmaceae Neo-Eboracensis s.n. / 1840 / "The leaf has a bitter taste—the decoction used for killing fleas by sprinkling it on the floor—and externally against 'pediculus morpio'. Also, with molasses, to make beer (instead of hops)"
S. AM. / R. Alany 69 / 1942 / "uso medicinal" / "Altamisa"

4506   *Pechuel-Loeschea leubnitziae* S.W. Africa / R. Seydel 3159 / 1962 / "Unangenehmer Geruch von Oeldruesen her. Ungern gefressen: macht bei Kuehen die Milch bitter"
S.W. Africa / R. Seydel 821 / 56 / "Vielfach Erreger aller gischer Hautausschlaege beim Menschen. Macht die Milch bitter" / "Bitterbusch"

4507   *Pectis angustifolia* var. *angustifolia* U.S.: Neb. / R. R. Weedon 4729 /
1968 / "Lemon-like smell"
U.S.: Neb. / W. J. Dress 10077 / 1969 / "lemon-scented"
U.S.: Colo. / T. S. Brandegee 309 / 1862 / " 'Smells like lemon' " /
" 'Lemon flower' "
Mexico / Cronquist & Fay 10768 / 1970 / "Plants lemon-scented"

4508   *P. schottii* Mexico / C. L. & A. L. Lundell 8009 / 1938 / "strong scented"

4509   *P. caymanensis* Cuba / J. A. Schafer 2784 / 1909 / "Aromatic herb"

4510   *Perezia dugesii* Mexico / R. McVaugh 22789 / 1965 / "flowers . . . spicily
fragrant"

4511   *Piptocarpha sprucei* Venezuela / J. A. Steyermark 61352 / 1945 / "birds eat
fruit" / " 'bejuco blanco' "

4512   *Piqueria peruviana* Ecuador / F. Prieto CP-22 / 1944 / "An infusion in hot
water of upper part of plant in flower used as a remedy in various skin infections; applied locally"

4513   *Plagiocheilus bogotensis* Ecuador / F. Prieto 2458 / 1945 / "An infusion of
the plant taken for backaches" / " 'Parlera (or perlera) sacha' "

4514   *Pluchea petiolata* U.S.: Tex. / E. Whitehouse 16606 / 1946 / "Plants with
strong, unpleasant odor"

4515   *P. dioscoridis* Tanganyika / R. E. S. Tanner 3339 / 1956 / "Zigua: For
stomachache, leaves boiled and the water drunk" / "Zigua: Nywenywe"
Tanganyika / R. E. S. Tanner 3570 / 1957 / "Zigua: For sharp stomach
pains, roots boiled and the water drunk" / "Zigua: Nywenywe"
Tanganyika / R. E. S. Tanner 3470 / 1957 / "Zigua: For stomache ache,
roots boiled and the water drunk" / "Zigua: Nywenywe"
Tanganyika / R. E. S. Tanner 3480 / 1957 / "Bondei: For sharp pains in
the stomach, roots boiled and the water drunk when cool" / "Bondei:
Mnywanywa"
Tanganyika / R. E. S. Tanner 3769 / 1957 / "Zigua: For swollen spleen,
roots boiled and the water drunk" / "Zigua: Mnyonwe"
Tanganyika / R. E. S. Tanner 1240 / 1953 / "Sukuma tribe: Roots pound-
ed, make into poultice to treat sores" / "Kisukuma: Nghungwambu gwa
muluguru"
Tanganyika / R. E. S. Tanner 623 / 1952 / "Reported to repel mosquitos" /
"Kisukuma: Mkungwambu"
Tanganyika / R. E. S. Tanner 3905 / 1957 / "Bondei—for gonorrhoea,
roots boiled and the water drunk" / "Bondei—Mnywanywa"

4516   *Polymnia fruticosa* Ecuador / W. H. Camp 1980 / 1945 / ". . . the leaves
are heated and tied onto the legs as a cure for rheumatism"

4517   *P. meridensis* Venezuela / L. Aristeguieta 2484 / 1956 / "medicinal" / " 'Es-
corzonera' "

4518   *Porophyllum pausodynum* Mexico / H. S. Gentry 4737 / 1939 / "Highly odorous"

4519   *P. pringlei* Mexico / R. McVaugh 22097 / 1962 / "Plant with strong sour odor"
Mexico / Y. Mexia 8815 / 1937 / "Extremely malodorous"

4520   *P. ruderale* var. *macrocephalum* Mexico / G. B. Hinton 2037 / 1932 / "sold in markets; eaten raw" / " 'Palo quelite' "
Mexico / Y. Mexia 8814 / 1937 / "Extremely malodorous"

4521   *P. ruderale* var. *ruderale* Ecuador / W. H. Camp 2342 / 1945 / "Used as an eye-wash for small children. The leaves ground and the water used to wash clothes; also used to make fiber" / " 'Penco blanco' "

4522   *Prenanthes alba* U.S.: N.Y. / C. T. Hastings s.n. / 1895 / "White Lettuce"

4523   *P. trifida* var. *trifida* U.S.: N.J. / H. N. Moldenke 10193 / 1937 / " 'Rat-tlesnake-root' "

4524   *Pseudoelephantopus spicatus* Cuba / Ex Herb. Collegii Pharmaciae Neo-Eboracensis s.n. / "used as a sudorific, like borage" / "lengua de vaca"

4525   *Pulicaria undulata* Egypt / M. Drar 112/1359 / 1932 / "Herb mixed with tea to give it a good flavour" / "Rabbúl"

4526   *Schkuhria pinnata* var. *pinnata* Ecuador / Camp & Prieto 2457 / 1945 / "Used to sweep out houses which are infested with fleas" / " 'Escoba de castilla' "

4527   *S. bonariensis* S. Africa / R. J. Rodin 3876 / 1948 / "pungent odor"

4528   *Sclerocarpus divaricatus* El Salvador / P. C. Standley 22089 / 1964 / "Remedy for fever" / " 'Calacote' "

4529   *Senecio cooperi* Costa Rica / Burger & Stolze 5710 / 1968 / "cut stem gives odor of coriander"

4530   *S. guadalajarensis* Mexico / no collector s.n. / no year / "Les jeunes tiges sont été employés . . . pour empoisonner les chiens errants" / "Clorincillo"

4531   *S. desiderabilis* Brazil / L. F. Pabst s.n. / 1969 / "Frutoplumoso usado por passaros para fazerem ninhos"

4532   *S. vaccinioides* Ecuador / W. H. Camp 2288 / 1945 / "The gente chew this plant when they have a tooth-ache" / " 'Cubilán' "

4533   *Sonchus exauriculatus* Tanganyika / R. E. S. Tanner 926 / 1952 / "Sukuma tribe: Leaves used as vegetable; roots boiled and water drunk to treat gonorrhoea" / "kikusuma: Msunga"
Tanganyika / R. E. S. Tanner 1339 / 1953 / "Sukuma tribe: Roots pounded, taken orally in water to treat syphilis; whole plant burned, ground up, eaten with food for ankylostomiasis" / "Kisukuma: Lusunga"

4534   *Sphaeranthus bullatus* Tanganyika / R. E. S. Tanner 2521 / 1956 / "aromatic"

4535   *S. suaveolens* AFR. / Lewalle 4315 / 1970 / "usage medicinal"

4536   *S. ukambensis* Tanganyika / R. E. S. Tanner 944 / 1952 / "Sukuma tribe: Roots chewed to treat dry cough" / "kisukuma: Igomb"

4537   *S. amaranthoides* Ceylon / Fosberg et al. 50761 / 1968 / "Very aromatic, mint-like odor"

4538   *Spilanthes americana* Venezuela / J. A. Steyermark 55313 / 1944 / "Boil plant for cutting temperatures and mashed juicy part for putting on tongue to cut out yaga (white stuff on tongue)" / " 'ya yi' "

4539   *S. leucantha* Ecuador / W. H. Camp 2688 / 1945 / "chewed when one has a toothache because of the 'sharp' flavor" / " 'Bontoncillo' "

4540   *Spilanthes* indet. Peru / Killip & Smith 23879 / 1929 / "Heads used for toothache"

4541   *S. acmella* Liberia / P. M. Daniel 97 / 1951 / " 'medicine' "

4542   *Stevia rhombifolia* Mexico / A. J. Sharp 441224 / 1944 / "fragrant when crushed"
Mexico / A. J. Sharp 441079 / 1944 / "fragrant, but of goats"
Mexico / G. B. Hinton 2277 / 1932 / "Uses: Mediane"

4543   *S. serrata* Mexico / A. J. Sharp 441278 / 1944 / "Strong unpleasant smell much like goats"

4544   *Tagetes erecta* China: Kwangtung / W. T. Tsang 21306 / "ill-smelling" / "Chau kok Fa"

4545   *Tanacetum senecionis* Himalaya / W. Koelz 680 / 1930 / "Strong odor"

4546   *T. tibeticum* Himalya / W. Koelz 2832 / 1931 / "plant scented"

4547   *Thelesperma longipes* Mexico / Stanford, Retherford & Northcraft 217 / 1941 / ". . . heavily grazed by goats . . . unpleasant odor"

4548   *Tithonia diversifolia* Tanganyika / R. E. S. Tanner 3483 / 1957 / "Bondei: For stomach ache in small children, roots boiled and water drunk" / "Bondei: Tugutu"

4549   *Unxia camphorata* Brit. Guiana / A. C. Smith 2175 / 1937 / "Plant used for tea"
Brazil / M. B. da Silva 6 / "Crushed leaves extracted with alcohol and used as 'smelling salts' for head aches" / "lerva cánfora"

4550   *Urbinella palmeri* Mexico / A. Cronquist 9538 / 1962 / "Plants with a mild sweet odor resembling that of celery"

4551   *Verbesina pinnatifida* Jamaica / G. R. Proctor 7434 / 1952 / " 'wild tobacco' "

552   *V. encelioides* Argentina / Di Lullo 39 / 1942 / "hierba medicinal"

553   *Vernonia patens* Panama / R. S. Williams 642 / 1908 / "Furnishes a bitter called 'salvia' "
Panama / G. P. Cooper 530 / 1928 / "Leaves used to cure insect bites by rubbing the swelling with a leaf" / " 'Sanalego' "

554   *V. canescens* Colombia / Pennell, Killip & Hazen 3 / 1922 / "Used medicinally" / " 'mano de tigre' "

555   *V. aemulans* Tanganyika / R. E. S. Tanner 2010 / 1955 / "Used as a vegetable" / "Kizigua: kunguchulu"

556   *V. brachycalyx* Tanganyika / R. E. S. Tanner 1641 / 1953 / "Sukuma tribe: Roots used to treat swollen gums" / "Kisukuma: Buding'wa mpimbi"

557   *V. cinerascens* Tanganyika / R. E. S. Tanner 1120 / 1952 / "Stems used for making beds" / "kisukuma: nghuwambu"
Liberia / P. V. Konneh 132 / 1951 / "used for Gonorrhea"
Tanganyika / R. E. S. Tanner 2950 / 1956 / "Leaves used as a vegetable" / "Zigua: kibavu cha Nguku"

558   *V. conferta* Sierra Leone / F. C. Deighton 1070 / 1928 / "The leaves and bark boiled and the hot liquid drunk for stomach ache. For children's worms, the bark is dried, ground and boiled and the liquid is mixed with ground benniseed and drunk" / " 'Nyina-Nyini' (Mende)"

559   *V. grantii* Tanganyika / R. E. S. Tanner 1217 / 1953 / "Sukuma tribe: Roots used in a children's skin disease, roots soaked and child washed in this water; roots used in preparation drunk to treat gonorrhoea" / "Kisukuma: Ilumba lya shimba"

560   *V. iodocalyx* Tanganyika R. E. S. Tanner 3607 / 1957 / "Bondei: For grinding pain in stomach, roots boiled and the water drunk" / "Bondei: Mhasha"
Tanganyika / R. E. S. Tanner 3938 / 1958 / "Sambaa: for stomach ache, roots boiled and the water drunk" / "Sambaa: mhasha"

561   *V. obionifolia* S.W. Africa / R. Seydel 3408 / 1965 / "Von Kleinvieh gefressen"

562   *V. oxyura* Tanganyika / R. E. S. Tanner 3190 / 1956 / "Zigua: for stomach ache and gonorrhoea, roots boiled in water and drunk" / "Zigua: Msugumbili"
Tanganyika / R. E. S. Tanner 2952 / 1956 / "Makua: leaves burnt and put on sores" / "Makua: Nakashe"

563   *V. pauciflora* Tanganyika / R. E. S. Tanner 1332 / 1953 / "Sukuma tribe: Leaves pounded, dried, cooked in tea or porridge to treat chest or lung pain" / "Kisukuma: Ndakunilwa"

564   *V. poskeana* Tanganyika / R. E. S. Tanner 4206 / 1959 / "Whole plant used in treating cough" / "Kizanaki: Busaru"

4565   *V. smithiana* Tanganyika / R. E. S. Tanner 3969 / 1958 / "Zigua: for
syphillis, plant burnt and applied with oil to affected parts"
Tanganyika / R. E. S. Tanner 3204 / 1956 / "Zigua: For 'white patches' on
skin, leaves burnt and ashes mixed with castor oil and rubbed on area" /
"Zigua: Mwezaluwala"

4566   *V. zanzibarensis* Tanganyika / R. E. S. Tanner 2883 / 1956 / "Bondei: For
stomach pains roots boiled and the water drunk" / "Bondei: Muuka"

4567   *V. andersonii* Hainan / K. S. Lung 311 / 1932 / "Tea made by boiling the
roots heals colds"

4568   *Wedelia trilobata* Guatemala / H. H. Bartlett 739 / 1905 / "Natives use it
for kidney trouble, and as a stomachic"
Peru / T. Plowman 2454 / 1969 / "flower used as liver medicine. Whole
plant is steeped as a tea and given for gonorrhea" / " 'yamu' "

4569   *W. menotriche* Tanganyika / R. E. S. Tanner 3112 / 1956 / "For snuff:
Roots ground up and dried" / "Zigua: Mwevute; Bondei: Pempeh"
Tanganyika / R. E. S. Tanner 3336 / 1956 / "Sambaa: For sore eyes,
sufferer sits over steam of plant boiling in water. For sore mouth, boiled
and the mouth rinsed" / "Sambaa: Mpembe"
Tanganyika / R. E. S. Tanner 2914 / 1956 / "Zigua: for stomach pains,
leaves and roots cooked and eaten" / "Zigua: Mvuti"
Tanganyika / R. E. S. Tanner 1940 / 1955 / "Roots used to treat stomach
ache: boiled and water drunk" / "kibondei: Pepe"
Tanganyika / R. E. S. Tanner 3747 / 1957 / "Bondei: To cure constipation, roots boiled and the water drunk" / "Zigua: Mvuti; Bondei: Mhanx-
anyoyo or Mpepe"

4570   *W. biflora* Guam / F. R. Fosberg 25338 / 1946 / "Sometimes used medicinally" / " 'masigsig' "

4571   *Asteracea* indet. Fr. Guiana / Fr. Guiana For. Service 7043 / 1935 / "Les
feuilles écrasées au pilon sont employées comme produit enivrant pour la
capture du poisson" / "WEIKOUNAME (Idiome Paramaka)"

Index to Families

Index to Genera

Index to Common Names

Index to Uses

Index to Families

Index to Genera

315

Index to Common Names

325

Index to Uses

NOTE: The *catalogue number* of each species, rather than the page on which it appears, is given in this index.

Analgesics, 285, 388, 628, 1907, 4413; back pain, 237, 1281, 2142, 4513; chest pain, 438, 627, 3265; headache, 438, 466, 512, 588, 914, 939, 1295, 1360, 1362, 1811, 1907, 2034, 2047, 2074, 2148, 2244, 2343, 2364, 2750, 2850, 3548, 3934, 4086, 4118, 4235, 4338, 4410, 4447, 4489, 4549; joints, 4127; muscles and ligaments, 1959, 2288; stomach, 512, 2431; taken externally, 22, 47, 222, 241, 534, 587, 1976, 2075, 2120, 2205, 2244, 2703, 2824, 3223, 3411, 3587, 3628, 3662, 3997, 4235; taken internally, 914, 1234, 3799, 4039, 4257, 4434. *See also* Anesthetics; Emollients

Anesthetics: local, 2392; toothache, 159, 453. *See also* Eye disorders; Oral cavity, toothache

Ankylostomiasis, *see* Vermifuges

Aromatics, 18, 19, 33, 34, 76, 77, 81, 95, 96, 141-44, 148, 152, 154-55, 157-58, 160, 166-69, 172, 180-81, 229, 242, 244, 246, 248, 258, 326, 356, 358, 369, 393, 396, 407-08, 417-19, 425, 427, 436, 440, 445, 448-50, 457, 461, 478, 483, 488, 605, 650, 658, 673, 677, 689, 713, 728, 753-54, 850, 937, 943, 945, 967, 972, 980, 988, 994, 1002, 1007, 1017, 1038-44, 1063-64, 1070-72, 1081-82, 1090, 1093, 1095-99, 1108, 1110, 1114, 1120-21, 1124-29, 1133, 1135-36, 1138-43, 1145, 1150, 1162-65, 1167-68, 1172, 1174-75, 1179, 1189, 1195-96, 1204-05, 1212, 1214-16, 1218-19, 1222, 1224-25, 1229, 1233, 1248, 1255, 1259, 1290-92, 1299, 1300, 1307-08, 1312, 1314, 1326, 1329, 1334, 1337, 1350, 1353, 1359-60, 1398, 1404, 1408-16, 1424, 1428, 1447, 1460, 1462, 1465, 1473, 1479, 1504, 1552, 1565-66, 1571-72, 1583-85, 1587-88, 1608, 1611-17, 1625-26, 1643, 1685, 1696, 1698-99, 1700, 1716, 1735, 1748, 1756, 1761, 1764, 1766, 1779, 1811, 1817, 1830, 1834, 1924-25, 1927, 1962, 2014, 2016-17, 2032, 2067, 2073, 2139, 2151, 2161, 2201-02, 2206, 2222, 2224, 2253-54, 2490, 2525, 2563, 2596, 2645, 2649, 2658-61, 2748, 2754, 2757, 2769, 2772, 2777, 2782, 2787, 2795, 2831, 2864, 2921, 2931, 2937, 2942, 2948, 2956, 2991, 3016, 3019, 3076, 3081, 3093, 3102, 3105-06, 3108, 3137, 3140, 3145, 3149-50, 3156, 3158, 3166, 3168, 3173, 3179, 3194, 3218, 3236, 3240, 3251-54, 3256-57, 3275, 3283, 3289, 3295, 3302, 3318-20, 3324, 3329, 3334, 3338, 3343, 3355, 3402-03, 3406, 3428, 3443, 3451, 3621-22, 3679-80, 3712, 3755-58, 3761, 3766, 3769, 3770-73, 3778, 3780, 3783, 4075, 4138-40, 4152, 4159-60, 4162-63, 4165, 4171, 4174-76, 4179-84, 4186-90, 4199, 4202, 4208-11, 4216, 4220-21, 4226, 4229, 4236, 4245-48, 4250-54, 4259, 4263, 4277, 4284, 4306, 4328, 4362, 4403, 4469, 4476, 4479, 4486-87, 4492-93, 4501-02, 4504, 4507-10, 4518, 4529, 4534, 4537, 4542, 4546, 4550; incense, 11, 162, 1106-07, 2697, 2933, 3297, 3450, 4446, 4472; perfumes, 161, 163, 471, 669, 1112, 1757, 2141-42, 2685, 2703, 2705, 3342, 3678, 4209, 4429, 4471. *See also* Attractants; Cosmetics; Oils

Arthritis, treatment, 2153

Astringents, 13, 330, 340, 823, 877, 1561, 1591, 1847, 2134, 2285, 2973, 3000, 3067, 3154

Attractants, 1778; fish bait, 231, 744, 1030, 2033, 3018, 4136; to insects, 112, 195, 232, 243, 249, 272, 281, 283, 308, 409, 413, 416, 429, 535-36, 614, 683, 705, 727, 743, 745, 821, 847, 924, 989, 1003, 1010, 1069, 1117, 1129, 1135, 1143, 1164, 1174, 1207, 1230, 1619, 1670, 1700, 1741, 1744, 1777, 1810, 1812, 1838, 1855, 1869, 2279, 2484, 2503, 2657,

357